T0339623

Applied Drought Modeling, Prediction, and Mitigation

Applied Drought Modeling, Prediction, and Mitigation

By

Zekâi Şen

King Abdulaziz University, Jeddah, Kingdom of Saudi Arabia

ELSEVIER

AMSTERDAM • BOSTON • HEIDELBERG • LONDON • NEW YORK • OXFORD
PARIS • SAN DIEGO • SAN FRANCISCO • SINGAPORE • SYDNEY • TOKYO

Elsevier
Radarweg 29, PO Box 211, 1000 AE Amsterdam, Netherlands
The Boulevard, Langford Lane, Kidlington, Oxford OX5 1GB, UK
225 Wyman Street, Waltham, MA 02451, USA

Notices
Knowledge and best practice in this field are constantly changing. As new research and experience broaden our understanding, changes in research methods, professional practices, or medical treatment may become necessary.

Practitioners and researchers must always rely on their own experience and knowledge in evaluating and using any information, methods, compounds, or experiments described herein. In using such information or methods they should be mindful of their own safety and the safety of others, including parties for whom they have a professional responsibility.

To the fullest extent of the law, neither the Publisher nor the authors, contributors, or editors, assume any liability for any injury and/or damage to persons or property as a matter of products liability, negligence or otherwise, or from any use or operation of any methods, products, instructions, or ideas contained in the material herein.

British Library Cataloguing in Publication Data
A catalogue record for this book is available from the British Library

Library of Congress Cataloging-in-Publication Data
A catalog record for this book is available from the Library of Congress

For information on all Elsevier publications
visit our website at http://store.elsevier.com/

ISBN: 978-0-12-802176-7

Table of Contents

Dedication

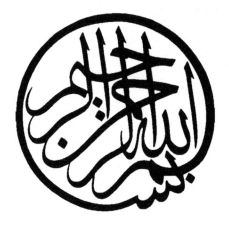

Noah phenomenon—(great flood, wet spell)

Joseph phenomenon—(drought, dry spell)

There is sensitive balance in nature as a sequence of dry and wet periods, which needs care for their preservations without destroying the balance in the environment. This book is dedicated to those who care for such a balance by logical, rational, scientific, and ethical applications for the sake of other living creatures' rights.

Preface

The sustainability of any society is dependent on different precious material resources such as water, energy, and technology, which must be traced by modern scientific research work outputs application so as not to meet with any restrictive, shortage, stress, or scarcity situations. Among the natural hazards phenomena, the most effective one for the long run is the continuation of dry spells (water demand deficiency), which occurs in the form of drought and leaves different imprints on the society at large. Droughts have a gradual "creeping" feature, with slow developments and prolonged effects on the daily activities of human life.

In general, settlers in humid regions have high confidence in their water resources supply and, therefore, they may not feel water scarcity impacts in time. Consequently, droughts may be more harmful in humid regions than arid regions. Accordingly, especially agricultural investments may be inflicted at maximum harm rates as a result of unexpected drought periods. Among the primary drought hazards are crop yield, animal husbandry, hydroelectric energy generation, reductions and decrease in industrial products, and navigation problems in low river flows. On the other hand, secondary effects include soil erosion, dust storms, forest fires, increase in plant diseases, insect hurdle, decrease in social and individual health, pollution concentration increase, deterioration in water quality, and so forth. In the literature simple drought descriptors are presented based on a single or few hydrometeorological variables such as rainfall, precipitation, solar irradiation, and wind speed. Rather than their individual applications, their joint assessments are given for the first time in this book by statistical ensemble averages and fuzzy inference system approaches. All of these indicate the importance of drought preparedness, early warning, proper drought modeling, and appropriate predictions. In this book the list of economic, environmental, and social drought impacts are explained in detail, and it gives the impression that there is no sector that may be safe from drought implications.

The most significant part of drought identification, assessment, and prediction studies is the modeling procedures that furnish the foundation for proper strategic planning, management, application, and implementation of output principles in a society prior the next drought occurrence. This book presents innovative drought modeling procedures by taking into consideration the inherent uncertainty feature in drought evolution. To account objectively for the uncertainty, probabilistic, statistical, stochastic, and fuzzy methodologies are employed with a set of simplifying assumptions. The necessary formulations and their quantitative applications through numerical solution approaches are presented for temporal and spatial drought durations, total end average deficits, and intensity with the necessary areal coverage extension formulations.

Drought events are explained in the last two chapters from the climate change and mitigation points of view with an emphasis on water resources supply and demand patterns, rainfall and runoff harvestings, groundwater recharge possibilities, and proper risk and hazard management points of view. As one of the mitigation procedures, weather modification and its application in Istanbul City, Turkey, is explained with some new formulations and it is recommended that at its present scientific level the weather modification (cloud seeding) procedures are not successfully applicable; therefore, cloud seeding must remain in the scientific research domain without practical, fruitful outputs. Different engineering structural drought combat procedures are explained and a list of recommendations for drought mitigation is provided.

The author has gained vast experience during a long stay as a staff member at the King Abdulaziz University Faculty of Earth Sciences and recently at the Faculty of Meteorology and Arid Lands, Excellency Center for Climate Change Research, Jeddah, Kingdom of Saudi Arabia. He became acquainted with different desertification, drought, groundwater recharge, water harvesting, and hydrogeological water management procedures and strategies and published numerous papers in top scientific journals. Another part of his extensive experience comes from meteorology, hydrology, and hydraulic studies at the Technical University of Istanbul, Turkey, in addition to the Turkish Water Foundation concerning the conjunctive and separate surface and groundwater resources under uncertainty principles and scientific modeling studies leading to predictions. His long experience for about 5 years in the workgroup of the Intergovernmental Panel on Climate Change (IPCC), as the freshwater resources chapter lead author provided a global picture and scientific views about possible climate change impacts on precious water resources including vulnerability, combat, and mitigation.

Most of the content of this book includes experience gained during the stay of the author in the Kingdom of Saudi Arabia; hence, he would like to extend his cordial appreciation to his colleagues at different faculties at the King Abdulaziz University and to its high-level administrators. The author would like to extend his appreciation to the Saudi Geological Survey (SGS), Jeddah, and Prince Sultan Research Center for Environment, Water, Desert, King Saud University, Riyadh, where he also gained experience. Similar gratefulness is also extended to the Turkish Water Foundation and Istanbul Technical University, and to those who made constructive suggestions during the preparation of this book.

I wrote several books in Turkish, English, and Arabic and many scientific papers, but nothing gives me the happiness as being at the service of people who seek scientific knowledge and information. Any fruitful impact of this book will make the author spiritually very content and happy. Finally, whatever my achievements, under their foundations is the patience and continuous support of my wife Fatma Şen, who deserves thanks from the bottom of my heart.

Zekâi Şen
Erenköy, Istanbul, Turkey
March 13, 2015

Chapter | ONE

Introduction

CHAPTER OUTLINE

1.1 GENERAL

Water is a major essential commodity for the survival of all living creatures. Life sustainability is not possible without it. Abundance of water brings comfort, whereas in its scarceness life becomes miserable. Human beings are dependent on water in almost every activity within the environment. If water is scarce or not available in sufficient quantities at a location, then human beings migrate to better water resources locations, which are riverbanks, lakes, seashores, oases, or shallow groundwater reservoirs. Evolution and development of any civilization has roots in water-related management activities. Such activities are the starts of social gatherings, cultures, and civilizations. The history of civilizations indicates that even in dry lands groundwater resources had dominant roles through shallow wells or natural springs. Any civilization is under the pressure of internal and external impacts and urges for food security, which cannot be achieved without water security. Water resources have been and still are under internal and external pressures. The foundation of any civilization

Applied Drought Modeling, Prediction, and Mitigation.

includes irrigation and agriculture, land use, seeding, and the quality control of products through technological developments, all of which drive the economic system of the society. Mismanagement of water resources, acid rains pollution, overexploitation, and other human activities play roles in the appearance, continuity, areal extent, and severity of droughts.

Even though the selection of settlement locations are made by humans, natural events such as droughts, floods, earthquakes, and others are among the external hazards that may affect societies at any time without preparedness against the final consequences. For instance, extreme water events such as droughts and floods should be managed in such a way that extra amounts of water should be stored in some way so as to be of benefit during future dry spells when the water supply may fall short of meeting the demand. Otherwise, the society may go through a water stress period until a suitable supply is either found from an engineering point of view or by the reoccurrence of abundant rainfall events. These days, water scarcity and stress increase day after day. Among the main reasons for water scarcity are the following points:

1. Increase in world population.
2. Burst in urbanization.
3. Increase in the needs of industrial production.
4. Differences in water distribution, movement, contamination, pollution, and deteriorations may result in undesirable ecological consequences.
5. During the last 25–30 years, due to global warming, greenhouse effects, and as a result of climate change, exploitable water resource quantities are bound to decrease in many parts of the world.

The most important effect of climate change on water resources is increase in the overall uncertainty associated with the management and supply of freshwater resources. Significant hydrological components such as storms, rainfall, stream flow, soil moisture, and evaporation are substantially random in their behavior and, accordingly, hydrologists or water specialists try to quantify them in terms of uncertain scientific methodologies; namely, probability, statistics, and at-large stochastic approaches ("see chapters: Temporal Drought Analysis and Modeling; Regional Drought Analysis and Modeling; Spatiotemporal Drought Analysis and Modeling") and, most recently, as chaotic and fuzzy systems ("see chapter: Basic Drought Indicators"). These scientific approaches provide predictions on the bases that the surrounding environmental effects and the climatic change are all stationary. Hence, classical approaches assume that the pattern of the local environmental and global climatic changes in the recent past will be repeated in the near future. It must not be forgotten at this stage that, certainly, the future pattern of climatic change and its consequences will not look like the past behaviors. It is, therefore, necessary to try and manage water supply systems with more care about the undesirable possible future extreme drought cases. This brings into the equation the concept of risk and management under a risky environment with probabilistic assessments and modeling ("see chapter: Climate Change, Droughts, and Water Resources").

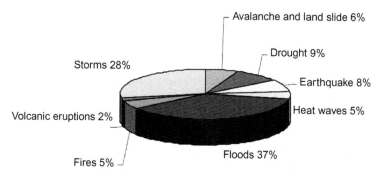

FIG. 1.1 Natural disaster percentages (WMO, 2005).

Water-related disasters (droughts, floods, hurricanes, typhoons, tsunamis) inflict a terrible toll on human life and property, far greater than earthquake damages (Fig. 1.1). About 90% of natural hazards are related to air, climate, and water.

Human activities should be planned in such a way that significant reduction of vulnerability can be made prior to drought occurrence ("see chapters: Climate Change, Droughts, and Water Resources; Drought Hazard Mitigation and Risk"). Nowadays, there is extensive knowledge, information, and capacity to disperse warnings even to the remotest places on the Earth, which help to alert people to take the necessary precautions against any natural disaster danger, in general, and drought effects, in particular.

On a global scale, very intensive and extensive drought distribution was observed during 1982–84. The most vulnerable parts of the world were West Africa, Sudan-Sahel, east and southeast Africa, southern and southeastern parts of Asia, the western Pacific and Australia, and southern parts of the United States.

Among the secondary effects of droughts are soil erosion, and consequent dust storms, forest fires, plant diseases, insect plagues, decrease of personal and public hygiene, increased concentration of pollutants, degradation of water quality, harmful effects on wildlife, and deterioration in the quality of the visual landscape. While floods, earthquakes, and cyclones are disasters associated with extreme events, droughts are the result of the low extremes such as unavailability of sufficient water. They seldom cause dramatic losses of human life except through famine. Generally, drought assessments at any point in a region can be achieved by taking into consideration time series records of the concerned variable. The first studies by Gumbel (1958) considered the probability of the lowest records during fixed periods. They are point wise instantaneous evaluations and, therefore, neither the drought coverage nor the areal extent can be modeled. Rather uncertain temporal and areal drought extensiveness must be modeled and predicted by quantitative methodologies such as probabilistic, statistical, stochastic, and (recently) fuzzy logic rule approaches (Şen, 2010). The first quantitative drought definition and studies by considering the threshold levels were due to Yevjevich (1967); later, various convenient applications were carried out by different authors (Downer et al., 1967; Llamas and Siddiqui, 1969; Saldarriaga and Yevjevich, 1970; Millian and Yevjevich, 1971; Guerrero-Salazar and Yevjevich, 1975; Şen, 1976, 1977, 1980a,b).

These methodologies provide quantitative gains in drought modeling and prediction; especially, initiation and continuation of agricultural drought triggers major problems between different water-dependent sectors. Uncertainty, lack of information, and ignorance about convenient methodological applications cause an increase in drought problems over time.

Consistent methodologies are not as easily available for drought modeling and prediction as for flood analysis. The most essential part of drought studies is its definition. Wilhite and Glantz (1985) have suggested that drought is dependent on different disciplines (meteorology, hydrology, agriculture, society) and, therefore, it needs different definitions. Tate and Gustard (2000), Demuth and Bakenhus (1994), and Dracup et al. (1980) summarized some of the most common drought definitions. They have also noted that there are confusions concerning different definitions, including "drought event" and "drought index" ("see chapter: Basic Drought Indicators"). Generally, a drought index implies a single number characterizing the general drought behavior at a measurement site, whereas a drought event definition is applied to select drought occurrences in a time series, including the beginning and the end of drought ("see chapter: Temporal Drought Analysis and Modeling"). The difference between these two time instances includes many drought characteristics such as drought duration, magnitude, intensity, and so on (Yevjevich, 1967; Şen, 1976, 1978). They continued to state that based on data availability climatic and regional drought characteristics require a suitable choice of definition. Beran and Rodier (1985) gave the most general drought definition as, "The chief characteristic of a drought is a decrease of water availability in a particular period over a particular area." Later, authors also provided similar definitions concerning different drought purposes (Allaby, 1998; Wilhite, 2000a; Boken et al., 2005; Tallaksen and van Lanen, 2004; Sheffield and Wood, 2006). Each definition is justified according to types of drought study in different sectors with different conclusions. For instance, in some seasons there may be enough precipitation events but mismanagement of water resources may lead to water shortages or water stress.

The most important motivation for this book is to provide primarily scientific, philosophical, logical, and linguistic information leading to necessary formulations, algorithms, and software for practical modeling, prediction, and applications in order to reduce the overall drought effect. The content has a wide range of practical, applicable, scientific, and mathematical models with examples and case studies for better appreciation of drought. Drought management cannot be without suitable and reliable models, which help to make future scenario predictions for the assessment of the worst and best drought mitigation strategies. Decision makers are then able to select the most suitable solution for their case among different scenarios. Risk-based drought preparedness plans and policies are always important for policy and other decision makers, which can be achieved objectively, if scientific models are available with reliable predictions ("see chapter: Climate Change, Droughts, and Water Resources"). The quantification of drought risk is possible only by means of mathematical models, which are given in practical detail in this book. Such models are among the most desired means of planning for predrought mitigation strategies, early warning systems and, accordingly, drought information dissemination among people prior to actual drought occurrence. The drought models also support information delivery systems, decision support tools to improve decision making, improved seasonal climate forecasts, drought planning, impact assessment methodologies, and for better and more effective monitoring, network planning, operation, and management.

The drought complexities and differences between various types are explained, modeled, and applied in the book. Special mathematical models are developed to address these differences as part of drought preparedness planning; otherwise, the differences may result in a failure of the mitigation and planning processes. The future behavior of drought can be assessed and evaluated through convenient mathematical models and their prediction capabilities in an applied manner, which are among the topics in this book. These models and their results may help to have a sustainable future against drought disasters.

The methodological explanations, models, and prediction approaches in this book help to bridge different sectors and, hence, the problem of drought effects can be reduced through modeling outputs. Otherwise, decisions taken by local and central administrators, managers, and politicians without scientific foundation may further bother the society and the consequences of drought may be worse than expectations.

1.2 HISTORICAL VIEW

Millions of years ago, during the Precambrian era, the climate had very a different structure than today. The average increase in the sun's mass was due to a steady increase in the sun's brightness as a result of the core pressure increase and more speedy energy consumption of temperature than nuclear reaction in the sun. According to the theoretical calculations, the sun had 30% more brightness than today. If this brightness was less at the beginning of the solar system then one could conclude that the world was colder, which might have caused the freezing of all water in the surface of the world (Sagan and Mullen, 1972). The calculations based on the constant atmospheric composition lead to the conclusion that about 2 million years ago, the world was frozen completely. This theory extends to the ice layers at Isu, Greenland, about 3.8 billion years ago. In each era, sedimentary layers are accumulated beneath the water bodies, which have been proven through the geological record studies (Walker et al., 2009).

The Paleolithic era continued until 12,000–10,000 BC and covers the geological Pleistocene era during which the climate had changed about seven times and at the beginning of this era, the world temperature had decreased. Later, there were three subsequent ice ages. Once more the temperature had decreased but the precipitation had increased. Expansion of the glaciers had affected plant and animal lives significantly. In the meantime, the climate belts had shifted to the south and the glaciers retreated toward the polar regions. Some species had increased whereas some others had decreased, but some others had extinguished. During the retreat, the glacier areas had converted to tundra, tundra to forests, and forests to deserts, which may be the main reason for oil reserves today. One can understand that the interglacier periods were hot with moist air, which is evident from the vegetation covers. For instance, in the eastern parts of the North American continent there were large forest areas. Pine trees had indicated reduction in the evaporation but in western areas of North America arid and semiarid regions had appeared frequently.

According to the first pollen records, the global temperature was only 1–2°C colder than today and the set of trees as forest had appeared during 10,200–9400 BC. On average, about 5°C lower temperatures than today had occurred during 9400–8300 BC. Later, very fast warming had prevailed and at the upper latitudes temperature was 1–2°C warmer than today. The start of cold fluctuation at upper latitudes reached to the maximum temperature prior to 7100 BC. After that time, the temperature at upper latitudes became partially warmer (Manley, 1953). In the meantime, in semiarid environments there were moisture fluctuations in pollen records. Even though the summer precipitation belt was comparatively moist in North America, Arizona and Mexico in the Colombia Plateau over an extensively large basin, there was a dry duration during 6000–2000 BC. Pollen records in tropical Africa indicated prevalence of one or more warm and moist media in the center of the southern Sahara, according to Mediterranean-type of vegetation. For more detailed information see Table 1.1.

Throughout history drought has been the companion of humanity. Over the years drought impacts have been felt in agriculture, water supply, industry, pollution control, energy, recreation, and a host of other activities related to water and society.

1.3 ATMOSPHERIC COMPOSITION AND DROUGHT

The significance of atmospheric composition is the absorption of some sun irradiation and its return to the Earth's surface so as to increase temperature. This is referred to as the greenhouse effect, leading to global warming. Some parts of high-frequency irradiation that is filtered through the ozone layer are reflected back from the clouds, deserts, and snow areas. Other parts are absorbed by the Earth's surface and reflected back in the form of low-frequency radiation. The ultimate source of energy for storm production is solar radiation. The balance of energy available for storm production may become manifested in three ways:

1. It provides the energy for increased evaporation over the ocean or other available moisture sources.
2. It provides the energy for development of horizontal air movement that can transport the moisture to the storm reception area.

TABLE 1.1 Historical Cycles of World Climate (Bronowski, 1978)

Date (year)	Region	Climate
9000–6000 BC	South Arizona	Hot and dry
7800–6800 BC	Europe	Cold and moist, ice layers 7000 BC left Switzerland at about 6840 BC
6800–5600 BC	North America and Europe	Cold and dry extinguishing mammals at Arizona
5600–2500 BC	Both hemispheres	Hot and moist
2500–500 BC	Northern Hemisphere	Generally hot and dry
500–M.S. 0 BC	Europe	Cold and moist
AD 330	America	Drought at southeastern
AD 600	Alaska	Ice advancement
AD 590–645	Near East and England	Cold winters are followed by extensive droughts
AD 673	Near East	Black Sea is frozen
AD 673–800	Mexico	Start of moist duration
AD 800–801	Near East	Black Sea is frozen
AD 829	Africa	Icing at Nile River
AD 900–1200	Iceland	Stagnancy at icing
AD 1000	Africa	Icing at Nile River
AD 1000–1100	Utah	Snow elevation is 300 m more than today
AD 1200	Alaska	Ice advancement
AD 1000–1215	America	Moisture in the west
AD 1220–1290	America	Drought in the west
AD 1226–1290	America	Extensive drought in the southeast
AD 1300–1330	America	Moisture in the west
AD 1500–1900	Europe	Generally cold and dry
AD 1570–1595	America	Between 1573 and 1593, dry northwestern America
AD 1880–1940	Both hemispheres	Increase of 11°C in winter temperature, 5.2 m reduction in ice lakes, 0% reduction in Artic ices, and 25% in Alp ices

3. It provides the energy for the vertical air movement that is essential over the storm reception area to lift the mass of moisture-laden air to enable it to cool and condense.

Droughts show themselves as one of the major climate change factors in the world. Apart from the atmospheric circulations, droughts also occur after volcanic eruptions and dust dispersion in the atmosphere, which may partially hinder the arrival of solar irradiation to the Earth's surface.

The global average temperature has increased between 0.5 and 1.3°C since 1856, when instrumental measurements were started. This implies global warming and consequent climate change impacts (IPCC, 2007). Such an increase is expected to give rise to important changes and events in the atmosphere. Hence, intensive natural events take place such as storms, tornadoes, above average temperature prevalence of air in winter season, and sea level rises, in addition to an increase in the intensity and frequency of droughts and floods at unexpected times and locations.

Precipitation deficiency in an area is the initial trigger for a possible extended period of dry spell, usually for a season, year, or several years. This deficiency gives rise to water or soil moisture shortages for various human activities (social, agricultural, or sectorial). Even though drought is considered as a rare random event, it is a normally recurrent event of climate. It may virtually occur in all climatic zones, but its characteristic features vary significantly from one region to another.

There is an optimum division in each area for each season to produce the physically possible maximum precipitation. If the horizontal airflow over the storm production area is above the optimum, moisture is then produced and transported rather quickly but the stored energy is exhausted sooner. Provided that the necessary lift is maintained, the storm may have great intensity with short duration. This may explain the nature of the depth–duration relationship for storms of physical maximum precipitation.

The water moisture and CO_2 concentrations in the atmosphere give rise to temperature increments. Increase in the sun's brightness causes primarily surface temperature increase and, therefore, dissociation accelerates leading to reduction in the CO_2 pressure, which decreases the effect of temperature rise and, hence, it cools again. During the long history of the sun, its brightness has increased and this feedback caused a slight increase in the temperature coupled with a significant decrease in the atmospheric CO_2 concentration (see Fig. 1.2). On the other hand, the heat source of the world within the significant components of the hydrological cycle and the atmospheric water moisture variations are presented in Figs. 1.3 and 1.4, respectively (Sırdaş, 2002).

Drought occurrences result mainly from variations inherent in the atmospheric circulation. It may depend also on such factors as the transport of volcanic ash and dust, which reduce the solar radiation reaching the Earth's surface. Aridity is a permanent climatic feature. In the driest zones, the variability of precipitation is the highest. Economic consequences of droughts are more important for humid regions because of people's unpreparedness for recurrent drought events and the large investments in agriculture, which may undergo big losses from droughts.

Drought does not mean a simple precipitation deficit only, but it is a result of humidity deficit with respect to long-term average humidity, which is the result of persistent unbalanced precipitation and evapotranspiration continuations.

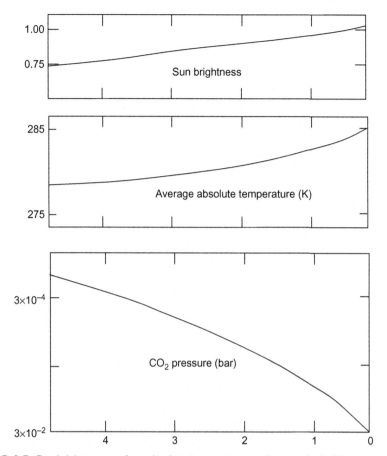

FIG. 1.2 Sun brightness, surface absolute temperature, and atmospheric CO_2 pressure.

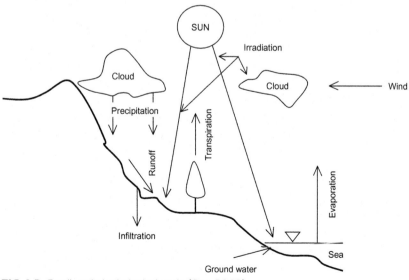

FIG. 1.3 Small-scale hydrological cycle (Şen, 2002).

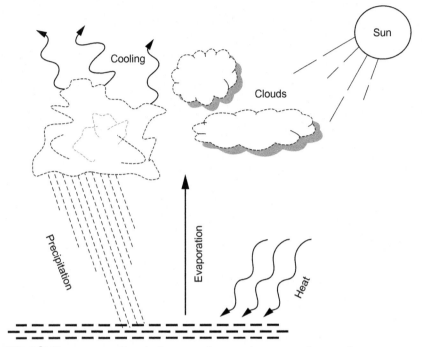

FIG. 1.4 Temperature effect in the atmosphere on the water moisture volume.

The causes of drought are not the same all the time. The same humidity deficit, depending on the time of the year, may lead to different consequences. It changes depending on the relative change of the present humidity level with respect to the previous level, air temperature, and wind conditions. One may not understand the drought effect from the meteorological records only, even though it may appear with lower precipitation occurrences than expectations. Possible deviations in the atmospheric circulation may give rise to meteorological droughts. Along the drought duration, a decrease in precipitation amounts is coupled with an increase in evaporation rates. Extreme atmospheric behaviors interact with ocean and, subsequently, meteorological and climate events affect landmasses according to droughts.

Droughts are manifestations of climatic fluctuations associated with large-scale anomalies in the planetary circulation of the atmosphere. They imply precipitation absence or weak precipitation occurrence for a long time over large areas. It is very difficult to identify and to clearly determine the onset as well as the termination point of a drought. It is a creeping phenomenon and its effects accumulate slowly and tend to persist over longer periods of time. Local and regional climate features are important in drought generation. This subsidence generates an adiabatic compression, which leads to an increase in temperature and, consequently, a reduction in the relative humidity. The subsidence further produces an inversion of temperature, which increases the static stability of the atmosphere and prevents the formation of sufficiently thick clouds to generate precipitation. Because at the start, the air is already dry, the relative humidity decreases further as the air mass subsides. Dry air at the start of subsidence becomes drier with the continuation of the subsidence. In such a dry environment, cloud formation is rather difficult and even after their formation, due to evaporation, the water droplets do not lead to rainfall generation. They arise in the precipitation deficit cases or during very light precipitation occurrences for long periods over large areas.

Droughts triggering mechanism in the atmosphere is not known definitely. Even today, one knows the consequences of droughts rather than their generation mechanisms. Among many drought impact factors are land use, environmental, hydrological, meteorological, and climate variability and variations. These factors are interactive in a chain reaction according to uncertain mechanisms; hence, establishment of reliable prediction models are not possible with high significance.

In a way, droughts are consequences of deviations in atmospheric events from long-term averages. Such regions are exposed to longer drought periods that are already extensive without sufficient rainfall or continuous reduction in its amount. This brings atmospheric subsidence over some areas that are subject to drought conditions. Hence, spatiotemporal drought behaviors become important for drought prediction ("see chapter: Spatiotemporal Drought Analysis and Modeling").

Droughts are not local, but temporarily expand over extensive areas as a result of high-pressure air movements (anticyclones, spiral movement of the air) and changes in high-pressure centers.

The major effect of droughts is large-scale atmospheric subsidence that may last for many years or local subsidence in mountainous regions. On the other hand, droughts are related to atmospheric complex movements with humid airflows that cause local precipitation. Clear sky and low humidity provide strong solar irradiation influx (on the average annually $200 \, \text{W/m}^2$) at many parts of the world, which cause increase in the soil temperature. Dry earth surfaces with high reflection properties (high albedo) and light colors give rise to significant losses as a result of this reflection. On such surfaces, long-wave irradiation reflections are intensive. Due to this reason, the real irradiation inputs in desert areas are comparatively lower as $80–90 \, \text{W/m}^2$.

Distribution of dry climates all over the world is related basically to subtropical regions where high-pressure centers are coupled with subsidence movements. These centers move toward polar regions in summer and toward the equator in winter. This situation generates three different drought belts as follows.
1. Desert core, where precipitation is either very little or nonexistent (approximately 20–30° latitude belts)
2. Tropical belt, where precipitation frequency is high in sunny seasons
3. Mediterranean climate belt or such belts that are oriented toward polar regions with rainy periods in winter

In all these belts, precipitation variability is high, but not all the parts of subtropical regions are dry; the subtropical high-pressure belt has divisions at several locations with abundant precipitation.

The African Sahel drought belt is under the effect of recent climate change with basic effects of fluctuations in high-pressure center locations. Statistical investigations of precipitation records indicate that in some parts of the dry belt, and especially in Sahel, continuation of abnormal humidity or the drought event has a definite trend (Nicholson, 1980, 1982; Hare, 1983). Droughts of 10 years or longer durations are more extensive in arid and semiarid regions and, generally, end with precipitation. Such a continuation implies that there is a feedback mechanism and natural water equilibrium works. In general, dry spells follow dry spells and humid (wet) spells follow wet spells. Finally, the normal situation is recovered and such feedback is important in drought mechanism modeling ("see chapters: Temporal Drought Analysis and Modeling; Regional Drought Analysis and Modeling"). The general circulation model (GCM), which takes into consideration atmosphere and ocean fluctuations, helps to predict surface climate in a computer environment through proper software. Such models also help to check the feedback mechanism function in the atmosphere. Dry earth surfaces, extreme plough of the surface, or unconscious agricultural implementations cause an increase in the feedback reflections (albedo). Many GCMs have shown that such feedback causes an increase in long-wave irradiation with cooling effects and also the subsidence events in the atmosphere, all of which give rise to an increase in precipitation amounts (Charney, 1975). Such a mechanism is referred to as the albedo feedback. Controllable land use is accepted as a certain method for microclimate regulations on the surface. Sufficient vegetation cover provides the key information for such a control. Production capacity reestablishment of a region is dependent on food and organic matter existence with high-filtration properties (Hare, 1983). Drought solutions necessitate climate-desertification work in more detail. In addition to GCM climatology, ecosystem simulation models and management techniques fill the gaps ("see chapters: Temporal Drought Analysis and Modeling; Regional Drought Analysis and Modeling; Climate Change, Droughts, and Water Resources").

Recent researchers show that the CO_2 concentration has increased twofold in the atmosphere. This triggers the greenhouse gas effect in an increasing manner. The natural hazards such as desertification and drought are thought to increase due to anthropogenic effects in recent years (IPCC, 2007). According to previous studies, carbon or oxygen measurements are good indicators for past temperature determinations. CO_2 concentrations indicate parallel behavior with temperature records at any time duration. CO_2 and CH_4 (methane) variations are observed depending on temperature fluctuations. Based on the examination of ice-drilling samples down to 2000 m depth at the Russian Vostok polar station, one can observe temperature fluctuations (Bender et al., 1997). Fig. 1.5 shows temperature variations since 1860 according to Centre for Climate Prediction and Research Institute. Since 1860, when the consumption of coal and fossil oil started, the CO_2 concentration has increased by 30% in the atmosphere; consequently, temperature increased by 0.6°C (IPCC, 2007). One can understand from this figure that at subtropical latitudes in winter seasons, there is more

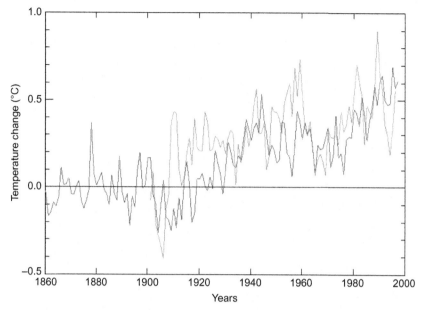

FIG. 1.5 The upper line indicates anthropogenic climate change affects, whereas the lower line is for measurements.

snow and consequent melting that gives rise to floods but in the summer seasons drought becomes more intensive (Baykan, 1994).

The atmospheric causes of droughts are not yet well understood scientifically within the general atmospheric circulation. For further research, it is necessary to couple atmospheric and oceanic circulations for better understanding of drought occurrences. Anomalies in sea surface temperature influence the flux of sensible heat and moisture at the atmosphere–ocean interface. Moisture and temperature together influence the subsequent latent heat release and determine the amount of precipitable water in the atmosphere. El Niño Southern Oscillation (ENSO) events are among the important factors that cause drought. According to Dilley and Heyman (1995), worldwide drought disasters double during the second year of an El Niño episode compared with all other years.

1.4 DROUGHT DEFINITIONS

Although the effect of drought is increasing steadily all over the world, unfortunately it is not yet very well understood, because the precise definition of drought is not available and its impacts are not assessed satisfactorily. Definitions are based on the basic career (hydrologic, agricultural, meteorological, geographical) or according to industrial, energy generation, water supply, navigation, and recreational regions. Roughly, droughts are defined as the temporary reduction in the rainfall, runoff, and soil moisture amounts and they are related to the climatology of the region. Dry climates, especially, are prone to drought effects due to the soil moisture deficiency and high variability in the rainfall occurrences as well as amounts.

A precise and universally accepted definition of drought is another problem that adds to the confusion about drought existence and degree of severity. Because drought is climate dependent, its definition should also consider the local and regional climatic features ("see chapter: Basic Drought Indicators"). On the other hand, drought implies different meanings and implications for a water manager, an agriculturalist, a hydroelectric power plant operator, and a wildlife biologist.

There are many and different definitions of droughts in the literature. However, it is defined generally as an extended period of rainfall deficit during which agricultural harvests are severely curtailed. Wilhite and Glantz (1985) have classified drought definitions as conceptual (ie, relatively vague (fuzzy) and operational), which is meant to provide specific guidance on aspects of drought onset, severity, and termination. Frequently asked questions for objective drought description are:

1. When is the starting time of drought?
2. How severe is the drought?
3. When is the termination time of the drought?

Droughts have a "creeping" feature, which is very gradual; hence, their developments are slow and have prolonged existence, sometimes over many years or even decades. Droughts are not confined by local topographic features or geological structures and they are more extensive over large areas. In practice, it is most often very difficult or impossible to tell the beginning of a drought; hence, their distinctions from human-induced desertification.

Among the main natural causes of drought are climate effects, but these are not the only ones. Mismanagement of water resources, acid rain pollution, over-exploitation, and many other man-made effects also contribute to the extent, appearance, continuity, and severity of droughts.

To cope with drought at early stages, short- and long-term drought predictions provide help to decision makers so they can respond to drought occurrences in a better and more precise manner. One should be careful about drought definitions as they are in given in various publications; they may need some modification to allow them to suitably present local and regional conditions. In most cases a crisp and a single definition of drought may not be helpful in attempting to solve the problem. Each drought definition must be specific to the region, application, and impact. In the identification of a drought, not only climatic factors but also water supply and demand patterns must be considered. Drought impacts are complex and vary regionally and temporally.

Drought severity is supported, apart from the precipitation deficiency, by high temperatures, winds, and low relative humidity. In many parts of the world, drought also relates to the timing, principally season of occurrence, delays in the start of the rainy season, occurrence of rains in relation to principal crop growth stages, the effectiveness of the precipitation such as the rainfall intensity, its areal extent, and the frequency of occurrence. All these effects indicate that each drought event is unique in its climatic characteristics, spatial extent, and impacts. It is not possible to find identical drought features and effects; therefore, researchers and decision makers should consider the local characteristics in any drought modeling, prediction, and combat in addition to generally valid formulations, procedures, and software applications. One should be careful about the existence of different types of information at different times and locations. It is possible that the climate forecasts may provide drought occurrence indication several months in advance, but social and economic indicators may gain prominence at the later stages when the drought or famine sets in.

Deficit is a term used for expressing that a quantity is less than the required level. Almost all social, economic, and natural events may have some context where deficit can be used to express unwanted situations. Likewise, drought is a term that is employed for reflecting the situations where hydrometeorological quantity such as water amounts may not satisfy a desired level of demand. Droughts usually exceed 1 year and may continue even for 8–10 years. Droughts cover more extensive areas than floods and their time occurrences have creeping phenomenon, which may not give the impression, initially, that there is already accumulating drought onset. They do not have distinctive reflections of the surface features and topography, but desertification phenomenon in humid regions may have a direct relation with drought events due to human activities. Irregular, irrational, and unmanageable water resources system design; unplanned population movements; industrialization; excessive use of agricultural lands; and abstraction of excessive groundwater resources and their water table declines cause dryness in the top fertile soil with loss of greenery, scenes leading to the start of desertification and its continuation may result in drought-triggering effects. Excessive grazing of cattle in pastoral lands, use of ineffective irrigation practices, deforestation, and nonconservation of enough and suitable soil may be counted among the factors that lead to doughtiness.

1.5 DROUGHTS, ARIDITY, AND DESERTIFICATION

Dryness is a common feature in the desert and steppe areas of the world, where there are water deficits throughout many years and centuries. The environmental life in such areas is adapted to water scarcity; therefore, only water deficit-bearing plants and animals can survive. Great deserts of the world are areas of dryness and aridity. The duration of dryness, steppe, and desert conditions are time continuous and have regional extensiveness. The desert areas occur in the northern and southern flanks of the tropical regions around the equator. Aridity is a continuous climate property, whereas drought is an extreme event that occurs during some time period. Rainfall variability is very high in arid and semiarid regions.

Droughts are among the rare events related to water availability, and people start to feel drought when there is not enough water. Drought periods may correspond to scarce rainfall events, excessive runoff exploitation, and groundwater abstraction more than safely rechargeable amounts from the rainfall directly and subsequent runoff indirectly. When drought comes, everybody is concerned; if it lasts, everybody tries to do his/her best to combat it but when it goes away everybody forgets except those who have been hurt (Yevjevich, et al., 1983). In general, humans realize droughts with the start of water shortages and stresses. With the appearance of drought everybody starts to worry and do everything in their power to handle it during its elongated duration, but after its disappearance all the stresses and pains are forgotten. In other words, traditional style of life restarts without any more worry until the next drought occurrence.

One of the dramatic long-term drought impacts, combined with human activities, is the degeneration of productive ecosystems into desert in the process called desertification. It is not exclusively a consequence of drought, but it may be accelerated by droughts through such phenomena as wind action in dry years, soil erosion during drought and postdrought periods, and particularly through human activities that are responsible for poor management of land, soil, crops, and herds. The desertification areas reflect more solar irradiation than the original land by causing changes in the thermal regime of the atmosphere that may tend to extend or intensify droughts.

Even though many societies have been subject to droughts for a long time, the effects of droughts cannot be assessed completely with the proper solutions. For instance, management of water, food, irrigation, and settlement issues continue to lack proper and sufficient planning where drought risk levels, necessary precautions, and early warnings are not established. Present categorizations are based on different activities such as meteorology, hydrology, agriculture, sociology, physiology, economy, and so forth. Socioeconomic impacts are not related to water supply only, but also to the water demand; hence, it is concerned with local population, and accordingly, with water resources management. Droughts trigger socioeconomic events and become sensible due to their temporal durations and spatial coverages.

The effects of drought are more harmful in humid regions than arid regions because settlers in humid regions are not ready for water scarcity impacts. Accordingly, agricultural investments face maximum harm as a result of drought periods. Among the primary economic hazards due to the drought impacts are crop yield, husbandry, hydroelectric energy generation, decrease in industrial products, and navigation problems in low river flows. On the other hand, secondary effects include soil erosion, dust storms, forest fires, increase in plant diseases, insect hurdle, decrease in social and individual health, pollution concentration increase, deterioration in water quality, and so on. Partial and continuous migrations appear due to droughts.

Generally, droughts are expected in arid and semiarid regions due to rainfall reduction, but the most devastating ones appear in humid regions with comparatively abundant precipitation. It is, therefore, necessary to distinguish between drought and aridity. For instance, 200 mm of precipitation may be enough for husbandry but wheat production needs at least 500 mm of precipitation; hence, 200 mm of precipitation implies dry conditions for the husbandry owners. The importance lies with the precipitation variability and fluctuations. In the assessment of droughts, generally regional extents are considered with climatic features ("see chapter: Spatiotemporal Drought Analysis and Modeling"). Additionally, the relative balance of water supply and demand is also important, especially at settlement areas. If water supply is less than demand, then a drought starts to appear and it affects many social, agricultural, and industrial activities. A continuation of persistent successive periods (months, seasons, years) with water supply less than demand constitutes extensive drought durations ("see chapter: Temporal Drought Analysis and Modeling"). Even though precipitation reduction plays a role as one of the input variables in the modeling of droughts, additionally water supply, demand, and conservation enter the modeling domain. For instance, precipitation does not provide direct water supply for plants; first, the soil moisture is recharged and then plant roots carry water inside the plant body. Likewise, precipitation does not provide direct water supply or irrigation water; instead, water is provided by rivers and groundwater resources. If a region is dependent on surface water resources, then the effect of drought is felt earlier than the areas where the water supply is heavily dependent on groundwater resources. A simple but effective rule that can be derived from this statement is that surface waters can be impounded behind surface reservoirs for short-term usages. However, they must be directed toward groundwater storage through natural and artificial recharge practices for use in the long run; especially during drought periods ("see chapter: Drought Hazard Mitigation and Risk"). Groundwater resources are primary alleviation sources during any drought or any other natural hazard cases such as earthquakes. Local hydrological cycle productivity possibilities must be enhanced for such activities. Conjunctive use of surface and groundwater resources comes into play for more beneficial consequences.

It is possible to deduce rationally that, although droughts may prevail at every corner of the world, there is a reverse relationship between annual average precipitation and drought impacts. Droughts are more intensive in arid and semiarid regions for two reasons:

1. Low precipitation has extreme variability, which means that compared to the average precipitation, the standard deviation from the mean is very big, showing the unreliability in the precipitation occurrences because the amounts vary from season to season and even from year to year.

2. In dry regions the length of drought is longer. In humid regions, drought durations may be persistent for a few or several months or years as a result of deviations from the average precipitation.

In arid and semiarid regions, accumulative effect of dryness starts to show its effects after some years and may continue for long durations of 10–15 years. For instance, drought that was effective from almost 1975 onward in Europe continued for almost 16 years. Such a situation in South Africa reached the level of famine and even worse cases existed in some areas.

The interaction between the land surface and the atmosphere is also a mechanism for changing the progression of prolonged droughts, which may cause desertification. For example, any change in albedo causes a decrease in vegetation cover and reduction in water vapor transfer associated with a decrease in soil moisture and surface water volumes.

1.6 DROUGHT IMPACTS

Droughts have interrelated sociopolitical, socioeconomic, and environmental impacts. Their effects are intricately related to the environmental economic and social fabric of a region or an entire nation. The consequences are population shifts and/or reduction, alteration of the social structure, great economic hardship, and significant environmental perturbations. In order to better understand the drought concept, the following four different events must be considered:

1. *Aridity:* It is a permanent natural condition and a stable climatic feature of a region.
2. *Drought:* It refers to a temporary feature of the climate or to unpredictable climatic changes.
3. *Water shortage:* It is understood mostly as a man-made phenomenon reflecting the concern with temporary and small area water deficiencies.
4. *Desertification:* It is a part of an alteration process in the ecological regime often associated with aridity and/or drought, but principally brought about by human activities, which change the surrounding environment to a significant degree.

Droughts represent a temporary imbalance in irradiative transfer including characteristics that are persistently lower than average precipitation with uncertain frequency, duration, and severity; have unpredictable consequences; represent overall diminished water resources; and have diminished average carrying capacity of the ecosystem. Among the drought effects are deterioration of farmland and rangeland, increase in wind erosion, reduction in natural flora and fauna, air quality deterioration, brush and pest infestations, and strained water supply. Responses to droughts necessitate resource regulation, rationing and/or recycling, and institutional measures.

From a meteorological point of view, drought appears when the precipitation amounts fall below the long-term averages in a persistent manner. From the hydrological side, it is equivalent to the water supply deficit in an area. Droughts are among the natural hazards such as volcanic eruptions, landslides, avalanches, tornadoes, floods, and earthquakes that affect more than 90% of the world population. Drought causes human deaths when it reaches the famine level with unreliable and insufficient food supply. According to worldwide predictions, each year about 70,000 km^2 area enters desertification classification. Possible social drought consequences may be noted as follows:

1. *World:* famine and collective deaths, international problems and conflicts, disruption in the social systems, deterioration, and severe health problems,

2. *Country:* health problems, food scarcity, increase in costs, foreign exchange losses, increasing subsidies to farmers by the government,
3. *Region:* deterioration of economic sectors and slackness, increases in unemployment, and social disturbances,
4. *Agriculture:* migration, bankruptcy, increase in debts, decrease in revenues, and dislocation,
5. *General view:* lengthy continuation of drought and water scarcity and its extension over large areas,
6. *Artificial droughts:* These occur as a result of various human activities such as wasting water resources without proper planning, systems, or rational management, and no conservation of currently available water resources' surpluses for future drought periods through proper impoundments, and
7. *Desertification:* This is a result of human–nature interaction in an unbalanced manner with consequent temporally lengthy and really extensive droughts. To prevent temporal and spatial expansions, droughts must be predicted in a scientific manner with necessary early warnings and precaution systems.

Furthermore, drought impacts can be classified according to economic, environmental, and social affairs as follows (Vlachos and Douglas, 1983):

Economic impacts:
1. Reduction in animal and meat production
2. Reduction of agricultural lands
3. Depletion of basic stocks
4. Restriction of grazing land extent
5. Water for animals may not be obtained and it may be costly
6. Expensiveness of animal feeding
7. Farm fires
8. Traditional crop yield reductions
9. Coverage of agricultural lands due to wind erosion and deposition
10. Insect invasion
11. Plant diseases
12. Weed effects on crops
13. Forest fires
14. Tree diseases
15. Deforestation and reduction in forest production
16. Endangering fish life
17. Defects affecting tourism
18. Industrial reduction
19. Unemployment increase
20. Increase of finance risks
21. Unbalance between income and expenses
22. Reduction in navigation facilities
23. Water transportation from far distances
24. Additional water resources storage development and incurred costs

Environmental impacts:
1. Damage on animal species
2. Drinking water deficit and/or lack
3. Disease occurrences
4. Damage to fish species
5. Damages to plant types
6. Deterioration in water quality and increased saltation
7. Air quality impact
8. Land appearance deterioration

Social impacts:
1. Protection of people from forest and farm fires
2. Low water flow that may affect health
3. Quality deterioration due to waste water
4. Unemployment
5. Difficulties for property owners
6. Reduction in savings
7. Pensions impact
8. Less establishment of small family farms
9. Uncertainty
10. Reduction in recreational areas
11. Dirty streets and cars
12. Water reuse in households
13. Amusement reductions
14. Increase in religious dependence

1.7 DROUGHT REGIONS

Among the precipitation-dependent droughts are continuous, seasonal, and irregular alternatives. The following four points help to understand droughts in a better manner.

1. *Deserts:* These are naturally arid regions with continuous water deficits. Less humidity, abundant solar irradiation, important temperature changes between day and night, and low annual rainfalls are characteristics of such regions. There are limited agricultural facilities and industrial developments are subject to limited water availabilities, rare human settlements, and insufficient food products. It is necessary to alleviate the existing conditions by various activities such as soil and irrigation improvements and water transportation solutions.

2. *Dry lands:* These areas appear due to natural effects with temporary water deficits such as semiarid regions. There are symptoms such as persistent rainfall occurrences with less than average rainfall amount, uncertain duration of rainless period, areal coverage and frequency, unpredictability of the events, decreasing water resources, and also steadily deteriorating bearing capacities. Desertification of agricultural lands, wind transportations, flora and fauna deteriorations, air quality reductions, and dust and sustainable

water resources maintenance signify such lands. In order to alleviate the situation, it is necessary to regulate resources in a rational manner, by putting in and enforcing restrictions and the establishment of risk management analysis for finding the most suitable and optimum solutions that are expected to provide the necessary early warnings and precautions.

3. *Water stresses:* These are temporary water imbalances due to human activities. Overexploitation of groundwater, reduction in surface water impoundments, insufficient surface and soil moisture contents, land use improvements, groundwater recharges, and different carryover capacities are among the most significant appearances. Changes in local hydrological systems, water quality, saltwater intrusion, and rivalry among water users are the impacts. Regulation of sources, rational usage, restriction enforcements, increase in the economic value of water, institutional restrictions, protections and conservations, innovations, and land use provide alternatives for desired precautions.

4. *Desertification:* It is a land degradation problem of major importance in the arid and semiarid regions of the world. Deterioration in soil and plant cover has adversely affected nearly 50% of the land areas as a result of human mismanagement of cultivation and rangelands. North America and Spain have the largest percentage of arid lands affected by desertification. Overgrazing and deforestation are responsible for most of the desertification of rangelands; cultivation practices inducing accelerated water and wind erosions are the most responsible factors in the rain-fed croplands. Improper water management leading to salinization is the cause of irrigated land deterioration. In addition to vegetation deterioration, erosion, and salinization, desertification effects can be seen in loss of soil fertility, compaction, and crusting. Urbanization, mining, and recreation have adverse effects on the land of the same kind as is seen on ranch, dry farming, and irrigated lands. Combating desertification can be done successfully by using already known techniques, provided that financial resources are available and the political will is effective ("see chapters: Climate Change, Droughts, and Water Resources; Drought Hazard Mitigation and Risk").

1.8 DROUGHT TYPES AND THEIR IMPACTS

Drought, as more than a simple lack of rainfall, follows a persistent moisture deficiency below long-term average conditions. Herein, the average implies that precipitation balances evapotranspiration in a given area during a certain time duration. However, depending on other drought-effective factors, similar moisture deficits may lead to different consequences at the certain time of year when they occur, preexisting soil moisture content, and other climatic factors such as temperature, wind, and relative humidity. Initially, drought is defined in terms of rainfall measurements that go beyond the meteorologist's vision. For instance, hydrologic drought occurs when surface water supplies steadily diminish during a dry spell. Continuation of successive dry spells, say monthly or yearly, may cause extensive drops in groundwater levels. Furthermore, agricultural drought occurs when a moisture shortage lasts long and hits hard enough negatively on cultivated crops. Soil conditions, groundwater levels, and specific characteristics of plants also come into play in this drought functional definition. A combined effect of all types of drought may cause undesired reflections in social life; hence, they are referred to as social droughts. Finally, any drought that may give rise to human starvation is a famine. The drought types are shown simply in Fig. 1.6. The most significant components involved in each drought type are given separately in Fig. 1.7. On the other hand, Fig. 1.8 presents the consequences that can appear after each drought type.

In general, for quantitative assessments of droughts, some long-term average value is determined to be "normal" and it will be referred to as the truncation or threshold level in this book. Any drought is a complicated function of a set of factors such as timing (principal season of occurrence most often in autumn or spring, delays in usual rainfall season start) and effectiveness (rainfall frequency and intensity). The occurrences of drought types take place in sequence. Naturally, all types start with the occurrence of meteorological drought, where the precipitation amounts are less than the normal level. In practical applications the normal level is considered as the arithmetic average of 30-years, and internationally it is adapted from 1960 to 1990. This standard normal duration provides an equal basis for the comparison of drought events all over the world. However, this period may be shifted to 1970–2000 or 1980–2010 and in the future to 1990–2020. However, it is advised in this book that if data are available one should then make comparisons for each 30-year period starting from 1960 onward.

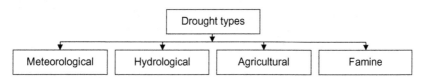

FIG. 1.6 Drought types.

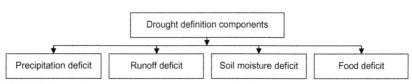

FIG. 1.7 Drought components.

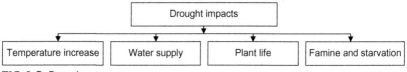

FIG. 1.8 Drought consequences.

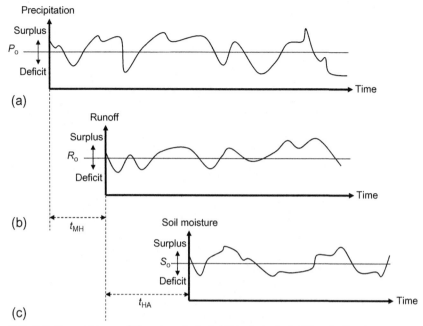

FIG. 1.9 Drought types: (A) meteorological, (B) hydrological, (C) agricultural.

Fig. 1.9 is a descriptive deposition of the relative situation of three drought types. The normal duration threshold values are P_o (R_o and S_o) for meteorological (hydrological and agricultural) drought. No need to say that all drought types originate from a deficiency of precipitation (Wilhite and Glantz, 1985), which spans an extended period of time. Following precipitation and after convenient antecedent conditions, the groundwater recharge takes place with an infiltration process and the soil saturates with water giving way to surface flow (runoff). If the amount of runoff is less than the threshold level, say R_o, then the hydrological drought starts to creep into the water resources scene. As it is obvious from Fig. 1.9 there is a time lag, t_{MH}, between the starts of meteorological and hydrological droughts. This time lag is very important for the early warning and preparedness activities prior to water shortage and stress events. Similarly, the time lag, t_{HA}, between the hydrological and agricultural drought starts may provide the necessary domain for adaptation prior to the vulnerability of agricultural drought.

Different aspects related to drought occurrences, their impacts, and necessary precautions for drought mitigation are presented in Fig. 1.10 together with adaptation, drought planning, and management alternatives.

In a drought context, most often each drought type is thought of as mutually exclusive, but they are mutually inclusive in a sequence shown in Fig. 1.11. This figure indicates that the evolution takes place from the left toward the right-hand side, but each successive pair has interference. As the meteorological drought subsides the hydrological drought takes over, and it is then followed by the

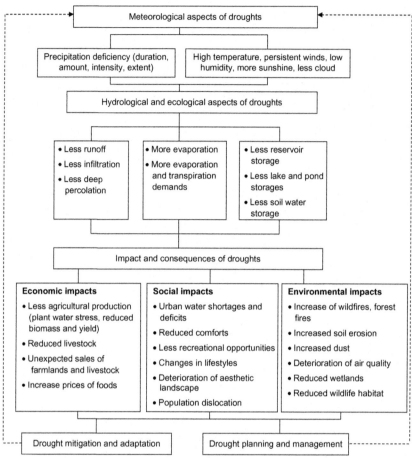

FIG. 1.10 Drought components and definitions (Salas and Pielke, 2002).

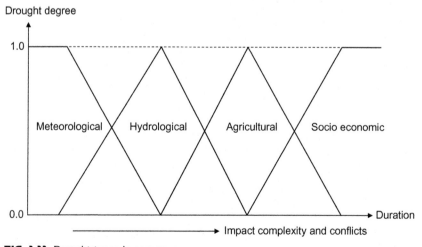

FIG. 1.11 Drought types in sequence.

agricultural drought, leading to the socioeconomical drought. Fig. 1.11 is the first indication for the fuzzy description of droughts, which will be dealt with in more detail in "chapter: Basic Drought Indicators".

1.8.1 Meteorological Drought

As mentioned earlier, meteorological droughts result from the stagnation or persistence of atmospheric high-pressure system over a region, which is typified by subsidence, clear skies, and low temperatures. Such an appearance in a region prevents normal atmospheric sequences of wet and dry weather progressions by the general atmospheric circulation.

Meteorological droughts are defined objectively by considering the rainfall temporal variations in any meteorology station (see Fig. 1.9A). As rainfall is a meteorological element and its generation depends in the background on the hydrological cycle and in the foreground on the formation of clouds and the condensation of water vapor in the clouds, these droughts can be considered as atmospherically related hazards. Rainfall deficits are the indicators of meteorological droughts. These deficits depend on many factors such as natural groundwater recharge, soil moisture sustenance, plant life support, and surface flow generation. Meteorological droughts are the initiators of other drought types; therefore, they have the least disaster potential. For instance, rainfall does not supply water to plants but the soil fed by rainfall does. Hence, even during rainfall deficit periods the soil fed by previous wet periods continues to supply plant water use. Likewise, rainfall does not supply water directly for irrigation, domestic, or industrial uses. Natural water sources are not rainfalls but rivers, lakes, and groundwater. This implies that in cases of insufficient rainfall, the demand for water can be obtained from other resources such as the surface flow, lakes, or groundwater reservoirs.

Meteorological drought starts with the decrease of successive rainfall amount (dry spells) below the long-term precipitation average in an area or location. The meteorological drought duration is also a location-dependent concept. A general definition based on precipitation amounts and duration states that, "drought is a period of more than some particular number of days with precipitation less than some specified small amount" (Great Britain Meteorological Office, 1951). The chosen specified small amount (threshold) is generally site and/or region specific, and depends on the problem under study. Care must be taken when definitions like this are applied to characterize and compare droughts in different regions. Depending on the meteorological features of each country, some countries take qualitative measurements for drought durations such as in England, for instance, where 15 consecutive rainless days are considered as the start of a drought period. More severe drought durations are considered to last more than 30 days. However, in arid regions such as the Kingdom of Saudi Arabia, any duration more than 2 years without rainfall is considered as the beginning of a drought period. In Bali, on the other hand, a 6-day duration of a rainless period means drought. In some areas drought duration is regarded as started if the rainfall amounts fall below 2.5 mm. Another approach is to consider the rainfall amount, and if 21 consecutive days have one-third of the historical rainfall, then a drought has begun. There are also other definitions that are based on the percentage of rainfall amount. For instance, if the annual rainfall is less than 60% or 70% of the actually measured rainfall amounts, then this year is considered as a dry year leading to a possible drought spell. At some other locations, 85% of the annual rainfall is taken as a threshold and such thresholds may be suggested according to the hydrometeorological convenience of the area.

Drought impacts are not important only in the planning and management of surface and subsurface storages, but also for meeting demand, if it falls short. All such definitions are subjective and do not help in prediction of any drought property for future precautions. On the other hand, lack of rainfall during the growing season is more damaging than any other period. Meteorological droughts appear as a result of rainfall reduction and there is no human inference in this phenomenon. These droughts do not have direct impacts on the economy or industry. For appreciation of meteorological drought, it is necessary to experience longer time duration than 1 or 2 years. It is necessary to distinguish such droughts from the climate change impacts that played a role in meteorological events over the last three decades. In some countries, as a relief from such droughts, weather modification, that is, cloud seeding methods, are employed. Up to now the efficiency of such a technique is not well documented ("see chapter: Drought Hazard Mitigation and Risk").

Meteorological droughts can be converted to numerical objective values through some formulas, the most famous and most widely used alternative being the Palmer Drought Intensity Indicator (Palmer, 1965; "see chapter: Basic Drought Indicators"). According to this criterion for drought classification in an area, it is necessary to have monthly and few or several years of precipitation and temperature records in addition to soil moisture balance. This shows objectively that a meteorological drought cannot be expressed by precipitation records only, but temperature consideration is also important.

1.8.2 Hydrological Drought

Hydrological droughts are more related to water demands and they appear whenever a marked reduction becomes appreciable in natural streamflow or groundwater levels, plus the depletion of water storage in dams and lakes for water supply. Hence, hydrological droughts are very important and significant for urban areas or industrialized regions as well as for agricultural activities. The main impact of hydrological droughts is on water resources systems. Hence, pollution of water resources also exacerbates the hydrological drought situation.

Hydrological drought is not attached to the precipitation decrease only, but additionally decreases in the surface flow and drops in the groundwater levels provide joint impacts. It is in effect in dry periods when water demand cannot be met sufficiently as a result of a continued dry weather condition. The drought duration, intensity, and frequency are all interrelated in any drought-stricken area (Sirdas and Şen, 2003).

Unfortunately, it is impossible to establish a direct relationship between precipitation amounts and the status of surface and subsurface water supplies in lakes, reservoirs, aquifers, and streams because these hydrological elements are used for multiple and different purposes such as irrigation, recreation, tourism, flood control, hydroelectric power production, domestic water supply, and environmental and ecosystem preservations. Time wise, a precipitation deficit uncertainty plays an additional role in prediction studies.

Most often, the definition of hydrological drought is dependent on surface water availability, but a base level is necessary also for the objective definition of these droughts (see Fig. 1.9B). For this purpose, there are many criteria based on either water demand, engineering design discharge, or statistical exceedence levels. For instance, in the case of water supply from a streamflow without any reservoir construction for water impoundment, the discharge continuity line is taken into consideration for the determination of basic level at 95% ("see chapter: Basic Drought Indicators"). By definition, drought flows are river discharges below this level, during which water abstractions must to be restricted or effluent discharges must be reduced, unless other additional sources are found.

According to the water balance calculations between water supply and demand levels, drought characteristic can be obtained temporally and spatially. In many countries, the hydrological drought impacts are on hydroelectric power generation and agricultural production. Any significant fall in the water level elevation, behind a hydroelectric power generation dam, is bound to cause reduction in the electricity supply, leading to socioeconomic consequences. On the other hand, hydrological drought affects the water quantity and quality deteriorations for irrigation purposes. Natural reduction in water quantity may imply quality deteriorations and reduction in dissolved oxygen demands, which restrict the use of water in agriculture.

To reduce the hydrologic drought impact effects, additional reservoirs can be constructed and at drought spell times impounded water can be transferred to areas needing water. For instance, in Istanbul City, the European side has more population but less rainfall than the Asian side; therefore, the drought impact appears frequently on the European side. In order to reduce this drought effect, extra water behind the Asian side dams is transferred beneath the Bosporus (see Fig. 1.12). Such engineering solutions have been in application all over the world since 1940.

The main watershed characteristics of major water supply reservoirs on both continents are given in Table 1.2 with the annual evaporation losses in Table 1.3.

During dry spell periods, evaporation losses are rather high in three of the major storage reservoirs, and therefore, during wet periods the water supply is supported primarily from these reservoirs so as to reduce overall water loss during 1 year. As a result of its location, in order to reduce the drought impact, Istanbul surface water supply reservoirs need four types of operation rules with increasing degree of complexity. These are

1. individual operation of reservoirs, hence, there are distinct operations for each of the surface water reservoirs considered,

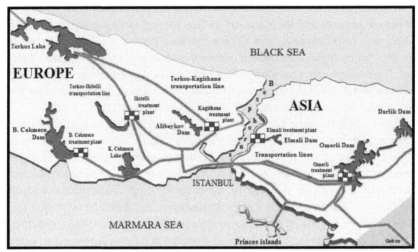

FIG. 1.12 Istanbul water distribution systems.

TABLE 1.2 Istanbul Watershed Characteristics

Continent	Watershed Name	Area (km^2)	Storage Volume ($\times 10^6$ m^3)
Europe	Terkos	619	142
	Alibeykoy	170	36
	B. Cekmece	629	100
Asia	Darlik	207	97
	Omerli	621	220
	Elmali	81	14
Total		2317	610

TABLE 1.3 Evaporation Losses

Reservoir Name	Area ($\times 10^6$ m^3)	Evaporation Height (m)		Evaporation Volume ($\times 10^6$ m^3)	
		Minimum	Maximum	Minimum	Maximum
Terkos	39	0.82	1.18	32	46
B. Cekmece	36	1.16	1.87	42	67
Omerli	23	1.22	2.01	28	46
Total		3.20	5.06	102	159

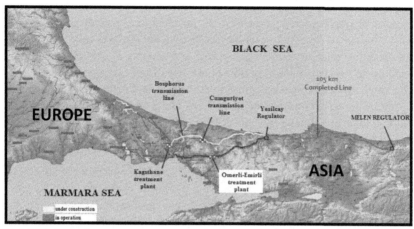

FIG. 1.13 Long-distance water transportation.

2. joint operation of the most significant two reservoirs on both sides of the city,
3. combined operation of all reservoirs on the continental basis for European and Asian sides of the city,
4. intercontinental operation of European and/or Asian side reservoirs. Management program and the practicing engineers and operators are handling the problem according to their professional know-how and in situ experience gains.

In the long run, the Istanbul's water demand is bound to increase; therefore, as shown in Fig. 1.13, water transmission lines are under construction from far distant locations (about 180 km away from Istanbul city) to transmit surface water, which otherwise flows into the Black Sea with no benefit. The local and central authorities try to augment the water resources supply facilities, but this cannot continue for sustainable water resources management and especially during drought impact periods.

Hydrological droughts are related to agricultural droughts by its effects on irrigation systems dependent on surface water and on river water quality, which is likely to be deteriorated with reduced levels of dissolved oxygen and the discharge of sediment-laden bottom water from reservoir storage. The standard solution for hydrological drought effects is the construction of surface or subsurface dams in addition to water distribution transmission lines for drought-stricken areas (Figs. 1.12 and 1.13).

Usually, hydrological drought is slow to develop and persists comparatively longer than meteorological drought. Groundwater, as one of the most significant components of the hydrological cycle, especially during drought periods, is a more conservative indicator for drought recovery. Effective groundwater resources exploitation depends on better subsurface hydrogeological conditions and understanding climatic influences. Streamflows are integrators of other indicators including soil moisture, groundwater, and precipitation. However, in many places their performances are influenced heavily by human activities because of unprecedented urbanization, development, and diversion works. For hydrological droughts, reservoir levels and storages are easy to measure, but operating rule curves may complicate assessments of drought conditions. Needless to say, experienced personnel are also of utmost importance; expects who know about the meteorological, hydrological, and agricultural activities in the area with verbal information that is very useful for successful prediction in addition to creating numerical models ("see chapters Temporal Drought Analysis and Modeling; Regional Drought Analysis and Modeling; Spatiotemporal Drought Analysis and Modeling").

Another solution or alleviation alternative for hydrologic drought offset is to recharge the groundwater resources during wet spells so that they can be withdrawn when needed. For instance, in arid regions flood or flash flood waters are injected into nearby aquifers so as to exploit them during drought periods ("see chapter: Climate Change, Droughts, and Water Resources"). The simplest solution is to identify, hydrogeologically, potential recharge locations at depression areas and direct flood waters toward such depressions, which provide temporary lakes for local users and, in the meantime, groundwater recharge can possibility take place (Şen, 2014).

Another possible short-term alleviation of drought impact is a request by the local and/or central administrators to the consumers to reduce their consumption rates and save each drop of water. This is possible at the supply and demand ends. Prior to the occurrence of drought, if the water manager expects a possible drought effect, then at times of emergency the water supply can be reduced by 10–15% physiologically without affecting consumers. If the aforementioned measures cannot alleviate the situation during the drought period, then water cuts are one of the last resorts. Needless to say, garden watering, car washes, and street cleaning by water can be forbidden until further notice.

1.8.3 Agricultural Drought

Agricultural droughts are mainly defined by considering the soil water and moisture availability to support crop growth instead of precipitation characteristics. Soil water and plant water uses are not directly related to precipitation only and its consequent infiltration into the soil, because infiltration rates vary according to antecedent moisture conditions, slope, soil type, and the intensity of the precipitation event. Soil water storage and transmission capacities vary from place to place in addition to temporal variations. Soils with low water-holding capacity are more prone to drought occurrences. Economic losses caused by droughts are mainly the reduction in the production of crops, cattle, industrial goods, waterway navigation, and hydropower.

Agriculture is a general plant growing activity for the sustenance and survival of human population on the earth. In their growing epochs, plants need a combination of solar energy from the sun, carbon dioxide from the atmosphere, and water from soil moisture. Today, at least two-thirds of all freshwater resources are used for irrigation, and consequent agricultural activities. Due to population growth, more water will be needed in the future for food security. On the one hand, rising economic and environmental problems might spur ways for increasing irrigable land coupled with extra pressure on the agricultural production from a growing population. On the other hand, the occurrence of even a slight drought condition in an agricultural area might result in large-scale hazardous effects.

Initially, agriculture may not cause significant human deaths but with its persistence migrations, social structure changes, heavy economic crises may take place with significant environmental problems. Soil moisture reduction to the level of insufficient root support is another factor for agricultural drought. Soil conditions, groundwater levels, and special characteristics of plants play a significant role in the occurrence of agricultural drought. Ecological drought is harmful to natural plants that could not be fed by irrigation. Dryness is dependent on continuous existence of dry months and years.

Comparison of crop yield amount in any year with any previous year's yield gives fuzzy impressions in drought classification as "very low," "low," "medium," "serious," and "harmful" (Şen, 2010). Specialists in agriculture benefit from such verbal categories in addition to weather and climate circumstances for defining various drought indicators ("see chapter: Basic Drought Indicators"). Measurements in one season for crop plantation are not sufficient to make predictions for long-term crop yields. Weather and climate predictions are also important in addition to crop yield properties for reliable crop plantation and subsequent crop yields. Agricultural drought and crop yield predictions are unsuccessful without long-term climate predictions (Boken and Şen, 2005).

Agricultural droughts are felt in all agricultural societies of the world including well-developed countries and in the societies where many people earn their livelihoods from agricultural production and activities. Plants need water in the soil for their growth and a shortage of this water supply results in an agricultural drought. Such water shortages may appear both in arable and postural soils of the world. It is possible to define agricultural drought if the soil moisture is insufficient to maintain average crop growth and yields; hence, direct assessment of agricultural droughts is possible by monitoring the soil moisture in agricultural lands instead of the water balance approach as with the Palmer Drought Severity Index (Palmer, 1965). The implications of agricultural droughts appear in many diverse sectors such as in the local and regional economy, social food self-sufficiency, and nutrition. Any reduction in the harvest is one of the indicators of the appearance of agricultural drought. For instance, wheat production in any region and year compared to the average crop product for the 5 previous years might indicate the existence of agricultural drought. The appearance of severe agricultural drought affects livestock, and their recovery is not possible within just a few years but, rather, on the average during 5 years or more. After an agricultural drought period, the society must bear the aftereffects for several future years.

When agricultural droughts emerge, the first reactions come from central governments and the subsidiaries start to take place as suggestions. The most important impact of the agricultural drought is on the most essential food production capacity of a region. By not being able to provide average plant water consumption, harvesting rates reduce. Agricultural drought affects many sectors outside the agriculture; hence, its impacts may appear as socioeconomic drought consequences more intensively than the hydrological drought.

Agricultural drought plays a significant role in the sustainability of a society. At the basis of agricultural drought lies food sufficiency and safety. These are fundamentals for the sustainability of a society. Unfortunately, today many societies, including the African continent, are on the verge of a final drought impact on their agricultural products. There are also societies that contend with famine. The reasons for these insufficiencies are scarcity and mismanagement of existing water resources, ignorance, and lack of information, in addition to population growth. On the other hand, in other parts of the world, there are food surpluses, and consumptive and extravagant styles of life. Agricultural drought includes deficiencies from the seeding to the harvesting stages. The soil moisture deficiency leads to earth dryness, which later turns into insufficient harvesting. The growth metabolism of plants gets affected by water deficiency; consequently, production becomes insufficient and even nonexistent due to drought effects. Plant water stress starts to end by the start of precipitation, but damages to productive agriculture and other such effects can be reduced by drought awareness, alertness, knowledge, and technological support.

In the simplest manner, agricultural drought can be defined also as a reduction in plant transpiration. Such a drought may emerge from time to time in any climate condition whether the region gets high or low precipitation. It is, therefore, necessary to separate continuous droughts in arid region climate conditions of the world from the others. One must not forget that agricultural droughts become more extensive areally after effective meteorological and hydrological droughts. The first indications of an agricultural drought are a reduction in precipitation and evapotranspiration.

In general, unconscious land use the upper basins, such as deforestation and changes in plant pattern, may worsen the surface water (runoff) quality, reduce infiltration and groundwater recharge possibilities, and hence, agricultural activities in the lower basins.

All of the aforementioned explanations point to the need for adaptation of threshold values for precipitation (meteorological drought), runoff (hydrological drought), and soil moisture (agricultural drought) calculations, modeling, and prediction studies (see Fig. 1.10). Any quantity lower than the adapted threshold value ("see chapter: Temporal Drought Analysis and Modeling") is the quantitative indicator of dry spells, and their persistent continuations lead to extended temporal and regional drought periods ("see chapter: Regional Drought Analysis and Modeling").

1.8.4 Socioeconomic Drought

NDPC (2000) explains socioeconomic drought as "Socio-economic [sic] definitions of drought associate the supply and demand of some economic good with elements of meteorological, hydrological, and agricultural droughts. It differs from the aforementioned types of drought because its occurrence depends on the time and space processes of supply and demand to identify or classify droughts. The supply of many economic goods, such as water, forage, food grains, fish, and hydro-electric [sic] power, depends on weather. Because of the natural variability of climate, water supply is ample in some years but unable to meet human and environmental needs in other years. Socio-economic [sic] drought occurs when the demand for an economic good exceeds supply as a result of a weather-related shortfall in water supply."

Drought impacts are complex, widespread, and have negative effects on societal activities. Due to several factors, societies and sectors become more vulnerable to droughts. Population increases, consumptive life styles, and urbanization require more water; if there is not a sufficient amount, then water stress and shortages take place. It is among the most significant duties of local and central administrators and decision makers to plan, operates and manage the available water resources in an optimum manner. In the meantime, researchers should direct their research activities toward the solution of such difficulties with the development of enhanced or innovative applied methodologies, if possible. If for such research studies, data availability, local knowledge, information, tools, and technological appliances are not available, then the local population cannot take measures to combat and mitigate drought effects. Extensive drought periods may give rise to local (among cities) and international (trans-boundary) water tensions that may end up with dangerous consequences for the environment, agricultural products, and societies.

Socioeconomic drought has social and economic consequences apart from the other drought types mentioned previously. It is well known that human activities are closely related to meteorological, hydrological, environmental, and agricultural events. Water supply and demand patterns, in particular, play a significant role in the occurrences of socioeconomic drought. Water supply and demand effects on the share of water and its distribution, grazing, and hydroelectric power generation, are also dependent on precipitation. During a socioeconomic drought, competition for water resources increases because different groups in the society, depending on their influence, start to ask for their priority in water supply. The policy makers are more concerned with a socio-economical drought than any other type. At times, depending on the regional features, drought consequences may not be very clear even for decision makers. As long as there is sufficient water storage (surface and subsurface) for the water supply, meteorological and hydrological droughts do not lead to socioeconomic drought problems. This is possible only in cases where demands are in balance with the availability of water supplies, even during low-precipitation seasons. Human-induced droughts may occur if water supply is comparatively less than the demand of the society. In periods of extensive droughts, the groundwater resources are overexploited and the limits of the aquifers may be reached if well-planned management procedures are not available. Whatever the situation, the decision makers should care for safe exploitation of the aquifers (Şen, 2014). The transfer of additional water resources from far distances is also possible, and in arid or semiarid regions desalination plants play the main role for water supply and demand balance.

1.8.5 Famine

Early civilizations were plagued by droughts and in some cases consequent famines. Agricultural drought is very important because of its implications for food, malnutrition, and consequently, possible famine. Extreme forms of agricultural droughts might lead to famine and mass deaths from starvation due to food insufficiency. In general, famine is defined as the "protracted total shortage of food" in a restricted geographical area, causing widespread disease and death from starvation (Dando, 1980). Famine is not a geographical but rather a cultural phenomenon. In many developed countries, severe agricultural droughts appear but with no deaths as a result of starvation, because there are stores of extra food supply from previous years or from adjacent areas without simultaneous drought impacts. Famines might also occur due to different effects such as war, social disturbances, infective diseases, or industrial explosions. They have occurred throughout human history, but today they tend to be associated with semiarid areas of subsistence or near subsistence in addition to drought-related crops.

At the most extreme situation of doughtiness famine may arise; hence, the necessary food for the sustenance of living creatures may not be produced. Wilhite and Glantz (1985) defined famine as "a process during which a sharp decline in nutritional status of at-risk population leads to sharp increases in mortality and morbidity, as well as to an increase in the total number of people at risk. It is not possible to say that the poor countries subject to drought have good nutrition and do not have health problems. As the drought is one of the extreme cases in the meteorology domain so is the famine as an extreme case of drought."

Dry areas, drought expansion, dry spells and their frequency of occurrences on daily, monthly, annual or more extensive bases are a combined result of different interrelationships. It is not sufficient to concentrate only on the meteorological and climatological features of the droughts; various other factors must be taken into consideration. Although there are many simple drought indicators, up to now there has not been a single drought indicator that includes a combination of physical and biological factors in the final decision (Gbeckor-Kove, 1989).

It is essential in the assessment of drought and desertification studies to consider precipitation areal extents ("see chapter: Spatiotemporal Drought Analysis and Modeling"). In arid and semiarid regions, pastures where cattle graze are subject to continuous threats from the weather and climate effects especially during dry periods. Subsequent dry durations cause severe drought durations that may cause desertification (Şen, 2013). For instance, Sudan, Somalia, and India had famine due to droughts in the beginning of the 1960s and toward the end of the 1970s. As a result of this hazard, thousands of people died in Africa and the effect shifted toward the east of the Great Sahara.

Ecosystems that have not been interfered with by human activities and stresses are more durable against natural hazards such as droughts. Such areas are under pressure during dry years, and especially, their attraction for agricultural activities poses additional questions. These natural events may be supported by water and wind erosion; hence, the rate of desertification increases. In summary, precipitation is the main limiting factor and, in semiarid regions, precipitation records are very important for vegetative cover, water resources, and the rate of desertification. To investigate temporal and spatial properties of drought there are many indicators that depend on precipitation records only, and more sophisticated ones such as the Palmer (1968) drought index that takes into consideration several hydrometeorological variables including precipitation ("see chapter: Basic Drought Indicators").

Famine is the most dangerous natural hazard and leads to starvation of living organisms. It appears especially in less-developed countries, and there is a need for international aid against its occurrence. Famine is the end point after the sequence of meteorological, hydrological, agricultural, and socioeconomic droughts. It is rather difficult to establish a direct relationship between drought and famine, because drought is a natural phenomenon but famine is a sociocultural event. All the drought types mentioned previously complete their dominance in a sequence and at the end famine might occur. If it is not possible to combat against famine effectively, then starvation becomes dominant in the region, leading to massive death tolls. There must be a balance between population growth and food supply. Similar to water resources deficiency, population increase (including migrations) in any region may lead to food deficiency.

Droughts vary greatly depending on the degree of water resources system and distribution development. Attached to any drought event, especially in the less-developed regions of the world, is food insufficiency due to restrictive agricultural activities, and consequent famine-related deaths in the human and living organism populations. Although famine is the most serious outcome of drought, it is not always directly related to drought. In areas of scarce water resources, and hence, limited agricultural products, the consequent insufficient food stokes malnutrition, which is the most widely spread disease in the world.

1.8.6 Water Shortages and Effects

Water stress appears during each drought period and affects water supply immediately. The first reactions come from the local government and municipalities and, in general, they look for plans and projects for better management of existing water resources and improvement of water supply. In the meantime, consumers try to minimize their water demands; hence, water saving activities start.

Water shortages occur as a result of long-duration nonrainy weather conditions. It also exists due to the failure to manage water resources in a rational manner. As a vital commodity for sustainability and survival, water is clean in nature and has cleaning properties.

Insufficiency of water in any region gives rise to different problems. For instance, during a drought, especially in households, one cannot perform cleaning operations and this may cause physiological effects on human beings. Epidemic diseases may occur if cleaning does not take place in communal places (schools, hospitals, quarters, etc.) in addition to spiritual imbalances. As a result of this, work efficiency is reduced at workplaces leading to the possibilities of economic losses. Even the tourism sector may be affected due to water shortages. Population increase brings to the front not only housing problems but also the lack of additional water supply sources, which may lead to water stress and then shortages.

During water shortages, the industrial sector cannot produce at full capacity and work output becomes lower, causing economic damage. Long-duration drought impact on water levels in dams brings with it the problem of insufficient hydroelectric power generation and its consequences in daily life, industry, and at every walk of life. Such a situation also affects irrigation possibilities and agricultural production in a negative manner. All these facts collectively impact negatively on the gross national profit and production of the region, in general, and the country, in particular.

It is necessary that sufficient water and water pressure be available at all times in fire hydrants. Water shortages endanger this need and, therefore, interfere with fighting fires, which inflict material losses in addition to spiritual and moral losses. In order to take precautions against drought consequences, the following points have both short- and long-range importance.

1. *Short-term adaptations:* With the start of drought, local and central administrators try to reduce the temporal and regional drought critical effects by adapting this approach. Among the alleviations are water cuts for short durations or reductions in water supply. Additionally, at locations under drought hazards, financial and food supports are provided to reduce the inflictions.

2. *Long-term adaptations:* Available water resources are planned according to scientific methodologies such as maintaining optimum water supply and distribution systems throughout the coming years by taking into consideration drought effects. These activities consider the water volume behind the surface dams and aquifers such that future drought possibilities can be reduced to a minimal level. For this purpose, not only one but several scenarios must be planned, and according to the actual situation, the most convenient one can be put into operation. Water supply and demand must be planned in balance over long periods. Among the solutions in the long run, water-resistant plants can be planted in anticipation of watering reductions. If possible, additional water resources possibilities (reservoirs, aquifers, lakes, water desalination) must be kept in reserve in case of emergencies due to extensive drought effects. One of the most important solutions in major cities is to reduce the leakage from water distribution networks.

Furthermore, short- and long-range adaptations must be kept in relation to each other in a harmonious manner. Drought policy must be executed simultaneously with drought planning. Governments must rely on scientific methodologies in order to alleviate drought hazards or reduce the final loss down to the absolute minimum.

In general, drought appears with unexpected water stresses and water deficits. Even though precipitation reduction is among the major factors, a decrease in soil moisture content, weakness in river flows, insufficient impoundments behind dams and their mismanagements all add up to an overall generation mechanism. A reduction in precipitation does not imply a decrease in water resources. It is necessary to consider water stresses not in an absolute sense but on a relative basis. For instance, relative considerations of recent water stresses in major cities imply that it will pass after some period of time. For this purpose, water transportation from remote regions, water conservation, water reuse, and finally, desalination facilities are among the technological remedy solutions.

1.9 SIGNIFICANT DROUGHT MITIGATION POINTS

In spite of all the inconveniences that droughts cause, they are not yet sufficiently understood in terms of characterization and impact assessment. There is no generally accepted definition of drought. Current definitions are based on different disciplines such as meteorology, hydrology, agriculture, geography, water resources development, water supply, industry, navigation, recreation, and others. From social and economic points of view, the definition of drought should consider not only water supply but also water demand. Hence, drought depends on the water use practices and the population. Drought appears when there are significant temporal or spatial water shortages in an area. If there is no question as to the definition of drought then everybody may think that they know what it is; however, if its explanation is required, then they cannot explain it to get a common agreement.

As mentioned earlier, it is not possible to identify the initiation and termination times of any drought. Accordingly, there is an endless discussion about the duration of any existing drought among the meteorologists, hydrologists, water specialists, and climatologists. Up to now, there has not been common agreement on the basis of objective drought definition and criteria among the specialists. It is, therefore, very important to define and, especially, to trace the drought behavior during its effective period by taking into consideration the views from different experts in various disciplines.

Drought is a creeping natural hazard that results from a deficiency of precipitation (runoff, soil moisture) from long-term average (over at least 30 years), which is referred to as a normal value. During such extended deficient periods over a season or longer periods, the water supply is insufficient to meet the demands of human activities and the environmental requirements for sustainable life. Although it is a natural hazard, it is accelerated by its impact on the local population and the environment. It has, therefore, a social dimension, which may penetrate into all human activities from agriculture and water supply to industrial works. Drought is regarded as a kind of climate departure from normal climatic conditions in an area. Glantz (1994) stated that drought is a normal part of climate, rather than a departure from normal climate. As such, it is complicated to control as a rare and random occurrence depending on many social, economic, and engineering aspects. Hence, its avoidance needs sound scientific approaches and methodologies, which should provide a firm basis for a drought combat policy and decision making process so as to minimize its hazardous consequences on humans and other living creatures ("see chapters: Climate Change, Droughts, and Water Resources; Drought Hazard Mitigation and Risk"). This perception has typically resulted in little effort being targeted toward those individuals, population groups, economic sectors, regions, and ecosystems most at risk (Wilhite, 2000a). Improved drought policies and preparedness plans that are proactive rather than reactive and that aim at reducing risk rather than responding to crisis is more cost-effective and can lead to more sustainable resource management and reduced interventions by government (Wilhite, 2000b). In general, hazard is used to describe the natural aspects of drought phenomenon and disaster expresses negative impacts on human and environmental situations.

Droughts are among the continuous threats to mankind in the future, as they were in the past. Apart from famine, they cause migrations, and hence, more demand on water resources. They curtail dramatically water supplies and food production by damaging agricultural systems, forests, wildlife, and especially soil. Unfortunately, there is no universal panacea and each country might experience drought impact some time in future, and accordingly, in addition to general drought curbing precautions, each country should have its own drought mitigation and combat program depending on local and regional climatological, geographical, and hydrological features. Water shortages are the primary causes and manifestations of drought anywhere in the world. Especially, in arid regions of the world, the drought duration seems almost endless. Vital water resources in the world are limited and their temporal and spatial distributions are very heterogeneous. Striving for economic growth leads to population increase, and hence, increasing reliance on water resources, and thereby, increase in the vulnerability to droughts. Droughts occur in any type of climate, whether arid or humid. Drought occurrences have extensive time and areal coverages, and accordingly, among their quantitative characteristics are drought duration, intensity, areal coverage, and vulnerability. In an agricultural sense, drought is considered as a soil moisture deficiency that has serious adverse effects on a community, which results generally in food production reduction.

Droughts have distinctive characteristics from other natural hazards in the sense that they occur slowly as a creeping phenomenon (Gillette, 1950). Hence, their effects accumulate slowly over a substantial period of time, where the onset and end of drought are difficult to determine, and scientists and policy makers often disagree on the bases (ie, criteria) for declaring an end to drought.

It starts sometime after the last rainy spell as a succession of rainless spells, which may be regarded as fine weather, especially in humid and semihumid areas. Long durations of rainless spells may not give the impression that there is an ongoing drought until the crops have withered and died. Another difficulty is in determining the end point of a drought because rainfall occurrence may not be the indication of drought end. One cannot be sure about the end duration unless there are good indications for this. The following points are among such indicators.

1. Return to normal precipitation and, if so, over what period of time does normal or above-normal precipitation need to be sustained for the drought to be declared officially over.
2. Consideration of precipitation deficits that emerge during the drought event need to be erased for the event to end.
3. Return of reservoir and groundwater level to normal or average conditions.
4. Impacts linger for a considerable time following the return of normal precipitation, so is the end of drought signaled by meteorological or climatological factors or by the diminishing negative human impact?

REFERENCES

Allaby, M., 1998. Oxford Dictionary of Plant Sciences. Oxford University Press, Oxford, UK.

Baykan, N.O., 1994. Büyük Menderes Havzası Kuraklık Eğilimleri (Büyük Menderes drought tendencies). In: Jeotermal Uygulamalar Sempozyumu'94. Pamukkale Üniversitesi, Denizli, pp. 383–397 (in Turkish).

Bender, M., Sowers, T., Brook, E., 1997. Gases in ice cores. Proc. Natl. Acad. Sci. 94 (16), 8343–8349.

Beran, M., Rodier, J.A., 1985. Hydrological aspects of drought. In: Studies and Reports in Hydrology, vol. 39. UNESCO-WMO, Paris, France.

Boken, V., Şen, Z., 2005. Techniques to predict agricultural droughts. In: Monitoring and Predicting Agricultural Droughts. Oxford University Press, USA, pp. 40–65 (Chapter 4).

Boken, V.K., Craqcknell, A.P., Heathcore, R.H., 2005. Monitoring and Predicting Agricultural Drought. A Global Study. Oxford Press, USA, 472 pp.

Bronowski, J., 1978. The Origins of Knowledge and Imagination. Yale University Press, Princeton.

Charney, J.G., 1975. Dynamics of deserts moisture index. Weatherwise 21, 156–161.

Dando, W.A., 1980. The Geography of Famine. Edward Arnold, London.

Demuth, S., Bakenhus, A., 1994. Hydrological drought—a literature review. Internal report of the Institute of Hydrology. University of Freiburg, Germany.

Dilley, M., Heyman, B.N., 1995. ENSO and disaster: droughts, floods and El Nino/Southern Oscillation warm events. Disasters 19, 181–193.

Downer, R., Siddiqui, M.M., Yevjevich, V., 1967. Applications of runs to hydrologic droughts. Hydrology paper 23, Colorado State University, Fort Collins, CO.

Dracup, A.J., Lee, K.S., Paulson, E.G., 1980. On the statistical characteristics of drought events. Water Resour. Res. 16, 289–296.

Gbeckor-Kove, N., 1989. Lectures on drought and desertification. In: Drought and Desertification. WMO/TDNo. 286World Meteorological Organization.

Gillette, H.P., 1950. A creeping drought under way. Water Sewage Works (March), 104–105.

Glantz, M.H. (Ed.), 1994. Usable science: food security, early warning, and El Niño. In: Proceedings of the Workshop held 31 October–3 November 1994 in Boulder, Colorado. National Center for Atmospheric Research, Boulder.

Great Britain Meteorological Office, 1951. The Meteorological Glossary. Chemical Publishing Co., New York, USA.

Guerrero-Salazar, P.L.A., Yevjevich, V., 1975. Analysis of drought characteristics by the theory of runs. Hydrology paper 80, Colorado State University, Fort Collins, CO.

Gumbel, E.J., 1958. Statistics of Extremes. Columbia University Press, New York, 375 pp.

Hare, F.K., 1983. Climate and Desertification: A Revised Analysis. WMO Publication, Geneva, 149 pp.

IPCC, 2007. The physical science basis. IPCC Working Group I fourth assessment report, Cambridge University Press, New York.

Llamas, J., Siddiqui, M.M., 1969. Runs of precipitation series. Hydrology paper 33, Colorado State University, Fort Collins, CO.

Manley, G., 1953. The mean temperature of Central England, 1968 to 1952. Q. J. Roy. Meteor. Soc. 79, 242–261.

Millan, J., Yevjevich, V., 1971. Probabilities of observed droughts. Hydrology paper 50, Colorado State University, Fort Collins, CO.

NDPC, 2000. Preparing for Drought in the 21st Century. National Drought Policy Commission.http://www.fsa.usda.gov/drought/finalreport/fullreport/pdf/reportfull.pdf.

Nicholson, S.E., 1980. The nature of rainfall fluctuations in subtropical West Africa. Mon. Weather Rev. 108, 473–487.

Nicholson, S.E., 1982. The Sahel: A Climatic Perspective. Club du Sahel (CILSS), Paris, 80 pp.

Palmer, W.C., 1965. Meteorological drought. US Weather Bureau research paper 45, p. 58.

Palmer, W.C., 1968. Keeping track of crop moisture conditions, nationwide: the new crop moisture index. Weatherwise 21, 156–161.

Sagan, C., Mullen, G., 1972. Earth and Mars: evolution of atmospheres and surface temperatures. Science 177 (4043), 52–56.

Saldarriaga, J., Yevjevich, V., 1970. Application of run-lengths to hydrologic time series. Hydrology paper 40, Colorado State University, Fort Collins, CO.

Salas, J.D., Pielke Sr., R.A., 2002. Stochastic characteristics and modeling of hydroclimatic processes. In: Potter, T., Colman, B. (Eds.), Handbook of Weather, Climate, and Water. John Wiley (Chapter 32).

Şen, Z., 1976. Wet and dry periods of annual flow series. J. Hydraul. Div., ASCE, Proc. Pap. 12497 102 (No. HY10), 1503–1514.

Şen, Z., 1977. Run-sums of annual flow series. J. Hydrol 35, 311–324.

Şen, Z., 1978. Autorun analysis of hydrological time series. J. Hydrol. 36, 75–85.

Şen, Z., 1980a. Statistical analysis of hydrologic critical droughts. J. Hydraul. Div., ASCE, Proc. Pap. 14 134 106 (No. HY1), 99–115.

Şen, Z., 1980b. Regional drought and flood frequency analysis: theoretical considerations. J. Hydrol. 46, 265–279.

Şen, Z., 2002. Su Bilimi Temel Konuları. (Basic Topics in Hydrology). Su Vakfı Yayınları (Turkish Water Foundation Publication), 227 pages.

Şen, Z., 2010. Fuzzy Logic and Hydrological Modeling. Taylor and Francis Group, CRC Press, New York, 340 pp.

Şen, Z., 2013. Desertification and climate change: Saudi Arabian case. Int. J. Global Warm. 5 (3), 270–281.

Şen, Z., 2014. Applied and Practical Hydrogeology. Elsevier, New York, 406 pp.

Sheffield, J., Wood, E.F., 2006. Characteristics of global and regional drought, 1950–2000: analysis of soil moisture data from off-line simulation of the terrestrial hydrologic cycle. J. Geophys. Res. Atmos. 112, D17115. http://dx.doi.org/10.1029/2006JD008288, 2007, 2036, 2039, 2040, 2050.

Sırdaş, S., 2002. Meteorolojik kuraklik modellemesi ve Türkiye uygulamasi (Meteorological drought modelling and application to Turkey. PhD Thesis, Istanbul Technical University, Istanbul, Turkey (in Turkish).

Sirdas, S., Şen, Z., 2003. Spatio-temporal drought analysis in the Trakya region, Turkey. J. Hydrolog. Sci. 48, 809–819.

Tallaksen, L.M., van Lanen, H.A.J. (Eds.), 2004. Hydrological Drought: Processes and Estimation Methods for Streamflow and Groundwater. In: Developments in Water Science, vol. 48. Elsevier Science BV, The Netherlands, 2035, 2036, 2039.

Tate, E.L., Gustard, A., 2000. Drought definition: a hydrological perspective. In: Vogt, J.V., Somma, F. (Eds.), Drought and Drought Mitigation in Europe. Kluwer Academic Publishers, The Netherlands, pp. 23–48.

Vlachos, E.L., Douglas, J., 1983. Drought impacts. In: Yevjevich, V., da Cunha, L., Vlachos, E. (Eds.), Coping with Droughts. Water Resources Publications, Littleton, CO.

Walker, M., et al., 2009. Formal definition and dating of the GSSP (Global Stratotype Section and Point) for the base of the Holocene using the Greenland NGRIP ice core, and selected auxiliary records. J. Quat. Sci. 24 (1), 3–17.

Wilhite, D.A., 2000a. Drought as a natural hazard: concepts and definitions. In: Wilhite, D.A. (Ed.), Drought: A Global Assessment. In: Keller, A.Z. (Ed.), Hazards and Disasters: A Series of Definitive Major WorksRoutledge Publishers, London, UK (Chapter 1).

Wilhite, D.A., 2000b. Preparing for drought: a methodology. In: Wilhite, D.A. (Ed.), Drought: A Global Assessment. Routledge Publishers, London, UK (Volume 2, Chapter 35).

Wilhite, D.A., Glantz, M.H., 1985. Understanding the drought phenomenon: the role of definitions. Water Int. 10, 111–120.

WMO, 2005. World Meteorological Organization Annual Report. WMO Publication.

Yevjevich, V., 1967. An objective approach to definition and investigation of continental hydrologic droughts. Hydrology paper 23, Colorado State University, Fort Collins, CO.

Yevjevich, V., Cunha, L.D., Vlachos, E., 1983. Coping with Droughts. Water Resources Publications, Littleton, CO, 417 pp.

Chapter TWO

Basic Drought Indicators

CHAPTER OUTLINE

2.1 GENERAL

The causes of major droughts begin with anomalies of less precipitation than normal in the weather or climate conditions that may affect hydropower generation and water demand for agricultural and industrial activities. Such natural phenomena can be assessed effectively if simple and practical

Applied Drought Modeling, Prediction, and Mitigation

formulations are available. Complicated models, procedures, and approaches may lead to finer resolutions of drought problems, but readily available direct verbal guidance or numerical indicators together with their corresponding verbal information are more desirable for practical applications. Droughts, among the most extreme natural events that affect society and the environment at large, can be dealt with initially via simple observations, calculations, and procedures.

In general, practical indicators describe the magnitude, duration, severity, or spatial coverage extent of drought. They are based on meteorological and hydrological variables, such as precipitation, temperature, streamflow, soil moisture, reservoir storage, and groundwater level. Several of them can be synthesized into a single indicator on a quantitative scale, often called a drought index. Although drought indices can provide ease of implementation, the scientific and operational meaning of an index value may raise questions, such as how each indicator is combined or weighted for an overall index derivation and how an arbitrary index value relates to hydrometeorological and statistical characteristics of droughts.

There are two types of drought indicators as described by Mawdsley et al. (1994); one is concerned with the environment and the other with water availability.

1. *Environmental indicators:* These are related to hydrometeorological and hydrological conditions and they help to measure the direct effect on the hydrological cycle. Among the most significant indicators are the deficits (dry spells) either in precipitation or streamflow or soil moisture or in their composition. They indicate the frequency of drought occurrences and various characteristics such as the drought duration and/or severity. The definition of these characteristics requires a threshold level, which is usually adopted as the arithmetic average or water demand level, but any other convenient level can also be adapted.

2. *Water resource indicators:* Generally, they measure severity in terms of the drought impact on water resources (surface or groundwater) that supply domestic, agricultural, and industrial uses, especially; and the impact on groundwater recharge, abstractions, and surface water levels behind the reservoirs. Water resources-related drought indicators are dependent not only on natural events, but also on the human interferences such as a steady increase in water demand versus limited supply facilities. Water resources mismanagement for water supply may also affect the drought conditions apart from a lack of rainfall or runoff. Water stresses and shortages are among the social dimensions of droughts due to mismanagement ("see chapter: Introduction").

Another definition of droughts posits two distinctive categories: conceptual droughts, where there are not specific formulations and procedures, and operational droughts that need treatment on a real-time basis including identification of the onset, severity, and termination of drought episodes ("see chapter:

Temporal Drought Analysis and Modeling"). In this context, the operational drought category is similar to the water resources indicators (Wilhite and Glantz, 1985).

The indices are useful tools for characterization and comparison of drought events regarding their timing and responses to a threshold level under circumstantial conditions. They provide information for drought management efforts provided that the operational aspects of droughts are known such as drought duration, severity, and frequency in addition to probabilistic, statistical, and stochastic features. Although there are numerous drought definitions in practical application, the most significant one is concerned with the water resources supply and demand balance. For instance, deficiency in demand may trigger a possible drought if the sequence and the intensity (amount) of the deficiencies are longer and greater than the response to general water resources. Because water resources, their exploitation, and consumption patterns differ from place to place, drought definition and calculation procedures take different forms as drought indicators. The most commonly encountered droughts are meteorological, hydrological, and agricultural types, all of which individually or collectively impact on society and economic channels; hence, the combined result appears in the form of a socioeconomic drought ("see chapter: Introduction"). There are various drought indicators (WMO, 1975a, b; Erinç, 1965; Ogallo, 1989; Rao, 1986; Al-Sefry et al., 2004) but each one has its pros and cons.

In many publications, including this book, when droughts are the main subject, quantitative and rather complex numerical methodologies and formulations come to the fore for their description. It is well understood that the models are main tools for predictions including droughts, but they need numerical data only. The models can be run provided a that numerical database is available, but unfortunately a verbal (linguistic) database is not considered in many applications. In fact, without linguistic information and a logical rule base one cannot develop meaningful models for prediction purposes (Şen, 2010). Therefore, it is emphasized in this book at various stages that linguistic data are very useful for reaching meaningful and effective conclusions, which are the primary requirements of administrators and decision makers. Drought combat, mitigation, and warnings cannot be effected without linguistic data. This chapter provides detailed information about current and practical drought indicators.

2.2 SIMPLE DROUGHT INDICATORS

Even at the initial stages of drought assessment, one should not depend directly on well-established and documented methodologies, techniques, or procedures; the preliminary reconnaissance study can be completed without involving scientific methodologies. The following simple considerations provide most of the essential and preliminary information about droughts.

1. Collect all types of information about the study area concerning drought and related aspects. These may include previous reports, papers, and gray literature in governmental offices about the area and its vicinity.
2. Conduct interviews with experienced local settlers about their views concerning past droughts and impacts. Although solutions are expected from the scientific and numerical model predictions, supplementing these with local verbal information helps to adjust the model outputs to convenient solutions for the study area.
3. Take groundwater-level measurements in observation wells, which may provide useful information about the droughts, provided that there is no interference from nearby pumping wells. In drought cases, the surface water resources are not replenished sufficiently by precipitation; consequently, administrators and decision makers seek support from groundwater resources. Properly managed groundwater resources provide support to alleviate, combat, and mitigate the drought spells (Şen, 2014a).
4. Review surface water impoundment-level records for useful information about the drought occurrences in a region. Reductions in surface water impoundment levels are among the first indicators of hydrological drought effectiveness.
5. Collect monthly rainfall data from the responsible authorities and examine them according to various simple descriptive and straightforward methods as follows:
    a. Have a look at the time series of the available rainfall records and try to interpret linguistically with simple rational and logical deductions and numerical calculations. For instance, in Fig. 2.1 the annual total rainfall record time series from the Istanbul, Turkey, Florya meteorology station (station number 17636) is presented with the mean annual rainfall amount of about 646 mm for a record duration from 1976 to 2006, inclusive. For definition of drought properties, it is first necessary to have a reference level, which is called the normal threshold level (in the statistical sense for 30-year period as explained in "chapter: Introduction") and it is considered herein simply as the mean annual rainfall amount.

    On the basis of the mean annual rainfall amount, the following information can be deduced from simple interpretations by visual examination of Fig. 2.1.
      i. There are seven drought durations on the mean annual rainfall threshold level.
      ii. The minimum duration is 1 year.
      iii. The maximum duration is observed as about 5 years during the 30-year duration of record.
      iv. The average drought duration is about 2.5 years, which can be rounded to 3 years to be on the safe side for drought mitigation studies in Istanbul City.

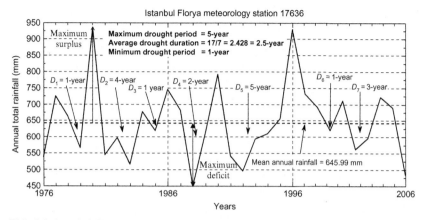

FIG. 2.1 Istanbul, Florya 17636 meteorology station time series.

    v. There is a maximum deficit during the 2-year drought period starting in 1988. Its amount is equal to $646 - 450 = 196$ mm.

    vi. The maximum surplus occurs in 1980 as $947 - 646 = 301$ mm.

    vii. It is also possible to determine the start and end times of any drought duration.

b. The histogram of the rainfall values may also provide some meaningful information about the drought possibility. The concept of a frequency diagram is a useful and simplistic procedure for interpretations. In Fig. 2.2, a frequency distribution diagram is given for the same rainfall data in Istanbul.

    One can deduce from Fig. 2.2 further useful information about droughts ("low"), floods ("high"), and especially for management ("medium") studies. The whole frequency diagram can be thought of in terms of three inclusive parts. Comparison of "low," "medium," and "high" rainfall amounts in two inclusive parts of the same time series yields useful information about the drought possibility in the location. In practical applications, the values in between are good indicators for water resources management. In the figure, it is obvious that drought possibility frequency is higher than flood possibility. Such frequency diagrams provide a basis for probabilities instead of possibilities through probability distribution functions (PDFs), as will be explained and used later in this and subsequent chapters.

c. In order to differentiate between the possibility of any change in the rainfall amounts, it is necessary first to standardize the given precipitation series, P_i ($i = 1, 2, \ldots, n$) where n is the number of measurements. In the case of Istanbul City, Florya station, there are $n = 30$ years of record. Precipitation statistical parameters, namely, arithmetic average, $\bar{P}$, and the standard deviation, S_P, do not have any effect on the

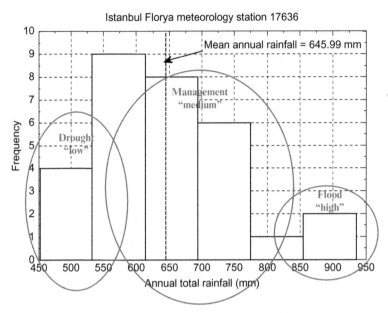

FIG. 2.2 Frequency diagram for Florya 17636 meteorology station.

time series serial structure; therefore, the standardized series, p_i, will have the same structural feature (wet and dry durations) as in the original time series. The standardization procedure is given in any statistics textbook, such as Benjamin and Cornell (1970), as

$$p_i = \frac{P_i - \bar{P}}{S_P} \tag{2.1}$$

In any drought study, it is also of interest whether there has been any decrease or increase in the "low," "medium," and "high" rainfall amounts in the meteorology station records. For this purpose, Şen (2012, 2014b) has suggested an innovative trend detection methodology that is very simple to understand and apply. The main question is whether there are significant changes in the three rainfall categories ("low," "medium," and "high"). In Fig. 2.3, the whole time series of 30-year duration is divided temporally into two halves (1976–90 and 1991–2006) and each half-record is sorted in ascending order and then they are plotted against each other. The scatter diagram in this figure indicates the relative positions of "low," "medium," and "high" rainfall amounts.

Theoretically, if the points lie on the 1:1 (45 degree) straight line, then there is no change and the measurements of each half are identical. However, this is never the case in practical applications and the points are expected to scatter around the 1:1 straight line. The upper

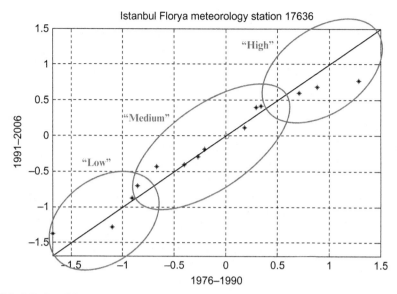

FIG. 2.3 Possible trends in Istanbul meteorology station rainfalls.

(lower) triangular area in Fig. 2.3 indicates the increments (decrements) of rainfall amounts. It is obvious from the figure that the "high" rainfall amounts have decreased during the 1991–2006 period compared to the 1976–90 period. However, all of the "low" and "medium" rainfall amounts do not show significant changes.

After the abovementioned explanations, one can conclude that so far as the water resources are concerned, there is not a significant drought effect in this meteorology station's rainfall amounts. This also implies that there is not a meteorological drought effect on the overall record duration, but as in Fig. 2.1, the average dry period has about a 3-year period; hence, the administrators and decision makers may take this point into consideration for providing water support during any possible drought duration in the future.

There is some disagreement about the definition of drought severity. In the literature of meteorological droughts, severity has been expressed through some indices. For instance, in the analyses of dry spells of monthly rainfall records in Australia, Foley (1957) suggested a drought severity index. The residual mass curve technique was used for developing this index. Other, more objective indices, such as the decile range (Gibbs and Maher, 1967) and the standardized indices (Gibbs, 1975), have also been introduced in subsequent years. The best known is the Palmer drought severity index (PDSI) (Palmer, 1965). Since its inception, the PDSI has evolved into numerous modified versions. For example, Karl (1986) has described a modified version known as the Palmer hydrological drought index (PHDI), which is used for water supply monitoring.

Another drought indicator parallel to the PDSI is the standardized precipitation index (SPI) (McKee, 1995). It has been shown that the values of the PHDI are highly sensitive to errors in the precipitation values, whereas the SPI has the potential to be a superior and yet a simpler index of drought severity than the PDSI (Guttman, 1998). The PDSI has been one of the most widely used indices in the United States, even though the SPI has advantages of statistical consistency and the ability to reflect both short- and long-term drought impacts (Hayes et al., 1999). An evaluation of common indicators, according to six criteria of performance, indicates strengths of the SPI and deciles over the PDSI (Keyantash and Dracup, 2002). Hydrological drought indicators relate to water system variables such as groundwater level, streamflow, reservoir storage, soil moisture, and snowpack. Hydrological indices include the surface water supply index (SWSI) (Shafer and Dezman, 1982) and the PHDI (Karl, 1986). The PHDI reflects longer-term hydrological impacts and the SWSI addresses some of the limitations of the Palmer indices by incorporating water supply information.

Most current drought severity indices are reviewed in this chapter. They are the Bhalme and Mooly drought index (BMDI) by Bhalme and Mooley (1980), crop moisture index (CMI) by Palmer (1968), Deciles (Gibbs and Maher, 1967), national rainfall index (NRI) by Gommes and Petrassi (1994), PDSI by Palmer (1965), percent normal (PN) suggested by Willeke et al. (1994), rainfall anomaly index (RAI) given by van Rooy (1965), reclamation drought index (RDI) by Weghorst (1996), SPI due to Mckee et al. (1993), SWSI presented by Shafer and Dezman (1982). On the other hand, are the soil moisture drought index (SMDI) suggested by Hollinger et al. (1993) and the crop-specific drought index (CSDI) by Meyer et al. (1993) and Meyer and Hubbard (1995) appeared after the CMI. Futhermore, the CSDI is divided into a corn drought index (Meyer et al., 1993) and a soybean drought index (Meyer and Hubbard, 1995). Except for these, the indices made by Penman (1948) and Thornthwaite (1948, 1963), have been used (Hayes et al., 1999). The characteristics of each index are summarized in Table 2.1. It is recognized that the PDSI is still the most widely used index, more advanced, and most scientific one. This chapter assesses the usefulness and pertinence of each index in the United States, explores their weak points, and finally, new indices are presented as a solution.

2.3 PALMER DROUGHT INDICATORS

Concerning drought conditions, there are two indicators related to the work of Palmer (1965). One of them helps to depict meteorological drought as the PDSI and the other is used for a description of hydrological droughts and it is referred to as the PHDI. The fact that PDSI gained so much attention and acceptance in the years following its development (Palmer, 1965), particularly in the United States, indicated that decision makers needed tools to monitor and respond to

TABLE 2.1 Various Drought Indices with Their Variables, Time Scales, and Concepts

Indicator	Factors	Time Scale	Main Concepts
PDSI	R, t, et, sm, rf	m	Based on moisture inflow, outflow, and storage
SWSI	P, sm	m	Similar to PDSI except sm factor
PN	r	m	Dividing actual r by the normal value
Deciles	r	m	Dividing the distribution of the occurrences over a long-term r record into sections each representing 10%
SPI	r	3m, 6m, 12m, 24m, 48m	Difference of r from the mean for a particular time and dividing it by the standard deviation
CMI	r, t	w	Like the PDSI except considering available moisture in top 1.6 m (5 ft) of soil profile
SMDI	sm	y	Summation of daily sm for a year
CSDI	ev	s	Summation of the value of calculated ev divided into possible ev during the growth of specific crops
RAI	r	m, y	R−r compared to arbitrary value +3 and −3, which is assigned to the mean of 10 extreme + and − anomalies of r
BMDI	r	m, y	Percent departure of r from the long-term mean

P, factors used in PDSI; r, precipitation; et, evapotranspiration; ev, evaporation; t, temperature; sm, soil moisture; rf, runoff; sn, snowpack; w, week; m, month; s, season; y, year; c, century; 3m, 3 months, etc.

drought events. Before the PDSI, most drought monitoring efforts used some representation of precipitation, but these were largely applicable to specific locations and not appropriate for many applications (eg, regional comparisons) (Heim, 2000). This indicator is preferable because it includes other meteorological variables in addition to rainfall (Smith et al., 1993). It is suggested by Palmer (1965) for measuring the soil moisture variability. It can make a comparison for standardized humidity conditions in an area on a monthly and weekly basis.

The PDSI and its assortment of companion indices were quickly accepted because they considered both supply and demand, even if (in retrospect) imperfectly. Palmer (1965) had attempted to develop a drought index that included a simplified two-layer soil model and a demand component affected by temperature (Heim, 2000). The index also attempted standardization for location and time, so that the values could be compared between different climate regimes. Historical calculations can easily be made, so comparisons through time at one spatial point are possible. The index provides a simple scale that decision makers and the public can associate with various levels of drought severity.

2.3.1 Palmer Drought Severity Index

This is the first comprehensive and the most frequently applied meteorological multivariable drought index in the United States (Palmer, 1965). It is based on the measurement of moisture supply departures from normal conditions, which is calculated from the water balance of a two-layer soil model using monthly mean precipitation and temperature data as well as the local available water content (AWC) of the soil.

As mentioned by Jacobi et al. (2013), in addition to more technical objections against PDSI use, there are also computational complexities and a lack of transparency. It has been already outlined by Alley (1984) that there are a multitude of computation methods, which are somewhat ambiguous procedures. Although the Palmer indices are widely applied within the United States, they have little acceptance elsewhere (Kogan, 1995).

In order to achieve an effective way of PDSI calculation it is necessary to consider the procedure at five consecutive and supportive steps or stages: hydrological, climatological, existing conditions, moisture anomaly index, and drought severity stages.

(1) *Hydrological stage:* This provides the computation of hydrological quantities by considering a set of assumptions and a water balance equation based on a long series (preferably, at least 30 years) of monthly precipitation records. For the surface layer, the water storage capacity is assumed as equal to 25 mm and the water content of the subsoil layer is dependent on the local soil properties. Potential evaporation, PE, is calculated according to the method by Thornthwaite (1948). The evapotranspiration, ET, losses from the soil is assumed to occur when the monthly precipitation, P, is smaller than PE, in which case one can calculate the surface layer evaporation loss, L_S, and the underlying layer loss, L_U, as follows:

$$L_S = \min\left[S_s, (\text{PE} - P)\right] \tag{2.2}$$

$$L_U = \frac{\left[(\text{PE} - P) - LS\right]S_u}{\text{AWC}} \quad \text{for}\,(L_u < S_u) \tag{2.3}$$

In these formulations, S_S and S_U are the available moisture at the beginning of the month in the surface and the underlying layers, respectively. Runoff is assumed to occur, if and only if, both layers are at moisture capacity.

On the other hand, if P is greater than PE then the recharge takes place and the potential recharge, PR, is given as the amount of soil moisture to hold the water capacity,

$$\text{PR} = \text{AWC} - (S_s + S_u) \tag{2.4}$$

where AWC is the available water content. Another hydrological component in the computations is the potential loss, PL, as the amount of moisture that could be lost from the soil by evapotranspiration along the zero precipitation months. Hence, its calculation can be achieved as follows:

$$PL = PL_s + PL_u \tag{2.5}$$

where each term on the right-hand side can be evaluated as

$$PL_s = \min[PE, S_s] \tag{2.6}$$

and

$$PL_u = \frac{[PE - PL_s]S_u}{AWC} \quad \text{for}(PL_u < S_u) \tag{2.7}$$

The last step in the hydrological computation stage is the runoff occurrence, RO, when $P > AWC$. The potential runoff occurrence, PRO, can be obtained by subtracting PR from AWC, which leads to the following equation:

$$PRO = AWC - PR = S_s + S_u \tag{2.8}$$

The last part of this expression can be appreciated from Eq. (2.4). The PDSI is sensitive to the AWC of a soil type. The two soil layers within the water balance computations are simplified and may not be accurately representative of a location. Thus, applying the index for a climate division may be too general. Snowfall, snow cover, and frozen ground are not included in the index. All precipitation is treated as rainfall, so that the timing of PDSI values may be inaccurate in the winter and spring months in regions where snow occurs. The natural lag between when precipitation falls and the resulting runoff is not considered. In addition, no runoff is allowed to take place in the model until the water capacity of the surface and subsurface soil layers is full, leading to an underestimation of runoff. The PDSI does not account for streamflow, lake and reservoir levels, and other longer-term hydrologic impacts of drought (Karl and Knight, 1985). Human impacts on the water balance, such as irrigation, are not considered.

In most cases the soil water content is not known. As a result the calculations are very sensitive to this factor. Its numerical values determine the drought intensity in addition to the start and ending times of wet or dry durations. However, it has very little scientific meaning. It is very sensitive to soil moisture, and therefore, provides general information for the study area. Furthermore, it provides a rather simple calculation method for the water balance between different soil elevations but cannot provide reliable information for regional works. Although this approach has been widely accepted, it is considered as an approximation only. The water balance model has been criticized on several grounds; for instance, soil moisture capacities of the two soil layers are independent of changes in vegetation.

(2) *Climatological stage:* This stage takes into consideration the long-term behavior feature of the PDSI computation in the study area. This long-term duration is adapted to the period of available historical precipitation record and helps to derive the moisture capacity of the lower layer in addition to the following four coefficients:

$$\alpha = \frac{\overline{ET}}{\overline{PE}} \tag{2.9}$$

$$\beta = \frac{\bar{R}}{\overline{PR}} \tag{2.10}$$

$$\gamma = \frac{\overline{RO}}{\overline{PRO}} \tag{2.11}$$

and

$$\delta = \frac{\bar{L}}{\overline{PL}} \tag{2.12}$$

These coefficients help to calculate the difference, d, between the precipitation under suitable climate conditions and real precipitation according to the following equation:

$$d = P - (\alpha PE + \beta PR + \gamma PRO - \delta PL) \tag{2.13}$$

(3) *Existing conditions:* This stage is a reflection of short-term drought features for reanalyzing the time series in order to calculate the moisture amount required for weather in each month. The climatically appropriate for existing conditions (CAFEC) value is computed. For example, the CAFEC value for ET_j for month j is

$$ET_j = \alpha_j PE_j \tag{2.14}$$

In this expression, PE_j is the potential evapotranspiration for the current month, j. Hence, the CAFEC precipitation value, P, is computed as

$$P = \alpha_j PE + \beta_j PR + \gamma_j PRO - \delta_j PL \tag{2.15}$$

(4) *Moisture anomaly index:* The deficit or surplus indicator for month, j, is the difference between the actual precipitation and the CAFEC precipitation, d, (Eq. 2.13), which is converted into moisture anomaly, Z, known as Z-index,

$$Z = K_j d \tag{2.16}$$

Herein, the weighting factor can be selected according to the following expression:

$$K = \frac{17.67\bar{K}}{\sum_{i=1}^{12} D_i \bar{K}}$$ (2.17)

where

$$\bar{K} = 1.5 \log \left[\frac{\frac{\overline{PE} + \bar{R} + \overline{RO}}{\bar{P} + \bar{L}} + 2.8}{D_i} \right] + 0.50$$ (2.18)

Herein, D_i is the arithmetic average of the absolute d values in month i. PDSI provides an opportunity to compare the weighting factors from one location to others.

(5) *Drought severity:* This stage includes determination of the drought severity after the Z-index time series analysis to develop criteria for the beginning and ending of drought periods. The values quantifying the intensity of drought and signaling the beginning and end of a drought or wet spell are arbitrarily selected based on Palmer's study of central Iowa and western Kansas and have little scientific meaning. During drought severity calculation, it is necessary, for each month, to calculate three intermediate indices X_1, X_2, and X_3 in addition to a probability factor. The beginning and the drought (wet period) termination can be defined in terms of the probability concerning the start and the end of the dry (wet) spell. A drought or wet spell is definitely over when this probability reaches or exceeds 100%, but the drought or wet spell is considered to have ended the first month when the probability becomes greater than 0% and then continue to remain above 0% until it reached 100%. During the period of "uncertainty" when an existing drought (wet period) may or may not be over (ie, when the probability is between 0% and 100%), the model computes the three intermediate indices X_1, X_2, and X_3. X_1 is the index value for an incipient wet spell, X_2 is the index value for an incipient drought, and X_3 is the index value for an established drought event or wet spell, respectively. All three intermediate indices are calculated using the following empirical expression:

$$X_j = 0.897X_{j-1} + \frac{Z_j}{3} \quad X_0 = 0$$ (2.19)

where Z_j represents the value of the moisture anomaly index or Z-index for the month, j. The Palmer's model selects the value of one of the intermediate indices and assigns it to PDSI depending on the value of the probability factor. For example, if the probability factor takes a value between 0 and 1, then PDSI takes the value of X_1; if the probability factor takes a value between 0 and -1, then

PDSI takes the value of X_2; and when the probability factor takes values larger than 1 or smaller than -1, then PDSI takes the value of X_3. The X_3 term responds much slower than PDSI to soil moisture changes and is an index for the long-term hydrologic moisture conditions known as PHDI. It should be noted that the Z-Index provides an indication of the persistence of the drought phenomenon, whereas PDSI denotes the drought severity.

The drought severity of ith month can be calculated by taking into consideration the summation of each dry month amounts.

$$X_i = \frac{Z}{2.691 + 0.309i} \qquad (2.20)$$

Alley (1984) has provided the calculation of ith month PDSI value from the previous month drought intensity, X_{i-1}, and the moisture anomaly, Z_i, for the current month as follows:

$$(PDSI)_i = 0.897(PDSI)_{i-1} + \frac{Z_i}{3} \qquad (2.21)$$

It can determine the start of drought after a delay of 7–8 months. All the basic terms of the water balance equation can be determined from the surface layer, including evapotranspiration, soil recharge, runoff, and moisture loss. In the calculations, human activities, such as irrigation, are not considered. PDSI values range roughly between -4.0 for "dry" and $+4.0$ for "wet" conditions. Table 2.2 provides the linguistic descriptions for "dry," "normal," and "wet" spells

TABLE 2.2 PDSI Moisture Status Categories (Palmer, 1965)

Status	Numerical	Linguistically
Dry spell	PDSI ≤ -4.0	"Extreme"
	$4.0 < PDSI \geq -3.0$	"Severe"
	$3.0 < PDSI \geq -2.0$	"Moderate"
	$2.0 < PDSI \geq -1.0$	"Mild"
	$1.0 < PDSI \geq -0.5$	"Incipient"
Normal	$0.5 < PDSI \leq 0.5$	"Near"
Wet spell	$0.5 < PDSI \leq 1.0$	"Incipient"
	$1.0 < PDSI \leq 2.0$	"Slightly"
	$2.0 < PDSI \leq 3.0$	"Moderately"
	$3.0 < PDSI \leq 4.0$	"Severe"
	PDSI ≥ 4.0	"Extreme"

corresponding to the numerical values calculated from the PDSI, PHDI, and Z-index calculations.

These values are a good match with the standard normal (Gaussian) PDF where after the two extreme values, namely, −4.0 and +4.0, there are rarely appreciable extreme values, as one can see in Fig. 2.4.

It is obvious from Fig. 2.4 that after "moderate" drought, "severe" and "extreme" drought occurrences have a very low probability of occurring. The following points are among the properties of PDSI.

1. It is a meteorological drought indicator and responds to abnormally long durations of dry and wet conditions.
2. As the conditions transit from dry to normal or wet, the drought calculations, according to PDSI hydrological impacts end on the lakes, water levels in the reservoirs, and on the other long-term effects (Karl and Knight, 1985).
3. It is calculated with precipitation and temperature data and regionally includes the water content of the soil.
4. It is possible to determine water balance basic terms from the inputs (like evaporation, soil moisture, and runoff and moisture losses from the surfaces).
5. Human impacts (like irrigation) are not incorporated in the PDSI calculations. Recent analysis methodologies along this line are explained by Alley (1984).
6. It includes also the drought duration. In this approach, along long-term averages, abnormal wet monthly value exerts a significant effect, which is not

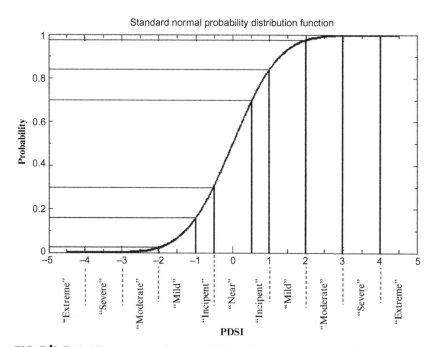

FIG. 2.4 Probability correspondences to Palmer drought severity index (PDSI).

desired at all. In a dry month series at times of nearly normal precipitation, the meaning of drought has not been understood properly.

PDSI can be applied to a station or to a set of stations in an area so as to deduce regional drought conditions. The PDSI's weaknesses and limitations have been identified over the years (Guttman, 1998; Guttman et al., 1992; Hayes et al., 1999). As pointed out by Alley (1984), there are considerable limitations and some of the most important ones can be summarized as follows:

1. The quantification of the intensity as well as the beginning and the end of a drought has little scientific meaning and they are subjectively selected.
2. The delay between the timing of precipitation and the resulting response of river flow and groundwater levels is not considered and may lag emerging drought by several months.
3. Its suitability is less favorable for mountainous areas and areas with frequent climatic extremes.
4. The sensitivity is rather high to the actual water content of a soil type.
5. The potential evapotranspiration is only approximated by applying the Thornthwaite (1948) method.
6. The PDSI is not designed for large topographic variations across a region and it does not account for snow accumulation and snowmelt.

On the other hand, Alley (1984) identified the following three primary benefits of the PDSI:

1. It provides decision makers with a measurement of the abnormality of recent weather events for a region and places current conditions in a historical perspective.
2. It also provides spatial and temporal representations of historical droughts. It has been widely used for a variety of applications across the United States. It is most effective at measuring impacts sensitive to soil moisture conditions, such as agriculture (Willeke et al., 1994). It has also been useful as a drought-monitoring tool and has been used to trigger actions associated with drought contingency plans (Willeke et al., 1994).
3. Finally, water managers find it useful to supplement PDSI values with PHDI values as a way to analyze additional hydrological information important to water management decisions.

PDSI is also useful for drought tracing, and its positive characteristics are determined by Alley (1984) as follows:

1. One finally obtains an indicator after the standardization of moisture conditions. It provides a comparison between the locations and months depending on the weight coefficients.
2. It also provides simplicity, as for the decision on the abnormality measurements of the weather conditions around the present time.
3. Additionally, it provides historical views among the continuous conditions regionally.
4. Past drought areal extents and occurrence times are also represented by this indicator.

The limitations of the PDSI involve its inability to fully characterize hydrologic, climatic, and geographical parameters and the variance in such parameters within river basins, either in the United States or in other countries (Alley, 1984; Karl and Knight, 1985).

2.3.2 Palmer Hydrological Drought Index

Recently, the PDSI has no longer been considered as a meteorological drought index but rather as a hydrologic drought index (PHDI), because moisture flow (precipitation), runoff, and storage are entered into the calculations but the long-term trend is not accounted for (Karl and Knight, 1985).

The PHDI is a modified version of the PDSI for assessing long-term moisture anomalies that affect streamflow, groundwater, and water storage. A primary difference between the two is in the calculation of drought termination, using a ratio of moisture received to moisture required to definitely terminate a drought. In the PDSI calculations a drought ends when the ratio exceeds 0. With the PHDI a drought does not end until the ratio reaches 100% (Karl, 1986; Karl et al., 1987). The PDSI and PHDI are calculated for climate divisions, typically on a monthly basis, with cumulative frequencies representing all months and all climate divisions (Karl, 1986). Table 2.3 indicates the ratio between the two indices and their corresponding linguistic descriptions as different categories with cumulative frequency percentages.

The values in this table vary spatially and temporally. Cumulative frequencies also vary depending on the region and time period under consideration (Karl et al., 1987; Nkemdirim and Weber, 1999; Soulé, 1992). For example, the category of "extreme drought," with the overall frequency of 4%, varies in frequency from less than 1% in Jan. in the Pacific Northwest to more than 10% in Jul. in the Midwest (Guttman et al., 1992; Karl et al., 1987). The Palmer indices and their water balance model have several limitations (Alley, 1984; Guttman et al., 1992; Karl, 1986; Karl and Knight, 1985). They are not particularly suitable for droughts associated with water management systems, because they exclude water storage, snowfall, and other supplies. They also do not consider human impacts on the water balance such as irrigation. The values for

TABLE 2.3 Drought Categorization

PDSI/PHDI	Category	Approximate Cumulative Frequency
0.00 to −1.49	"Near normal"	28% to 50%
1.50 to 2.99	"Mild to moderate"	11% to 27%
3.00 to −3.99	"Severe"	5% to 10%
−4.00 or less	"Extreme"	<4%

determining the severity, the beginning and end of a drought, were arbitrarily selected based on Palmer's studies of central Iowa and Kansas in the United States. The methodology used to normalize the values is only weakly justified on a physical or statistical basis. For instance, for climatic regions with a large interannual precipitation variation, the statistical measure of "normal" is less meaningful than other measures, such as the range, median, or mode of the precipitation PDF (Wilhite and Glantz, 1985). The indices are based on departures from climate normal, with no consideration of precipitation variability, so they tend not to perform well in regions with extreme variability in rainfall or runoff (Smith et al., 1993).

2.4 SURFACE WATER SUPPLY INDEX

The SWSI drought indicator was suggested by Shafer and Dezman (1982). It is designed to be an indicator of surface water conditions, and Hayes et al. (1999) described the index as "mountain water dependent," in which snowpack is a major component. There are four inputs for the SWSI: snow accumulation, precipitation, runoff, and water reservoir storage. Depending on the season, it is useful for calculating snow accumulation in winter seasons, precipitation, and water storage in the reservoir. During summer seasons in the SWSI equation, snow accumulation is replaced by runoff. The application of the SWSI resembles the PDSI, with incorporation of hydrological and climatological features into a single index value. Each component has a monthly weight assigned to it depending on its typical contribution to the surface water within that basin, and these weighted components are summed to determine a SWSI value representing the entire basin. The SWSI is also centered on zero and has a range between −4.0 for dry and +4.0 for wet extreme conditions. Due to various activities such as flow diversions or new reservoirs, the SWSI has limitations, which must be adjusted by considerations of changes in the weight of each component. Heddinghaus and Sabol (1991) indicated that it is difficult to maintain a homogeneous time series of the SWSI index. On the other hand, extreme events also cause a problem if the events are beyond the historical time series; then the index will need to be reevaluated to include these events.

Calculations are made on a monthly basis, where each variable is divided by its long-term arithmetic average for normalization and through a frequency analysis the nonexceedence probabilities are obtained. According to each component contribution to surface water (runoff) in a given drainage basin, the nonexceedence probabilities are multiplied by weighting coefficient. The summation of these weighted values concentrates on zero by subtraction of 0.5. On the other hand, to reduce into the same variation domain (between −4.2 and +4.2) with PDSI the final values are divided by 12. Shafer and Dezman (1982) gave the following equation and the calculation of SWSI is achieved on monthly basis as

$$\text{SWSI} = \frac{aP_S + bP_P + cP_R + dP_{St} - 50}{12} \qquad (2.22)$$

where P_S, P_P, P_R, and P_{St} are nonexceedence probabilities of snow, rainfall, runoff, and storage values, respectively; and a, b, c, and d are snow, rainfall, runoff, and reservoir storage coefficients, respectively. Their addition should be equal to 1, because they are percentage weights $(a+b+c+d=1)$.

Monthly data for each component are analyzed according to probabilities of occurrence, combined into an overall index, and weighted according to their relative contributions to surface water in the basin. A modified SWSI provides stronger statistical foundations to the index, with drought categories and cumulative frequencies similar to the PDSI classification shown in Table 2.4 (Garen, 1993).

If each hydrologic area is heavily influenced by snowpack, then SWSI provides water supply conditions for each area. If there are changes either in data or water supply sources, then recalculation is necessary to account for changes in the frequency distributions and the weights of each component. It is possible that there may be discontinuity in any station, which cannot be included in the calculations, and therefore, a new station must be added to the system with its frequency features. Because the calculations of SWSI are specific for each basin, the comparison among the basins or regions is not possible (Doesken et al., 1991).

The SWSI scale has been proposed by Shafer and Dezman (1982) to further develop the PDSI by considering the moisture conditions. In the PDSI for

TABLE 2.4 Drought Categories According to SWSI

Status	Numerical	Linguistically
Dry spell	PDSI ≤ -4.2	"Extreme"
	$4.2 < $ PDSI ≥ -3.0	"Severe"
	$3.0 < $ PDSI ≥ -2.0	"Moderate"
	$2.0 < $ PDSI ≥ -1.0	"Mild"
	$1.0 < $ PDSI ≥ -0.5	"Incipient"
Normal	$0.5 < $ PDSI ≤ 0.5	"Near"
Wet spell	$0.5 < $ PDSI ≤ 1.0	"Incipient"
	$1.0 < $ PDSI ≤ 2.0	"Slightly"
	$2.0 < $ PDSI ≤ 3.0	"Moderately"
	$3.0 < $ PDSI ≤ 4.2	"Severe"
	PDSI ≥ 4.2	"Extreme"

homogeneous region soil moisture, the algorithm is adjusted and "mountain water dependence" as well as "mountain snow accumulation" are also considered among the main factors. Garen (1993) modified the original SWSI procedure to incorporate water supply forecasts during the winter season. SWSI takes into consideration not only flat surface regions but also the sloppy regions with soil moisture without snow melt. It has been in use along with the PDSI with the following benefits:

1. It provides ease of data and calculation facilities against the surface water resources representative measurements.
2. It provides a convenient model, and therefore, it is in practical use.

In spite of these benefits, in the practical use of SWSI, there are also drawbacks, some of which are mentioned in the following list:

1. It has limitations in practical uses. There is a need for determining the effect on the system with addition of a new station into the system and for a new frequency analysis.
2. It is necessary to develop new algorithms for each component that will be included in the SWSI, such as the usual water level variations in the channel bed, runoff separation, or as a new water storage reservoir. It is, therefore, difficult to calculate this indicator from a single time series.
3. If the events are based on the historical time series then there may appear problems at extreme values and the indicator needs re-evaluation of the events by use of the frequency analysis.
4. Rather short and continuous hydrometeorological observation series existence affects the methodology.
5. Over extensive areas, due to significant differences from one location to another, the weights show significant differences.
6. In case of missing measurements at a station and, additionally, a few variables that may not have any measurement at all, a new frequency analysis may be needed.
7. In cases of a newly constructed dam or any reservoir in the region for each water balance case, the weights must be readjusted.
8. If there is not a priory extreme value record, then the PDF must be allocated to each one.
9. Because the SWSI has a special value for each drainage basin, it does not allow comparison with other basins.

2.5 PERCENT OF NORMAL INDICATOR

The percent of normal indicator (PNI) is one of the simplest measurements based on the rainfall record at a location. Its application is straightforward through the analysis of using percent of normal, which is very effective for a single case such as location or season. Its calculation is given as division of actual precipitation, P_a, by normal precipitation, P_n, which is accepted worldwide as at least a 30-year mean (normal period as explained in "chapter: Introduction"), and finally, multiplying by 100%.

$$PNI = 100\frac{P_a}{P_n} \qquad (2.23)$$

This expression may yield different results depending on the season and location, because the PDF of many precipitation measurements are skewed, where the mean value is not equal to the mode or median values as in the normal (Gaussian) PDF case. As the PNI is dependent on the mean value, such discrepancies are expected after each calculation. For instance, for a positively (negatively) skewed precipitation PDF, the median is less (more) than the mean, so below-average (normal) precipitation occurrences are more likely than above-normal precipitation. In general, monthly rainfalls abide with a logarithmic-normal PDF, which is positively skewed.

The PNI is quite effective for comparing a single region or season but easily misunderstood because "normal" is a mathematical construct that does not necessarily correspond with expected weather patterns. In general, precipitation measurements on monthly or seasonal scales do not abide with a normal PDF, but the use of PNI implies a normal population PDF where the mean and median are equal to each other. The actual precipitation measurements tend to accumulate around the median rather than the mean value. Consequently, if one expects average (normal) precipitation on any day, she or he will usually get a value that is below the average. On the other hand, as the normal value depends on time and location, one cannot compare the frequency of the departures from normal between time periods or locations. This makes it difficult to link a particular value of a departure with a specific impact occurring as a result. Therefore, mitigating the risks of drought based on the departures from normal is not a useful decision-making tool when used alone. The use of PNI can make it difficult to link a value of a departure with a specific impact occurring as a result of the departure, and thus to design appropriate drought mitigation and response measures (Willeke et al., 1994).

The most effective case is possible if there are records for at least 30 years, which is regarded internationally as normal period. One can then take the arithmetic average of a 30-year record and adopt it as a base indicator, which has been already applied in Fig. 2.1. The application can be done for various basic time units such as days, weeks, months, and years. The benefits of such an approach are as follows:

1. It provides better results at small-size areas and for short durations.
2. It can be understood and applied simply and varies depending on the location and season.

On the other hand, among the drawbacks of this indicator one can count the following points:

1. Use of the arithmetic average is convenient in the case of almost symmetrical rainfall PDFs such as the normal (Gaussian) PDF. Otherwise,

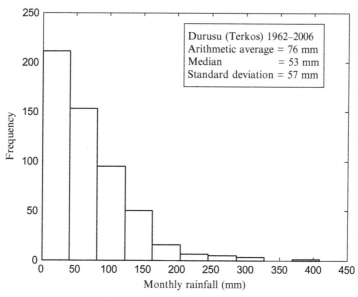

FIG. 2.5 Istanbul monthly rainfalls.

instead of mean the median value must be employed. Only in the symmetrical PDFs such as normal PDF, do the arithmetic average, median, and mode values coincide. However, in practical applications, most often automatically, the arithmetic average is taken into consideration without considering the shape of the PDF, and this causes significant errors in the end products.

2. Seasonal and monthly rainfalls do not abide with normal (Gaussian) PDF; especially in arid and semiarid regions and even in subtropical humid regions, monthly rainfall PDFs are in the form of exponential, logarithmic normal, or Gamma PDF forms. For instance, Fig. 2.5 presents the long-term monthly rainfall frequency distribution function in Istanbul, Turkey. Herein, the average value is always greater than the median value,

3. In the definition of this indicator the term "normal" is for statistical convenience and it has nothing to do with meteorology or climate.

It is obvious that the monthly rainfall PDF is not symmetric; hence, in the future, low rainfall amounts are expected to occur more frequently than "medium" or "high" rainfall amounts.

Example 2.1 PNI Calculation

As an application of this indicator, Seyhan River flow records from the southern part of Turkey are given in Table 2.5 with corresponding PNI values and classes as surplus (S) or deficit (D).

TABLE 2.5 Seyhan River Flows and PNI Values

Year	Discharge (m³/s)	PNI	Class	Year	Discharge (m³/s)	PNI	Class
1940	3106.11	291.74	S	1966	1484.42	139.43	S
1941	1316.72	123.67	S	1967	1020.12	95.82	D
1942	1245.76	117.01	S	1968	1212.79	113.91	S
1943	1077.13	101.17	S	1969	1225.61	115.12	S
1944	1265.1	118.83	S	1970	1278.91	120.12	S
1945	1156.15	108.59	S	1971	903.69	84.88	D
1946	1339.03	125.77	S	1972	761.42	71.52	D
1947	1070.57	100.55	S	1973	745.89	70.06	D
1948	1119	105.10	S	1974	680.28	63.90	D
1949	771.64	72.48	D	1975	1144.62	107.51	S
1950	635.69	59.71	D	1976	940.1	88.30	D
1951	1022.28	96.02	D	1977	921.99	86.60	D
1952	953.72	89.58	D	1978	1195.85	112.32	S
1953	1757.94	165.12	S	1979	1166.36	109.55	S
1954	847.04	79.56	D	1980	1019.35	95.74	D
1955	986.82	92.69	D	1981	1202.05	112.90	S
1956	854.8	80.29	D	1982	1279.55	120.18	S
1957	641.11	60.22	D	1983	835.24	78.45	D
1958	1147.77	107.81	S	1984	1180.63	110.89	S
1959	883.01	82.94	D	1985	837.6	78.67	D
1960	847.44	79.60	D	1986	937.62	88.07	D
1961	879.8	82.64	D	1987	871.21	81.83	D
1962	866.94	81.43	D	1988	1015.95	95.42	D
1963	1196.56	112.39	S	1989	779.67	73.23	D
1964	660.8	62.07	D	1990	838.4	78.75	D
1965	1169.7	109.87	S				

The arithmetic average, standard deviation, and median values are 1064.67, 370.57, and 1019.35 m³/s, respectively. Because these are annual flows, their underlying PDF is expected to be normal (Gaussian) as shown in Fig. 2.6 with the most suitable normal frequency distribution function.

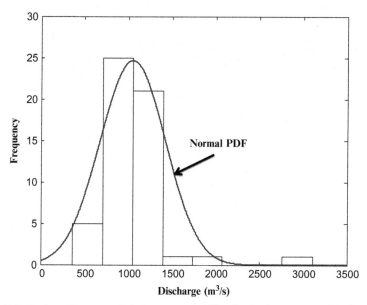

FIG. 2.6 Seyhan River annual discharges probability distribution function (PDF).

PNI values are calculated according to Eq. (2.23) and presented in the same table. A classification is also provided by considering that if $PNI > 100$ then "surplus," S, otherwise for $PNI < 100$ "deficit", D, states prevail. PNI indicates that the most severe drought year was in 1950 because it had the minimum $PNI = 59.71$. From the same table, one can identify that the 1985–90 period has the longest drought duration.

2.6 DECILE INDICATOR

The limitations in the use of PNI can be avoided to a certain extent by the Decile indicator (DI) approach, where the long-term precipitation record is divided into tenths of percentiles, called deciles. The lowest 20% is "extremely below normal"; next lowest 20% is "below normal"; middle 20% is "near normal"; next highest 20% is "above normal"; and highest 20% is "extremely above normal" (Gibbs and Maher, 1967). Fig. 2.7 indicates these five different domains on a standard normal PDF.

The DI was selected over the PDSI for the Australian Drought Watch System for simplicity, consistency, and understandability (Smith et al., 1993). For the effective and accurate calculation of the DI procedure, long climatological records (30-year duration) are needed with consistent observation. With long-term rainfall records at hand, first of all they are sorted in an ascending order; hence, the cumulative PDF is derived empirically and even, if necessary, a theoretical PDF can be matched. The vertical cumulative probability axis is then divided into equal length subintervals as 0–0.1; 0.1–0.2; …; 0.9–1.0, which

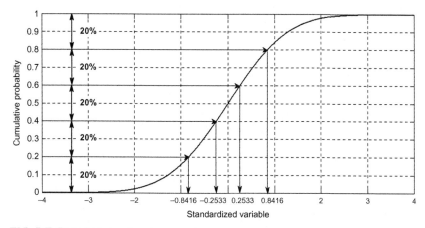

FIG. 2.7 Drought index domains.

TABLE 2.6 Dry and Wet Period Classification

Initial Conditions	Numerical	Linguistically
1–2	<20%	"Extremely below normal"
3–4	Near to 20%	"Below normal"
5–6	Next to 20%	"Near normal"
7–8	Above 20%	"Above normal"
9–10	>20%	"Extremely above normal"

are deciles (see Fig. 2.7). The first decile implies that the precipitation amount cannot be more than 10%, the second decile implies precipitation amount occurrence between 10% and 20%, and so on. It is also possible to combine two successive deciles; hence, there remain five classes as in Fig. 2.7. The classification for DI is given in Table 2.6.

Example 2.2 DI Calculation

For the application of DI, 30-year rainfall amounts are considered at one of the water supply reservoirs on the European side of Istanbul, Turkey, which is named Terkos Lake (see Fig. 1.12). The cumulative PDF of these data are calculated according to the following steps:

1. The data are first sorted from the smallest to the biggest in ascending order.
2. Each one of the ordered rainfall is attached with an integer rank number, m. Hence, the smallest rainfall has its rank as 1, whereas the longest has the number of sample, n, which is equal to 30 in this example.
3. The exceedence probability of the n ranked flows is calculated according to the following formula:

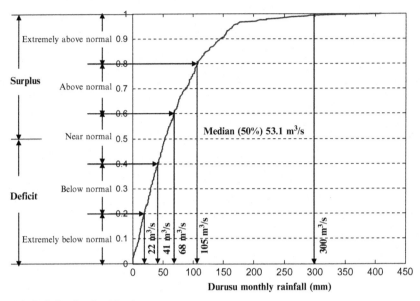

FIG. 2.8 Decile classification.

$$p_m = \frac{m}{n+1} \tag{2.24}$$

4. The plot of the ordered rainfall against their probability of exceedances gives rise to the cumulative PDF as shown in Fig. 2.8.

In Fig. 2.8, on the vertical axis the first lower (upper) five initial conditions as surplus (deficit) are shown and later the classifications are made according to Table 2.6. From the cumulative PDF, it is possible to find the limits of each decile on the horizontal axis. For instance, values that are far below the normal constitute very dry conditions and the rainfall amounts vary between 0 and 22 m³/s. Other subsets can be grouped in the same manner according to Table 2.6.

2.7 CROP MOISTURE INDEX

As a Palmer indicator derivative, the CMI reflects moisture supply in the short term across major crop-producing regions. It helps to identify potential agricultural droughts. It is not useful for long-term drought monitoring. The CMI uses a meteorological approach developed by Palmer (1968) to monitor week-to-week crop conditions. In comparison to the PDSI, which monitors long-term meteorological wet and dry spells, the CMI is designed to evaluate short-term moisture conditions across major crop-producing regions. It is based on weekly mean temperature and total precipitation within a climate division, and incorporates the CMI value from the previous week. The CMI responds rapidly to changing conditions, and it is weighted by location and time, so weekly maps, if available, can be used to compare moisture conditions at different locations.

The CMI responses to variability conditions swiftly and it is dependent on the closeness and time weights. It is calculated widely on a weekly basis and used for the comparison of moisture conditions in different regions. In order to show moisture conditions during a CMI precipitation in any drought period, CMI value interpretation may be used. Another characteristic of CMI is during the tracing of long-term drought, if its typical value is close to zero for the start and ending of drought in each season. Generally, such a limiting value prevents tracing moisture conditions outside the seasonal development. In general, drought duration exceeds a few years. The CMI is not suitable at the beginning of special plant growing season for seed development. The values of this indicator vary between −4 and +4. "extreme," "medium," "near normal," "above normal," and "abundant" water supply cases corresponding in sequence to −4, −2, 0, +2, and +4 CMI values.

2.8 ERINÇ DROUGHT INDICATOR

Erinç (1965) has developed a simple drought indicator for Turkey to describe the drought problem, for arid/humid area and duration separation purposes. His indicator considers as input variables precipitation and the maximum temperature. In his formulation, monthly average precipitation, $\bar{P}$ (mm), and maximum monthly temperatures, $T_{\max}$ (°C), values are used as follows:

$$\text{EDI} = \frac{\bar{P}}{T_{\max}} \tag{2.25}$$

Because evapotranspiration loss is very high with high monthly average maximum temperature, values less than 0°C are not considered in the calculations. Hence, unreliable effects on precipitation are overlooked, as for the case of ineffective evapotranspiration monthly temperature averages during freezing months. On the contrary, it is possible to take into account the remnant effects of snow and ice from the previous months in the current month calculations. The Erinç drought indicator (EDI) is classified by considering its results with the plant development expansion areas (see Table 2.7).

It is possible to use EDI for any duration or season. In order to determine which precipitation effective class is valid, it is necessary to multiply the EDI values by different weights depending on the duration. For instance, the 1-year indicator is multiplied by 12, the 2-month indicator by 6, the 3-month indicator by 4, the 4-month indicator by 3, and the 6-month indicator by 2.

2.9 WATER BALANCE INDICATORS

There is a common belief that the most useful drought indicators take into account water balance type of equations that adjust the difference between water supply and demand (Ogallo, 1989; Gbeckor-Kove, 1989). By taking into

TABLE 2.7 Erinç Classification

Numerical	Linguistically	Vegetation Cover
<8	"Completely dry"	Desert
8–15	"Dry"	Desertification
15–23	"Semidry"	Arid
23–40	"Semihumid"	Forest
40–55	"Humid"	Moist forest
>55	"Very humid"	Very moist forest

consideration the balance between water supply and demand, it is possible to trace drought and its calculations. For instance, a common water balance formulation for plant-soil medium is given as follows:

$$P = R + D + E_T + S + W \tag{2.26}$$

where, P, R, D, E_T, S, and W are precipitation, runoff, drainage outside the plant root layer, evapotranspiration, soil water storage change, and water accumulation within the plant body. In cases of intensive drought durations P, R, and D are close to zero; hence, the balance between E_T, S, and W shows the intensity of the drought. As already explained in Section 2.2, for evaluation of weekly soil moisture conditions, PDSI is a good example for water balance equation. This indicator is dependent on the number of months, recharge, water loss, and water balance factors. However, it is necessary to pinpoint at this stage that the most real drought indicators are the ones derived from the water balance models, but in general, there are many limiting factors that hinder the use of such approaches, including difficulties in accurate model parameter estimations. In many places, water balance input parameter estimations become unreliable or insufficient depending on the precipitation network insufficiency (WMO, 1983).

2.10 KÖPPEN DROUGHT INDICATOR AND MODIFICATIONS

The areal limitation of droughts is very important ("see chapters: Regional Drought Analysis and Modeling; Spatio-Temporal Drought Analysis and Modeling") and, in general, it is related to microclimatological features. There have been many trials in the past in order to show the precipitation effectiveness on various climatic conditions through different simple formulation approaches. The majority of these approaches depend on the relationships among precipitation amount, regime, and temperature. The first of such studies was done by

Köppen (1918), who determined the dry region boundaries according to the following formulation:

$$P = 2T + 20 \qquad (2.27)$$

where P is the annual precipitation amount in cm and T is the annual temperature in °C. This equation does not consider the precipitation pattern and, accordingly, it was modified by the same author in 1922 as follows:

At dry (B) and moist (A, B, C) climate boundary:

$$\text{Areas with winter precipitation}: P = T + 22$$
$$\text{Areas with summer precipitation}: P = T + 45 \qquad (2.28)$$
$$\text{Areas with complete precipitation}: P = T + 35$$

These were further modified by the same author in 1936 and the following equations are advised:

At dry (BW) and semidry (BS) climate boundary:

$$\text{Areas with winter precipitation}: P = T$$
$$\text{Areas with summer precipitation}: P = T + 14 \qquad (2.29)$$
$$\text{Areas with complete precipitation}: P = T + 7$$

Semidry (BS) and moist (A, B, D) climate boundary:

$$\text{Areas with winter precipitation}: P = 2T$$
$$\text{Areas with summer precipitation}: P = 2(T + 14) \qquad (2.30)$$
$$\text{Areas with complete precipitation}: P = 2(T + 7)$$

If in the most dry summer month, the precipitation amount is less than one-third of the winter precipitation and also less than 40 mm, then the station is counted as in the "winter precipitation" class. If the average precipitation amount during the most precipitation summer months is more than one-tenth of the most dry winter month precipitation amount, then the station falls within the "summer precipitation" classification. All cases outside these two are considered in the "complete precipitation" class.

2.11 MARTONNE DROUGHT INDICATOR

In climatological research, another frequently used approach was proposed by De Martonne (1926) as a drought indicator, Martonne drought indicator (MDI).

$$\text{MDI} = \frac{\bar{P}}{\bar{T} + 10} \qquad (2.31)$$

TABLE 2.8 Classification

Numerical	Linguistically
$DI_M < 10$	"Desert"
$10 < DI_M < 20$	"Semidesert"
$20 < DI_M < 30$	"Semimoist"
$DI_M > 30$	"Moist"

where $\bar{P}$, is the annual average precipitation in mm, and $\bar{T}$ is the annual average temperature in °C. With the application of this expression the bigger the indicator values, the higher the precipitation effect; that is, the climate is moist proportionally. The classification according to this indicator is given in Table 2.8. This formulation can also be used on a monthly basis.

On the other hand, according to Thornthwaite (1948) the effect of precipitation in any station can be calculated as follows:

$$M_I = \frac{100S - 60D}{PET} \qquad (2.32)$$

where M_I, S (D), and PET are the moisture indicator, the summation of monthly water surplus (deficit) over 1 year, and annual potential evapotranspiration, respectively. In moist (dry) regions its value is positive (negative). Places where this indicator has a value close to zero are among the moist and dry regions. Table 2.9 shows the moisture indicator classification related to the climate.

TABLE 2.9 Thorthwaite Classification

Numerical	Linguistically	Climate
>100	Completely moist	Moist
80–100	Moist	
60–80	Moist	
40–60	Moist	
20–40	Moist	
0–20	Semimoist	
(−20) to 0	Arid to semimoist	Arid
(−40) to (−20)	Semiarid	
< (−40)	Completely arid (desert)	

TABLE 2.10 Frequency Classification

Numerical	Linguistically	Decile Ranges
>90	"Extremely above the average"	10
80–90	"Significantly above the average"	9
70–80	"Above the average"	8
30–70	"Average"	4–7
20–30	"Below the average"	3
10–20	"Significantly below the average"	2
<10	"Extremely below the average"	1

A classification according to precipitation occurrence frequency has been in use since 1969 as developed by Gibbs and Maher (1967). In this approach, the mean is used to indicate the appearance of precipitation in percentages between the 30th and 70th percentiles. The first decile indicates either completely dry or wet conditions and the tenth decile indicates complete humidity condition. The categorization is presented in Table 2.10.

2.12 BUDYKO–LETTAU DROUGHT RATIO INDICATOR

In recent years, the Budyko–Lettau drought ratio indicator (DR_{BL}) is among the most frequently used indicators for identification of those areas that are bound toward desertification in arid and semiarid regions. The indicator that is dependent on the solar irradiation as presented by Budyko (1958) has been used by Lettau and Lettau (1969) for the DR_{BL} according to the following equation:

$$DR_{BL} = \frac{R}{L\bar{P}} \tag{2.33}$$

where R is the solar irradiation rate, L is the latent heat for water evaporation, and $\bar{P}$ is the annual average precipitation. DR_{BL} is the ratio of the net irradiation budget of the Earth's surface to the evaporative annual average precipitation amount. In the DR_{BL} calculation, annual amounts are used and heat transportation by wind is not taken into consideration (Hare, 1983). Budyko (1958, 1974) has suggested the classification in Table 2.11 for his categorization.

Additionally, Budyko has defined many dry ecosystems with plants as desert. However, Hare (1983) has adjusted Budyko's DR_{BL} limitations according to the plant growth within desertification. On the other hand, the main interest has been accumulated around 2–7 DR_{BL}, which represents land regions. These lands correspond to regions where desertification effects have significant dimensions. Hare (1983) indicated the numerical and linguistical relationships as in Table 2.12.

TABLE 2.11 Budyko Classification

Numerical	Linguistically
>3.4	"Desert"
2.3–3.4	"Semidesert"
1.1–2.3	"Arid"

TABLE 2.12 Aridity Belts

Numerical	Linguistically
>10	Complete desert. Apart from the groundwater resources or interconnected well systems, the impact of desertification is not important because of the settlements
7–10	Edge of desert. Desertification is important at belts where there are benefits from ranch animals
2–7	Semiarid belt. There is extensive desertification because of wide ranch animal husbandry and tillage of the soil
<2	Semimoist climate. Desertification is not significant

2.13 ARIDITY INDEX (AI)

Dryness is a common feature in the desert and arid areas of the world where there have been water deficits throughout many years and centuries. The environmental life in such areas is adapted to water scarcity, and therefore, only water deficit-bearing plants and animals can survive in these areas. Great deserts of the world are areas of dryness and aridity. The duration of dryness, arid, and desert conditions are time continuous and have regional extensiveness. The desert areas occur in the northern and southern flanks of the tropical regions around the equator.

Climate variability also has an important role in the drought occurrence in an area. In arid and semiarid regions there are aridity effects (Al-Sefry et al., 2004). For instance, in the Arabian Peninsula, the average annual rainfall amounts are less than 100 mm in coastal areas, less than 50 mm in desert areas, and about 500 mm in mountainous regions; hence, there is a distinctive regional variability in arid and semiarid regions, whereas in humid areas the spatial distribution has comparatively less variability. These areas are subject to intensive solar irradiation, and therefore, evaporation and evapotranspiration rates are very high, which cause continuous water loss. Additionally, rain gages density is very scarce and any study based on records in arid regions may be questioned for the possibility of the representativeness of any statistical quantity in the area.

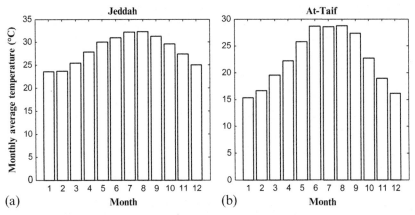

FIG. 2.9 Average monthly temperatures: (A) Jeddah, (B) At-Taif.

Another climate variable is temperature, which is a continuous variable and there are also significant temperature differences between various areas. For instance, in the Arabian Peninsula, the lowest temperature degrees are in Jan., with variability between 15 and 23°C, and the maximum temperature degrees are in Jun. within a range from 28 to 32°C. Average monthly temperature variations are given in Fig. 2.9 for two locations; namely, Jeddah and At-Taif within the Kingdom of Saudi Arabia.

The Jeddah station lies on the coastal area of the Red Sea at an elevation of about 10 m above mean sea level, whereas At-Taif station is located towards the eastern direction from the Jeddah station at about 150 km but at an elevation more than 2100 m above mean sea level. The morphological difference between these two stations makes the distinctive differences both in rainfall and temperature records. It is possible to identify in the whole year two periods with low temperature degrees from Dec. to Feb., and a high temperature period between Jun. and Aug.

A common property of arid and semiarid regions is insufficiency of rainfall; hence, dependence on groundwater resources. In these areas, the planning and strategic conservations have top priority among water resources activities especially for agriculture. In any strategic planning, the two most significant variables are rainfall and temperature. Aridity index, AI, is based on these two variables, where the ratio of average monthly rainfall, $\bar{R}$, to average monthly temperature, $\bar{T}$, is (Al-Sefry et al., 2004)

$$A_i = \frac{\bar{R}}{\bar{T}} \qquad (2.34)$$

In an arid region, generally, the AI definition has the least value equal to 0 and this corresponds to the case of no rainfall.

Example 2.3 AI Calculation

The application is performed for the Jeddah and At-Taif stations according to Eq. (2.34), where rainfall is shown in mm and temperature in °C units; accordingly, the dimension of the AI is mm/°C. The monthly aridity index for each station is shown in Fig. 2.10.

Commonly, AI value varies around 1, and in Fig. 2.10 its value approaches 2 for the Arabian Peninsula. The higher the AI, the higher the amount of rainfall during low temperatures. High AI provides groundwater recharge possibility. Jeddah, which is in the coastal area of Saudi Arabia next to the Red Sea, has AI less than 1, which is not affected significantly from the eastern Mediterranean frontal rainfall events. On the contrary, At-Taif, which is also close to the Red Sea but at an elevation more than 2100 m above mean sea level receives rainfall especially due to the monsoon effects from the Indian Ocean in Apr. and eastern Mediterranean air movements in Nov.

2.14 STANDARDIZED PRECIPITATION INDEX

In 1993, a group of scientists at Colorado State University (McKee et al., 1993) developed a new drought index, the SPI. Extensive studies showed that the PDSI is highly correlated (typically $r > 0.90$) with precipitation at certain time scales (almost always 3–12 months), and therefore, temperature added little supplementary information. Although based on precipitation alone the SPI was designed to address many of the weaknesses associated with the PDSI and intended to provide a direct answer to the questions most commonly posed by water managers. The different time scales are designed to reflect the impacts of precipitation deficits on different water resources. For instance, soil moisture conditions respond to precipitation anomalies on a relatively short scale, whereas groundwater, streamflow, and reservoir storage reflect longer-term precipitation anomalies. The SPI relies on a long-term precipitation record, typically at least for 30-year normal duration as explained in "chapter:

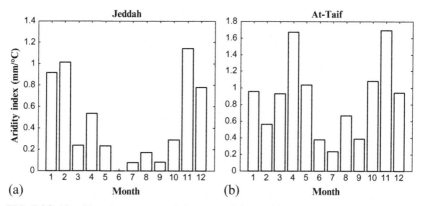

FIG. 2.10 Monthly aridity indexes: (A) Jeddah, (B) At-Taif.

TABLE 2.13 SPI Classifications

Numerical	Linguistically	Time Category (%)
0.00 to (0.99)	"Light"	34.1
(−1.00) to (−1.49)	"Medium"	9.2
(−1.50) to (−1.99)	"Intense"	4.4
< (−2.00)	"Extreme" ("Very intensive")	2.3

Introduction". This record is fitted to a PDF, such as the Gamma, logarithmic-normal, or Pearson III PDFs, so that a percentile on the fitted distribution corresponds to the same percentile on a normal (Gaussian) PDF (Panofsky and Brier, 1958). Percentile is then associated with a Z score for the standard Gaussian PDF and the Z score is the value of the SPI. The categories of the SPI, according to McKee et al. (1993), are as in Table 2.13.

The main advantage is its standardization so that the values represent the same probability of occurrence, regardless of time period, location, and climate. Equal categorical intervals have differing occurrence probabilities. For instance, the probability differential between an SPI of −1.0 and −1.5 is 9.1% as "moderate" drought; between −1.5 and −2.0, the probability differential is 4.4% as "severe" drought, even though both represent an index differential of 0.5. SPI has been designed for understanding precipitation reductions at multiple time scales. Different time scales reflect the convenience of water resources to drought conditions. Soil moisture conditions respond relatively to short-term precipitation anomalies. According to the time average duration, SPI value provides the deviation of the precipitation, P_i, amount at time instant, i, from the long term average, $\bar{P}$, and the division of this difference to the standard deviation, S_P. This is the same procedure in the statistics literature for the standardization of a given time series, which is here the precipitation measurements as $P_1, P_2, \ldots, P_i, \ldots, P_n$, where n is the number of data. The SPI, which is shown as p_i, is defined according to Eq. (2.1) for simple drought interpretations. Such a standardization procedure provides the following benefits:

1. The arithmetic average of the standardized sequence, $\bar{SPI}_i$, is zero.
2. Its standard deviation is equal to 1.
3. The standard series has randomly distributed positive and negative deviations, where positives (negatives) imply wet (dry) spells. The sequence of consecutive negative values shows dry periods (drought duration) and positive wet durations.
4. The standardized sequence does not have any dimension.

The partition of standard SPI normal PDF is shown in Fig. 2.11.

In this manner, the SPI is useful for calculating duration of the drought, its total deficit, and its intensity. SPI shows the relationship neither with regional drought nor with other meteorological variables in a standard manner. Its results

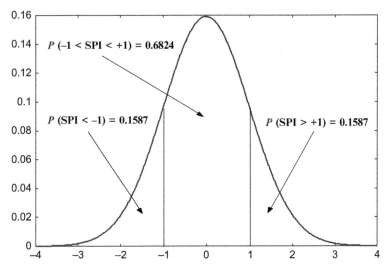

FIG. 2.11 Standardized precipitation index (SPI) standard normal PDF (*P* indicates probability).

provide only the time variability of the drought and their properties. In practical efforts, the general application of the SPI requires that the precipitation records are in accordance with the normal PDF, but especially for precipitation records less than annual duration, their distribution does not obey the normal (Gaussian) PDF. The SPI yields the shape of the precipitation data distribution at hand, precipitation percentages, and the total deficit during any drought duration. On the basis of SPI, drought can be defined as the sequence of successive dry spells, where the standardized precipitation value is less than zero and even smaller than −1. On the other hand, any drought starts with the fall of SPI to less than zero level and ends with the appearance of a value more than zero level. Drought intensity can be defined according to the classification given in Table 2.13.

If the time duration is as short as 3 or 6 months, then SPI yields many values above and below the zero level. During the basic time durations of 12, 24, and 48 months, SPI responds rather slowly to the changes in the precipitation. In Table 2.14 different cumulative probability values are given corresponding to the SPI values.

From the SPI structure, researchers in any part of the world can identify the significance of drought at any definite time scale, or wet period anomalies can also be determined. For the first time Thom (1958) suggested that the most convenient PDF for precipitation records is the Gamma PDF (Şen and Al-Jadid, 1999). This PDF has positive *x* values only and its mathematical formulation is given as follows:

$$f(x) = \frac{1}{\beta^{\alpha} \Gamma(\alpha)} x^{\alpha-1} e^{-x/\beta} \tag{2.35}$$

TABLE 2.14 SPI and Probability Values

Numerical	Cumulative Probability
−3.0	0.0014
−2.5	0.0062
−2.0	0.0228
−1.5	0.0668
−1.0	0.1587
−0.5	0.3085
0.0	0.5000
+0.5	0.6915
+1.0	0.8413
+1.5	0.9332
+2.0	0.9772
+2.5	0.9938
+3.0	0.9986

Here, $\alpha > 0$ and $\beta > 0$ are shape and scale parameters, respectively. $\Gamma(x)$ is the Gamma function given as

$$\Gamma(\alpha) = \int_0^\infty y^{\alpha-1} e^{-y} \mathrm{d}y \tag{2.36}$$

This PDF has most frequently right (positive) skewness as shown in Fig. 2.12.

SPI depends on matching Gamma PDF to a given precipitation series. In the meantime, for each station with α and β parameters in addition to certain time duration such as 3-month, 12-month, 48-month, and so on, one can make predictions. The estimation of these PDF parameters can be achieved through the maximum likelihood principle according to the following formulations (Thom, 1966):

$$\hat{\alpha} = \frac{1}{4A}\left(1 + \sqrt{1 + \frac{4A}{3}}\right) \tag{2.37}$$

$$\hat{\beta} = \frac{\bar{x}}{\hat{\alpha}} \tag{2.38}$$

and

$$A = \ln(\bar{x}) - \frac{\sum \ln(x)}{n} \tag{2.39}$$

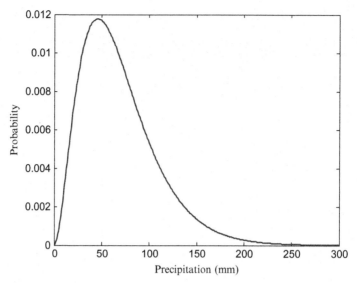

FIG. 2.12 Gamma PDF.

where n indicates the number of precipitation. After the calculation of these parameters for a certain time scale, they are used for the cumulative distribution function (CDF) of observed precipitation as follows:

$$G(x) = \int_0^x g(x)dx = \frac{1}{\hat{\beta}^{\hat{\alpha}}\Gamma(\hat{\alpha})}\int_0^x x^{\hat{\alpha}-1}e^{-x/\hat{\beta}}dx \qquad (2.40)$$

or $t = x/\beta$ transformation leads to

$$G(x) = \frac{1}{\Gamma(\hat{\alpha})}\int_0^x t^{\hat{\alpha}}e^{-t}dt \qquad (2.41)$$

Thom (1958) has given incomplete Gamma PDF, $G(x)$, values in the form of a table. This function is unknown for any x. Any precipitation distribution includes also zero precipitation values; in this case the CDF can be written as follows:

$$H(x) = q + (1-q)G(x) \qquad (2.42)$$

Herein, q indicates the probability of precipitationless duration. If a sequence has length, n, with m zero precipitation number then one can calculate $q = m/n$. The CDF, $H(x)$, with its arithmetic average equal to 0 and variance 1, is converted to standard normal (Gaussian) CDF, Z-score, which are known as the SPI values. Panofsky and Brier (1958) have presented the way to convert any PDF, say Gamma PDF, to a standard normal PDF.

Example 2.4 SPI Calculation

In the following sequence the SPI methodology has been applied to the monthly precipitation amounts at Terkos Station, Istanbul, Turkey, starting from 1962 onward. Such a comprehensive study has been achieved by Sirdas and Şen (2003) for the European side of Turkey (see Fig. 1.9). Herein, 1-, 3-, 6-, 12-, 24-, and 48-month duration successive precipitation arithmetic averages are considered. Naturally, the first duration (1-month) data are direct monthly precipitation values. The software written through the MATLAB® program by the author has been explained step-by-step as follows:

1. For the sorted sequence of the given rainfall series the corresponding probabilities are calculated from Eq. (2.24). The CDF, which is denoted by x, is plotted as in Fig. 2.13A. Hence, the precipitation variation interval is read from the horizontal axis; for instance, $0 \leq x \leq 450$. This variation domain is digitized in the MATLAB software by steps 1, increments as a statement, $x = 0:1:450$.
2. The precipitation data less than 1-year duration is assumed to abide with the Gamma PDF.

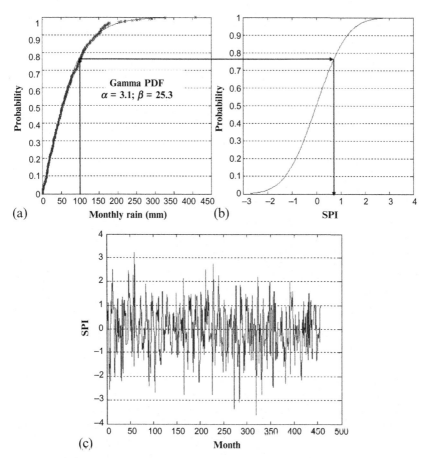

FIG. 2.13 (A) Monthly observation and theoretical Gamma cumulative distribution function (CDF), (B) SPI CDF, (C) SPI time variation.

3. The data at hand is denoted shortly in the MATLAB software as "**data**" and by using **gamfit (data)** statement; α and β parameters are obtained for the Gamma PDF. These parameters are given in Fig. 2.13A.
4. The most suitable theoretical Gamma CDF is fitted to the data in Fig. 2.13A. The convenient MATLAB statement for this procedure is $y =$ **gamcdf(x,α,β)**, where α and β parameter numerical values are substituted. Herein, y is the nonexceedence probability of given x precipitation amount.
5. To overlay the theoretical CDF over the data in Fig. 2.13A, first of all **hold on** statement and then the **plot(x,y)** is executed. Hence, the smooth curve of Gamma CDF is obtained as in Fig. 2.13A. It is now possible to visualize how close this theoretical curve is to the data scatter points.
6. Now, it is the time to convert this Gamma CDF into a standard normal PDF for SPI analysis. For this purpose, first of all it is necessary to calculate the corresponding Gamma CDF nonexceedence probability, p, for each **data** value. The probability can be calculated simply by the statement $p =$ **gamcdf (data,α,β)**, where α and β parameters are the numerical values calculated at step 3.
7. These calculated probability values should also be valid for the standard normal (Gaussian) PDF. The probability and the corresponding SPI values are calculated by the **SPI = norminv(p,0,1)**. The resulting normal (Gaussian) PDF is given in Fig. 2.13B.
8. To check the standardization of the SPI values, if the MATLAB statement **plot(SPI,p)** is executed, then the normal (Gaussian) CDF as SPI is given in Fig. 2.13B.
9. Finally, to show the time variation of the SPI values, MATLAB statement **plot(SPI)** is executed leading to Fig. 2.13C, which provides a basis for the classifications in Table 2.13 by using the values on the vertical axis.

All these steps are repeated for 3-, 6-, 12-, 24-, and 48-month durations in a similar manner and the resulting graphs are given in Figs. 2.14–2.18, respectively.

Comparison of Figs. 2.14–2.18 indicates that as the duration increases there is increase in the drought effect. For instance, it is possible to find at -1 level how many durations are successively less than this value. This hides in it all the features including length, intensity, and frequency of a drought period. The calculation of these parameters is presented in "chapter: Temporal Drought Analysis and Modeling, Section 3.4".

2.14.1 Pros and Cons of SPI

SPI provides early warning for different time scales, helps to calculate the drought, and it is simpler to use than PDSI. However, depending on the initial values and information, the conclusions and interpretations may differ significantly. SPI considers the low precipitation values, groundwater, reservoir and soil moisture storages, in addition to streamflow and snow melt. Some of the benefits from the use of SPI are as follows:
1. It is precipitation dependent only.
2. Its calculation is simple.
3. It is related to probability only.
4. It calculates precipitation deficiency during the wet and dry durations.
5. It starts to calculate percentage of the values based on the average of the first time scale.
6. It has a normal PDF, and therefore, it can trace the wet period as it does for the dry period.

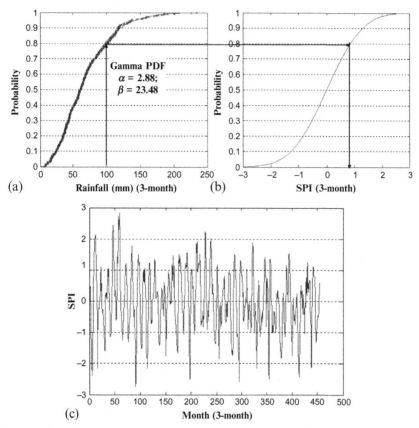

FIG. 2.14 (A) Three-month observation and theoretical Gamma CDF, (B) SPI CDF, (C) SPI time variation.

7. It can be calculated for different variables such as snow accumulation, water impoundments, streamflow, and soil moisture and groundwater storage.

In Eq. (2.1) the standard deviation of the data in a given region varies according to the homogeneity within the region. In homogeneous areas (either very wet or dry) the standard deviation does not vary from one location to another. In this case the fluctuations are rather small around the mean value. In homogeneous regions the standard deviation is small, and accordingly, the deviations from the mean are also small; therefore, the SPI values will also be small. On the contrary, in a normally wet region the standard deviation will have comparatively higher value and, accordingly, deviations will be different and, consequently, the SPI values will have significant deviations at different scales.

In the SPI applications by Sırdaş (2002) the most significant problem was that SPI does not provide sufficient information at regions of low or intensive drought. McKee et al. (1993) provided only four domains for the SPI classification, as in Table 2.13. The classifications are proposed on the basis that the

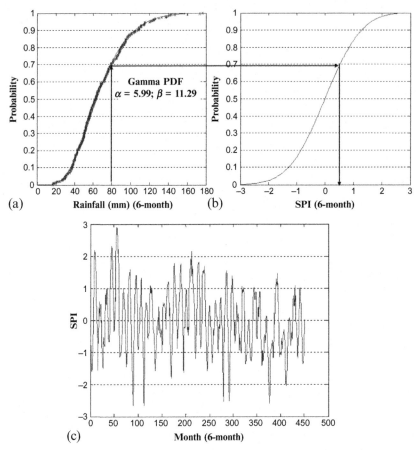

FIG. 2.15 (A) Six-month observation and theoretical Gamma CDF, (B) SPI CDF, (C) SPI time variation.

standard deviation is equal to 1. This has been considered as constant for all the stations. It has been observed that very dry regions, due to homogeneity in the data, appear as "low," "dry" but "low" and "medium" dry regions, and may appear as "dry" regions (Sirdas and Şen, 2003; Şahin and Sırdaş, 2002).

Even with the development of the SPI and the SWSI, four major limitations to drought monitoring remain, as mentioned by Wilhite and Pulwarty (2005) and enumerated in the following items.

1. *Temporal frequency of data collection:* Most changes are slow, but drought status can change appreciably in the course of a day (eg, with heavy rain or snow), or over a few days to a week (eg, spells of high heat, low humidity, high winds, significant evaporation or sublimation of snowpack, sensitive phonologic stages, shallow soil moisture depletion by high evapotranspiration demand, and so forth). Thus, daily updates represent the right frequency of new information. In many instances, data are measured and collected only

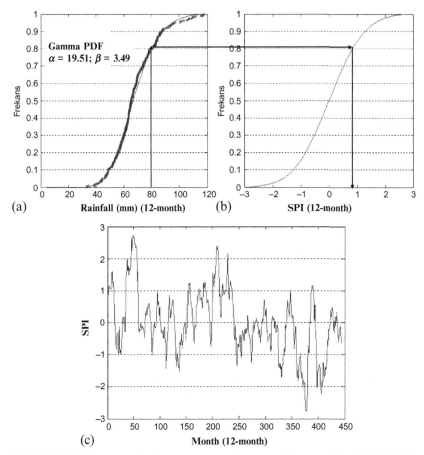

FIG. 2.16 (A) Twelve-month observation and theoretical Gamma CDF, (B) SPI CDF, (C) SPI time variation.

on a monthly, or sometimes weekly, basis and are often unavailable for several days or weeks because of manually intensive processing methods. Concerns about the quality of near–real-time data and the quality control process involved in avoiding usable recent data have also contributed to these temporal limitations.

2. *Spatial resolution:* In most cases spatial resolution has been at a coarse, regional scale. Much of the climate information is organized by climate divisions, which may fail to provide the required spatial detail of drought conditions needed by decision makers, especially where topographic gradients predominate.

3. Use of a single indicator or index to represent the diversity and complexity of drought conditions and impacts: Decision makers need multiple indicators to understand the spatial pattern and temporal time scales within and between regions.

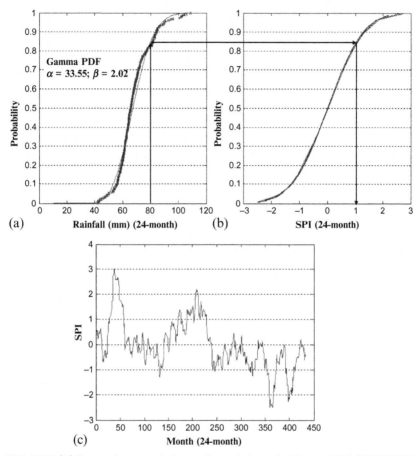

FIG. 2.17 (A) Twenty-four-month observation and theoretical Gamma CDF, (B) SPI CDF, (C) SPI time variation.

4. Lack of reliable drought forecasting products: To respond effectively, decision makers need to anticipate the development and cessation of a drought event and its progression.

2.15 TYPICAL PROBLEMS WITH INDICATORS AND TRIGGERS

In practical applications, single indicators provide inadequacy for decision makers and, therefore, multiple indicators and triggers must also be used as supportive tools. Often, indicator scales may be incomparable and trigger values may be statistically inconsistent. Comparison of the three indices illustrates common problems with indicators and triggers in drought plans. These problems exist not only for values of indices such as SPI, PDSI/PHDI, and SWSI, but also for values of indicators such as total monthly precipitation,

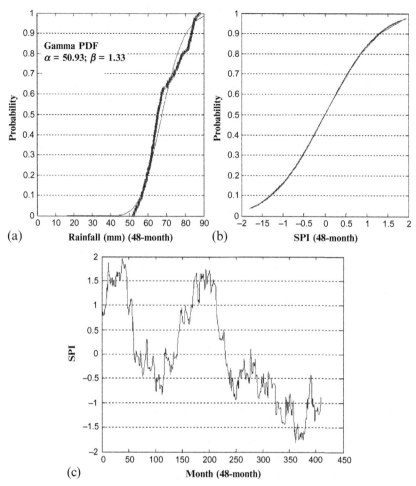

FIG. 2.18 (A) Forty-eight-month observation and theoretical Gamma CDF, (B) SPI CDF, (C) SPI time variation.

average monthly streamflow, and average monthly reservoir levels for different reasons.

There is no consistency between different drought indicators, although many of them are based on the CDF, but their classification percentages are significantly different from each other. One can notice that "severe drought" occurs 4.4% of the time for the SPI, 5% for the PDSI/PHDI, but 12% for the SWSI with the same verbal description. For instance, for SPI "moderate" drought category implies that $-1.49 < SPI < -1.00$; any value between these two limits has the same description. This is very restrictive, has crisp meaning for the drought category, and one cannot make meaningful interpretations. However, logically the closer the SPI value is to -1.00 (-1.49) the more it should have "mild" ("severe") ingredients. If the index value is on one of the boundary values, say, -1.50, then what does it mean? There is confusion as to the meaningful and objective conclusion because a value of -1.5 represents a cumulative probability of 6.7% for the SPI, approximately 27% for the PDSI/PHDI, and 32% for the SWSI. Another drawback is that the indicator value changes according to frequencies, which depend on record duration and location. It is not even possible to depend on more than one indicator, because confusion arises due to different values. This also implies that an effective temporal or spatial drought management cannot be planed, operated, and maintained in a drought-prone area. As mentioned earlier in this chapter (Section 2.2), linguistic information gathering and field reconnaissance studies may help toward more meaningful understanding and better solutions. For this purpose, as explained by Şen (2010), fuzzy drought classification helps the decision makers (see Section 2.18).

2.16 PERCENTILES FOR DROUGHT INDICATORS AND TRIGGERS

As mentioned earlier, different drought indicators provide results that are inconsistent, but there is a statistical way to bring them all together in a statistically consistent manner through the PDF concept. For this purpose, each drought indicator value is considered completely independent from each other. This provides an opportunity to treat all the results collectively through a convenient PDF. This approach provides a consistent and equitable basis for evaluation, ease of interpretation, and application to water management decisions by relating indicator values to familiar concepts such as return periods and occurrence probabilities ("see chapter: Climate Change, Droughts and Water Resources"). Indicators and indices can be transformed to percentiles by fitting a CDF to the data (such as a Gamma or Pearson III PDFs for precipitation) or by developing an empirical CDF using ranking algorithms, plotting positions, or other CDF estimators (Harter, 1994). The drought plan triggers are then based on percentiles instead of single indicator or index values.

To develop an empirical CDF, various indicator values are sorted into ascending order and then each value is attached with an empirical probability according to Eq. (2.24). The plot of the indicator values against the empirical probabilities provides a scatter diagram similar to Fig. 2.8. A Gamma, logarithmic-normal, or Pearson type III CDF can be matched to the scatter of points theoretically, and in future drought calculations and predictions, this theoretical CDF can be employed for planning purposes.

Example 2.5 Statistical Combination of Drought Indices

In an area, usage of different drought indicators for the same data source may lead to different numerical values and, accordingly, the verbal interpretations may also become different. Table 2.15 provides different drought index values given both numerically and linguistically.

TABLE 2.15 Various Drought Indicators

Drought Indicator	Numerical	Linguistically
PDSI	2.53	"Moderate"
PHDI	1.90	"Mild"
Z-index	3.11	"Severe"
SWSI	2.35	"Moderate"
SPI	2.92	"Moderate"
Average	2.56	"Moderate"

It is obvious from this table that out of five indices, three yield "Moderate" and other two "Mild" and "Severe" conditions. It is well known that none of these values are exact; hence, they can be treated as independent random variables. Each one represents a sample from the universal set of drought indices. They can be treated by Eq. (2.24) and by considering the positive values as the reflection of negatives and arranging them in ascending order, it is possible to obtain a CDF value for each drought index. Their scatter plot is given in Fig. 2.19.

On the other hand, one can fit a theoretical PDF to the scatter points for obtaining the universal (population) behavior of the drought indices. Herein, the theoretical Gamma PDF is adapted as has been already used for SPI calculations, and the necessary background for this PDF is already given in Section 2.14. The parameters of the most suitable Gamma CDF and its plot are given on the same figure. One can observe that there is a good match between the scatter and theoretical Gamma CDF.

Now, the main question is what to adapt for the final decision as the drought index. In order to decide on this, it is first necessary to depend on a certain level

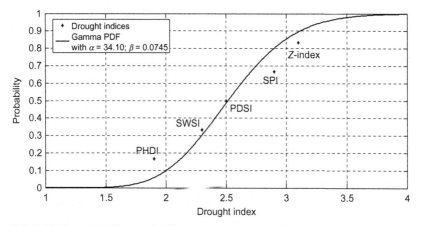

FIG. 2.19 Drought indices probability.

of risk. A set of risks can be chosen as in the following table, where reliability is the complementary of the risk. The values in the last column can be obtained in the MATLAB program through the statement of **gaminv([0.99 0.95 0.90 0.85],34.10,0.0745)**.

Risk (%)	Reliability (%)	Composite Drought Index
1	99	3.66
5	95	3.30
10	90	3.11
15	85	2.99

More detailed information about drought risk calculations is given in "chapter: Climate Change, Droughts and Water Resources". One can see that the first three risk levels fall within the "severe" drought classification and the last one into the "moderate" class.

2.17 TRIPLE DROUGHT INDICATOR

In the SPI calculations only the rainfall amounts are taken into consideration at a station without its relationship with other meteorological variables such as temperature, humidity, evaporation, solar irradiation, wind speed, and sunshine duration. It is well known that each one of these variables has some role in the occurrence of rainfall events. It is, therefore, logical to think about the relative effects of a set of these variables on the rainfall amounts and also on the SPI values. In order to achieve a concise relationship visually, one can plot a map between three of these meteorological factors provided that one of them is the standardized rainfall amounts or SPI values.

Temperature and closely related evaporation and the air humidity are also drought effective variables. Şen (2008) has incorporated precipitation temperature and the air humidity in drought analysis, and he proposed the triple drought indicator (TDI). On the Cartesian coordinate system the horizontal and vertical axes are reserved for humidity and temperature records, respectively, and on the perpendicular axis the temperature–humidity plane SPI values are considered. This provides scatter of points on the temperature–humidity plane with SPI values attached to each scatter point. This is a triple axes system, which constitutes the basis of the TDI procedure. From a mathematical point of view, in this triple axes system temperature and humidity are independent variables on which SPI values are dependent. The following points indicate the benefit from such a TDI:

1. It provides the pattern of precipitation variation based on the temperature and humidity.
2. It is possible to understand where there are deficits (wet periods) or surpluses (wet periods) according to temperature and humidity predictions.
3. According to temperature and humidity predictions, one can know the average and the standard deviation of the precipitation.

4. For any given value of air humidity, it is possible to trace the precipitation variation with temperature.
5. Similarly, for any given temperature value one can obtain the change of precipitation with air humidity.
6. On the basis of temperature and humidity values their minimum (maximum) precipitation ranges can be determined. These areas are the locations of precipitation deficits (surpluses).
7. Likewise, one can determine the minimum precipitation area and the corresponding temperature and humidity variation ranges can be calculated. These areas show how the real drought impacts vary according to precipitation, temperature, and air humidity variations.
8. For a given temperature–humidity couple the precipitation amount can be predicted for farmers. Additionally, the dry (wet) parts and their amounts can be identified.
9. Given a temperature (humidity) level, air humidity (temperature) variation can be determined.

The spatiotemporal drought behaviors were investigated by Sirdas and Şen (2003) for Turkey, where the drought period, magnitude, and SPI values are presented to depict the relationships between drought duration and magnitude.

Example 2.6 TDI Application

The application of the above explained TDI approach is presented for seven locations from different climate regions of Turkey (Fig. 2.20). In these stations there are precipitation, temperature, and humidity records from 1930 to 1990 and their features are presented in Table 2.16

Each one of these stations has a different meteorological regime; hence, SPI variations. The respective SPI values can be represented similar to the ones given in Section 2.14. However, in the following sequence, instead of SPI time series,

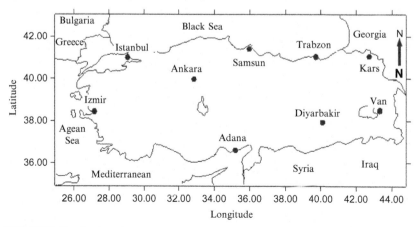

FIG. 2.20 Station locations.

TABLE 2.16 Different Climate Zones in Turkey

Location	Latitude	Longitude	Elevation (m)	Climate Type
Istanbul	4058	2905	33	Maritime–Continental
Ankara	3957	3253	891	Continental (steppe)
Adana	3659	3521	27	Maritime (Mediterranean Sea)
Diyarbakır	3754	4014	677	Continental (semiarid)
Trabzon	4100	3943	30	Maritime (Black Sea)
Van	3827	4319	1661	Continental
İzmir	3826	2710	29	Maritime (Aegean Sea)

SPI variations related jointly to temperature and air humidity changes are presented with interpretations under the light of classifications in Table 2.13. In all stations, only 1-month SPI values are used, but the reader can construct similar TDI maps for 3-, 6-, 12-, 24-, and 48-month SPI values, again based on the temperature and air humidity joint variability.

(1) Istanbul is located at the northwestern part of Turkey, where Aegean Sea and Black Sea maritime climate type is altered due to the effects of continental air movement from the Balkans (southeastern Europe). As a result, winter seasons are cold and rainy whereas summer seasons are rather warm with long-term sunshine durations and high humidity. A TDI map for Istanbul is given in Fig. 2.21, where zero SPI line indicates the separation boundary between dry and wet spells.

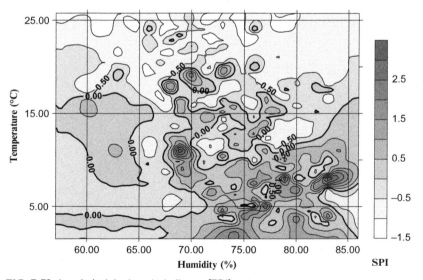

FIG. 2.21 Istanbul triple drought indicator (TDI).

In Fig. 2.21, it is obvious that the drought spells occur at high temperatures irrespective of the air humidity. However, the effect of the air humidity appears more at "low" and "medium" values, whereas at "high" temperature and air humidity the drought effect is relatively weaker. The SPI islands surrounded by bold thick lines are distributed throughout the temperature and humidity ranges. On the other hand, for any given air humidity value dry SPI values are almost around the 15°C temperature. Additionally, relatively "low" air humidity and in "high" temperature regions droughts are dominant, while rather "medium" and "low" temperatures are associated with wet conditions.

(2) Ankara is located at the central part of the Anatolian Peninsula and it is subject to dry air movement resulting in a continental climate regime. In the north, the Black Sea and in the south, the Mediterranean Sea-born evaporation sources with the wind effects leave their moisture contents along the hills of mountains that are parallel to the Black Sea and the Mediterranean Sea coasts and, therefore, Central Anatolia including the capital, Ankara, has low humidity with dominance of dry air. Compared to the Istanbul case the SPI of Ankara in Fig. 2.22 indicates low air temperatures with 85% moisture content and more stagnant dry conditions.

In Fig. 2.22 the lowest SPI values appear in the very low humidity region at any temperature. It is possible to interpret from this figure that as if there is a fuzzy broken line that separates the most effective dry (drought) and wet spell regions based on the temperature–air humidity domain. One can conclude that the weather and meteorological prediction is more reliable at Ankara than Istanbul, where there are comparatively more randomly distributed patches of dry spells in Fig. 2.21.

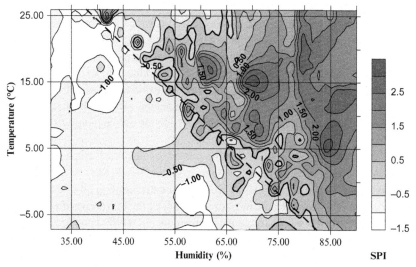

FIG. 2.22 Ankara TDI.

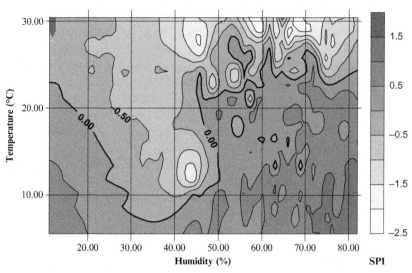

FIG. 2.23 Adana TDI.

(3) Adana has its typical Mediterranean Sea-born maritime climate regime in southern Turkey, where the winter seasons are cool and rather short but summer seasons have high temperature with high moisture. In Fig. 2.23 a dominant continuous zero SPI bold line separates the temperature–humidity variation area into two very distinctive parts. This zero line can be adapted as a discriminant line for possible future drought occurrence traces. Almost at 50% humidity level there is a sudden temperature increase. "High" moisture values result in negative SPI drought parts at "high" temperatures more than about 25°C. The most effective temperature range that falls within the 30% to 50% range of humidity causes dry conditions at even "low" temperatures.

(4) Diyarbakır is located in the southeastern province of Turkey in a semi-arid region, where dry and relatively low humidity range and especially in summer months there are long-duration sun irradiations. Fig. 2.24 is very similar to Adana TDI. Again, a continuous zero line separates approximately the dry and wet spell temperature–humidity domain. Dry SPI values fall within the humidity range from 40% to 60%. Such a shift is possible due to semiocean effect with penetration of Mediterranean Sea air into the region as a result of depression morphology. Approximately below 20°C, outside the above mentioned humidity range, there is no drought appearance.

(5) Trabzon region is located at the pattern of Eastern European air mass destination after moisture-laden passage over the Black Sea. At this station the sunshine duration is rather short. Due to its location among all other stations, this is the one that receives the most precipitation. Fig. 2.25 shows drought, dry (wet) regions' interference around the Trabzon location based on the SPI values with temperature and air humidity foundation.

The SPI pattern in this figure explains unsteady air movements with changes in weather conditions and in such cases reliable drought predictions are not

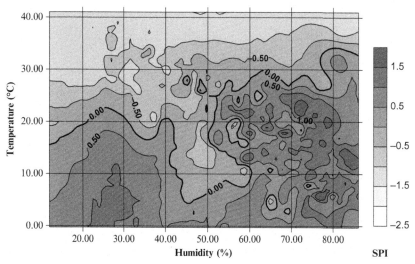

FIG. 2.24 Diyarbakır TDI.

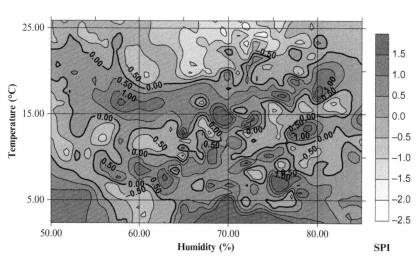

FIG. 2.25 Trabzon TDI.

possible. There is a set of zero lines that separates the wet and dry spells, but it is not possible to identify stable region on the temperature–humidity domain.

(6) Van region falls in the most eastern part of Turkey, where there are high mountains and dry air masses that reach the region with short sunshine duration and in winter months there are snowfalls. The SPI map of this region is given in Fig. 2.26, which has its own special pattern compared with all the previous TDIs.

White patch domination (according to Table 2.13 "light drought") is greater than any other station and this area concentrates especially at "high" temperatures coupled with "low" and "medium" humidity in addition to "high" temperature

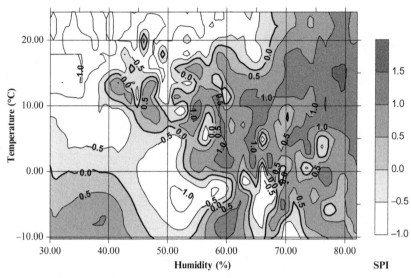

FIG. 2.26 Van TDI.

and "low" humidity. Prior to anything, whatever the temperature degree on average there is no drought at relative humidity values more than 60%.

(7) Finally, in the west of Turkey, along the Aegean Sea coastal area, İzmir station has average sunshine duration, but also has high relative humidity. Fig. 2.27 again has a characteristic SPI pattern. The droughts arise at any humidity level reaching to 60% in cases of more than 15°C temperature. The relative humidity increments above this level causes temperature increases but decreases in drought effect.

2.18 FUZZY LOGIC APPROACH

In almost all of the drought indicators, first a numerical value is obtained and then the entrance of this numerical value to a convenient table gave linguistic information as "mild," "moderate," "severe," "extreme," and similar words as drought categories. The information content of each such fuzzy word differs to a certain extent between the specialists and the users. Hence, there is fuzzy content between the numerical values and their verbal descriptions. In order to alleviate such differences in the meaning and convert them to meaningful numerical drought degrees, fuzzy logic principles can be used as first suggested by Zadeh (1968) and later in a complete book on fuzzy concepts in hydro science by Şen (2010).

In Section 2.16 the confusion about the drought indicators has been solved through a statistical consistency methodology by considering probability attachments as percentiles, but in this section the fuzzy categorical words in various drought indicator tables are combined through fuzzy membership functions (fuzzy sets) for each word in a logical order. Of course, in the drought indicator

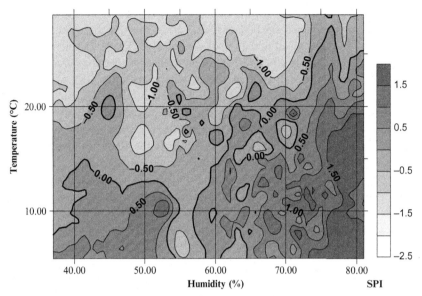

FIG. 2.27 Izmir TDI.

tables (Tables 2.2–2.4, etc.) each word is ordered according to the numerical drought indicator values.

It is rather clear that there is a wide gap between what the given tables are attempting to say (such as dry or drier-than-normal conditions) and the message received by users (definitely dry conditions, perhaps the worst drought in living memory). Some of this gap arose from confusion about the use of fuzzy terms such as "wet," "dry," "nearly dry," and so on. It appears that users and forecasters interpret such fuzzy words in slightly different ways. In cases of drought periods the forecasts and media releases intend to indicate that dry conditions are more probable than wet conditions. Many users, however, interpreted "near dry" as almost certainly dry or certainly not wet. In general, climate forecasters may be more knowledgeable about the basis and accuracy of drought predictions and their potential values than the average individual user such as a farmer. Some decision makers may dismiss the potential value of predictions for decision making due to uncertainty about the accuracy of the forecasts and confusion arising from the terminology.

Murphy (1988, 1993) noted that forecasts must reflect uncertainty in order to satisfy the basic maxim of forecasting—that a forecast should always correspond to a forecaster's best judgment. When the forecasts are expressed in probabilistic terms, they become more logical, because as explained in "chapter: Introduction", the atmosphere is not deterministic. The vague recognition that information matters has not led to agreement on when, how, and under what conditions it influences the behavior of administrators and decision makers. Despite the vast and growing array of institutions involved in collecting, analyzing, and disseminating information potentially relevant to global scale through local governance, the understanding of the role that these "information institutions" play remains limited (Keohane and Nye, 1989). Despite this limitation, some notions are emerging. In recent decades, experience of a decreasing trend of annual precipitation or a higher frequency of drought events may alert the local people for potential threat of climate change. Decision making under uncertainty becomes a very important and critical issue. Unfortunately, little consideration has been given to the real possibility that in climate change, impacts may be too conservative or underestimate the degree of change in the frequency and severity of extreme events for some locations.

There are many different definitions of droughts in the literature. However, it is defined generally as an extended period of rainfall deficit during which agricultural harvests are severely curtailed. The most important effect of climate change on water resources will be a great increase in the overall uncertainty associated with the management and supply of freshwater resources. Significant hydrological components such as storms, rainfall, streamflow, soil moisture, and evaporation, all effecting agricultural activities, are substantially random in their behavior; accordingly, any hydrologist or agriculturalist will regard them in terms of uncertain scientific methodologies, namely, statistics, probability, and at-large stochastic and most recently as fuzzy systems. All these scientific approaches provide future estimations on the basis that the surrounding environmental effects and the climatic change are all stationary. However, in the recent past, the patterns of the local environmental and global climatic changes are expected to repeat in the near future. It is possible to quantify the uncertainties with some probabilistic formulations under restrictive assumptions (Sections 2.6 and 2.16). It must not be forgotten at this stage that certainly the future pattern of the climatic change and its consequences will not look like past behaviors; therefore, it is necessary to try and manage water supply systems with more care about the undesirable extreme drought cases in the future. This brings into focus the concept of risk and management under the risky environment ("see chapter: Climate Change, Droughts and Water Resources") with possibility (fuzzy) and probabilistic assessments and modeling.

Wilhite and Glantz (1985) have classified drought definitions as conceptual, that is, relatively vague (fuzzy) and operational, which is meant to provide specific guidance on aspects of onset, severity, and termination. The latter descriptions are among the objective quantities that are frequently asked about in any drought study: When is the starting time of the drought? How severe is the drought? And when is the termination time of the drought? In fact, the time difference between the termination and starting instances shows the duration of drought.

Although many areas all over the world are prone to drought risks, unfortunately the existence of drought and its appearance have been vague, imprecise, uncertain, and most of the time fuzzy, not only for laypersons but also for the experts in the area. It is, therefore, difficult to find a unique definition for droughts and consequently, the current definitions are rather based on expert and professional concepts. In order to alleviate the case, most often droughts are classified right from the beginning according to their attachment of economic consequences and physical causes. Even though droughts inflict economic losses, they are social phenomena as well.

Example 2.7 Fuzzy Logic Description of Drought

In Example 2.5, various drought indices values are considered as independent random variables and, accordingly, the probability approach is used based on the numerical values. Herein, the drought values are assumed as fuzzy variables as already given linguistically in different tables throughout this chapter. There are fuzzy subsets such as "extreme," "severe," and so forth. It is possible to represent these subsets collectively in a fuzzy universal drought index domain through membership functions, which are adapted in this work as triangular shapes. The reflections of drought classification from Table 2.2 as fuzzy subsets are given in Fig. 2.28. In this figure, the reader who is not familiar with fuzzy modeling can appreciate that neighbor classifications have overlaps; that is, they are mutually inclusive.

The vertical axis shows the membership degrees and its domain of variation is from 0 to 1, inclusive. The template in Fig. 2.28 provides the chance for separating the membership degrees or percentages of any numerical drought index value contribution into two neighbor drought classifications. As an example herein, two of the drought indices values are taken into consideration from Table 2.15 as PHDI $= -1.90$ and SPI $= -2.92$. For the fuzzy inference, the simplest procedure includes the following points:

1. Enter the drought index value from the horizontal axis.
2. Draw a vertical line from the location of the drought value (broken line).
3. Identify the intersection points on the two neighbor fuzzy subsets.
4. Read linguistically the names of both subsets.
5. Draw horizontal lines from the intersection points until they cut the vertical axis for membership degrees.
6. Read the two values corresponding to these points.
7. Describe the drought verbally in two parts.

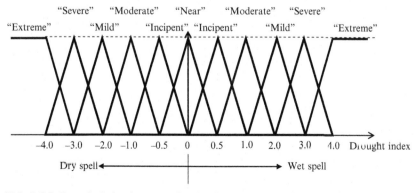

FIG. 2.28 Drought index fuzzy membership functions.

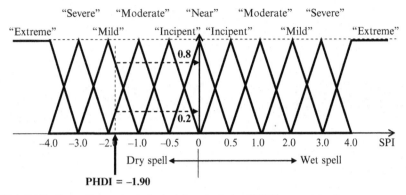

FIG. 2.29 Fuzzy Palmer hydrological drought index (PHDI) categorizations.

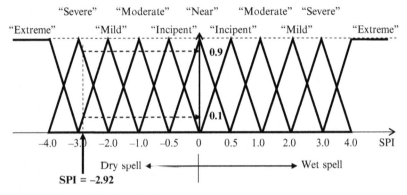

FIG. 2.30 Fuzzy SPI categorizations.

The application of the aforementioned procedure is shown in Fig. 2.29 for PHDI. It is obvious that the PHDI value reaches to "moderate" and "mild" fuzzy drought classes with 0.2 and 0.8 membership degrees, respectively. This is tantamount to saying that the "moderate" drought is bound to occur at a 20% level whereas the "mild" alternative may occur at an 80% level.

The application of the similar procedural steps to SPI drought indices is shown in Fig. 2.30.

In case of SPI, the drought classification shifts toward a more extreme domain. Its −2.92 value meets "mild" and "severe" fuzzy sets at 0.1 and 0.9 membership values that correspond to 10% and 90% drought occurrences, respectively.

2.19 CONTINUITY CURVE

The continuity curve (CC) method provides information about the hydrometeorological quantity, Q, by considering the maximum and minimum quantities and others in between. Especially based on the use of daily quantities, the CC is equivalent to the subtraction of the CDF from 1, where one can obtain the exceedence time percentage for any given threshold (truncation level). Generally, 70% and 90% (Q_{70} and Q_{90}) exceedence percentages are used for low-quantity conditions. For instance, Q_{95} implies the quantity value corresponding to 95% exceedence. Q_{95} and Q_{90} flow values are frequently used in most of the academic studies and applications and in different countries as a quantitative indicator.

In the assessment of droughts, management of reservoirs, hydroelectric power generation, and water resources management studies the CC is employed for proper design determination. In order to produce CC, the time series is truncated at a set of different truncation levels between the maximum and minimum values and then the corresponding time percentages for exceedence are calculated. In Fig. 2.31 the hydrometeorological time series is cut at Q_0 level.

After the truncation one can obtain big ($Q > Q_0$) and small ($Q < Q_0$) series. In the meantime, they are also referred to as the water surplus and deficit, respectively. They are very important in drought or flood calculations as has been explained in detail by Şen (2005, 2008). If the discharge number bigger than Q_0 is n_b, its exceedence percentage during the whole past record can be calculated as $p_b = n_b/n$, where n is the number of observations. For any given truncation level, there is a percentage (probability) value. Logically, an increase in the truncation level causes a decrease in the number of bigger quantities and, accordingly, p_b percentages also decrease. Hence, there is an inverse relationship between the truncation level and its percentage as in Fig. 2.32, which is a nonlinear relationship. At the extreme cases, maximum (minimum) quantity value, the percentage is equal to 0 (1).

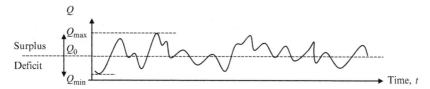

FIG. 2.31 Truncation for continuity curve (CC).

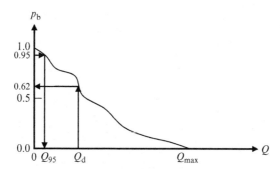

FIG. 2.32 Continuity curve.

For any given discharge value on the horizontal axis, there is a corresponding percentage value of the vertical axis in Fig. 2.32. Provided that the CC is obtained from more than 30-year data, it is considered as a reliable tool for quantity exceedence probability calculations. For instance, the CC is useful to determine the design discharge for hydroelectric power generation discharge from a river without any storage structure. It is also useful for irrigation, sedimentation, and water contamination preliminary studies.

The CC can be used for comparison between different drainage basins. For this purpose, different CCs are plotted on the same graph and then their arithmetic averages are taken for a representative CC in the drainage basin. Later, different arithmetic average CCs are drawn on the same scale Cartesian coordinate system and this provides visual and then quantitative comparisons between the drainage basins. It cannot provide predictions for very low (smaller than the minimum discharge) and for very high (bigger than the maximum discharge) discharges. To alleviate this drawback, it is necessary to fit the most suitable CDF to the available CC, as already done in Section 2.6. The reliability of a water system in the area requires that the necessary discharge is available for at least 180 days (approximately at 0.5 probability level). For this purpose, one can enter the vertical axis with 0.50 values and then find the corresponding discharge value on the horizontal axis (see Fig. 2.32).

As for the surface waters, preparation of the CC is convenient, if it is constructed by considering water year, which starts from Oct. 1 each year and ends on Sep. 30. It is also possible to construct the CC on a monthly basis for determination of low-flow months. One can also incorporate the snow melt amounts in the CC. It is recommended for practical uses that the CCs should be updated after each 5-year period.

Example 2.8 CC Calculation

If there is Q m^3/s discharge for hydroelectric power generation at a location along a river, then by use of the CC one can determine what the percentage is and the total times for proper energy generation. Provided that the exceedence probability, p_b, is known then the total time for the electricity generation can be calculated as $t_b = p_b T$, where T is the record duration. For a given design discharge, Q_d, in Fig. 2.32 the corresponding exceedence time percentage is 62%. Its multiplication by the record duration, say $T = 50$ years, yields the duration as $50 \times 0.62 = 31$ years over which the discharges are more than truncation discharge (design discharge). The subtraction of this calculated duration from total duration, T, gives the time duration during which one cannot generate energy.

REFERENCES

Alley, W.M., 1984. The Palmer drought severity index: limitations and assumptions. J. Clim. Appl. Meteorol. 23, 1100–1109.

Al-Sefry, S.A., Şen, Z., Al-Ghamdi, S.A., Al-Ashi, W.A., Al-Bardi, W.A., 2004. Strategic ground water storage of Wadi Fatimah, Makkah region. Technical report SGS-TR-2003-2. Saudi Geological Survey, Jeddah, 168 pp.

Benjamin, J.R., Cornell, C.A., 1970. Probability, Statistics and Decisions for Civil Engineers. McGraw-Hill, New York, 684 pp.

Bhalme, H.N., Mooley, D.A., 1980. Large-scale droughts/floods and monsoon circulation. Mon. Weather Rev. 108, 1197–1211.

Budyko, M.I., 1958. The Heat Balance of the Earth's Surface (N.A. Stepanova, Trans.). U.S. Department of Commerce, Washington, DC, 259 pp.

Budyko, M.I., 1974. Climate and Life. Academic Press, New York, 508 pp.

De Martonne, E., 1926. Areisme et indice artidite. C. R. Acad. Sci. Paris 182, 1395–1398.

Doesken, N.J., McKee, T.B., Kleist, J., 1991. Development of a surface water supply index for the western United States. Climatology report number 91-3. Colorado State University, Fort Collins, Colorado.

Erinç, S., 1965. Yağış müessiriyeti üzerine bir deneme ve yeni bir indis (On the precipitation effect and a new indice). İstanbul Üniversitesi Coğrafya Enstitüsü, no: 41, İstanbul (in Turkish).

Foley, J.C., 1957. Droughts in Australia. Review of records from earliest years of settlement to 1955. Bulletin no. 47. Bureau of Meteorology, Commonwealth of Australia, Melbourne, Australia.

Garen, D.C., 1993. Revised surface-water supply index for Western United States. J. Water Resour. Plan. Manag. 119 (4), 437–454.

Gbeckor-Kove, N., 1989. Drought and desertification. WMO/TD-No. 286, WCAP No. 7. World Meteorological Organization, Geneva, pp. 41–73.

Gibbs, W.J., 1975. Drought, its definition, delineation and effects. Drought special environmental report no. 5. WMO/TD-No. 193, WCP No. 134. World Meteorological Organization, Geneva.

Gibbs, W.J., Maher, J.V., 1967. Rainfall deciles as drought indicators. Bureau of meteorology bulletin no. 48. Commonwealth of Australia, Melbourne.

Gommes, R., Petrassi, F., 1994. Rainfall variability and drought in Sub-Saharan Africa since 1960. Agrometeorology series working paper no. 9. Food and Agriculture Organization, Rome, Italy.

Guttman, N.B., 1998. Comparing the palmer drought index and the standardized precipitation index. J. Am. Water Resour. Assoc. 34 (1), 113–121.

Guttman, N.B., Wallis, J.R., Hosking, J.R.M., 1992. Spatial comparability of the Palmer drought severity index. Water Resour. Bull. 28, 1111–1119.

Hare, F.K., 1983. Climate and Desertification. A Revised Analysis. WMO Publication, Geneva, 149 pp.

Harter, T., 1994. Unconditional and Conditional Simulation of Flow and Transport in Heterogeneous, Variably Saturated Porous Media (Ph.D. dissertation). The University of Arizona.

Hayes, M., Wilhite, D.A., Svoboda, M., Vanyarkho, O., 1999. Monitoring the 1996 drought using the standardized precipitation index. Bull. Am. Meteorol. Soc. 80 (3), 429–438.

Heddinghaus, T.R., Sabol, P., 1991. A review of the Palmer drought severity index and where do we go from here? In: Proceedings of the 7th Conference on Applied Climatology. American Meteorological Society, Boston, pp. 242–246.

Heim, Jr., R.R., 2000. Drought indices: a review. In: Wilhite, D.A., (Ed.), Drought, Volume 1: A Global Assessment. Routledge, London, pp. 159–167 (Chapter 11).

Hollinger, S.E., Isard, S.A., Welford, M.R., 1993. A new soil moisture drought index for predicting crop yields. In: Preprints Eighth Conference on Applied Climatology, American Meteorological Society, Anaheim (CA), 17–22 January 1993. AMS, pp. 187–190.

Jacobi, J., Perrone, D., Duncan, L.L., Hornberger, G., 2013. A tool for calculating the Palmer drought indices. Water Resour. Res. 49, 6086–6089.

Karl, T.R., 1986. Sensitivity of the Palmer drought severity index and Palmer's Z-index to their calibration coefficients including potential evapotranspiration. J. Clim. Appl. Meteorol. 25, 77–86.

Karl, T.R., Knight, R.W., 1985. Atlas of Monthly Palmer Hydrological Drought Indices (1931–1983) for the Contiguous United States. Historical Climatology Series 3–7. National Climatic Data Center, Asheville, NC.

Karl, T., Quinlan, F., Ezell, D.S., 1987. Drought termination and amelioration: its climatological probability. J. Clim. Appl. Meteorol. 26, 1198–1209.

Keohane, R., Nye, J., 1989. Power and Independence, second ed. Scott, Foresman and Company, Glenview, IL.

Keyantash, J., Dracup, J.A., 2002. The quantification of drought: an evaluation of drought indices. Bull. Am. Meteorol. Soc. 83, 1167–1180.

Kogan, F.N., 1995. Droughts of the late 1980s in the United States as derived from NOAA polar-orbiting satellite data. Bull. Am. Meteorol. Soc. 76 (5), 655–668.

Köppen, W., 1918. Klassifikation der Klimate nach temperature, niederschlag und jahresverlauf. Petermann's Mitteilungen 64, 193–230, 243–248.

Lettau, H.H., Lettau, K., 1969. Shortwave radiation climatonomy. Tellus 21, 208–222.

Mawdsley, J., Petts, G., Walker, S., 1994. Assessment of drought severity. British Hydrological Society, Occasional paper 3.

McKee, T.B., 1995. Drought monitoring with multiple time scales. In: Preprints, Ninth Conference on Applied Climatology, Dallas, TX. American Meteorological Society, Anaheim (CA), pp. 187–190.

McKee, T.B., Doesken, N.J., Kleist, J., 1993. The relationship of drought frequency and duration to time scale. In: Preprints Eighth Conference on Applied Climatology. American Meteorological Society, Anaheim (CA), 17-22 January 1993. AMS, pp. 179–184.

Meyer, S.J., Hubbard, K.G., 1995. Extending the Crop Specific Drought Index to soybean. In: Preprints Ninth Conference on Applied Climatology, American Meteorological Society, Dallas (TX). American Meteorological Society, pp. 258–259.

Meyer, S.J., Hubbard, K.G., Wilhite, D.A., 1993. A crop-specific drought index for corn: I. Model development and validation. Agron. J. 86, 388–395.

Murphy, J.M., 1988. The impact of ensemble forecasts on predictability. Q. J. R. Meteorol. Soc. 114, 463–493.

Murphy, A.H., 1993. What is a good forecast? An essay on the nature of goodness in weather forecasting. Weather Forecast. 8, 281–293.

Nkemdirim, L., Weber, L., 1999. Comparison between the droughts of the 1930s and the 1980s in the Southern Prairies of Canada. J. Clim. 12, 2434–2450.

Ogallo, L.J., 1989. Drought and desertification (Report of the CCL rapporteur on drought and desertification in warn climates to the tenth session of the commision for climatology, 1989, Lisbon). WCAP. No. 7. WMO/TD-No. 1-40. World Meteorological Organization, Geneva.

Palmer, W.C., 1965. Meteorological drought. U.S. Weather Bureau research paper no. 45, p. 58.

Palmer, W.C., 1968. Keeping track of crop moisture conditions, nationwide: the new crop moisture index. Weatherwise 21, 156–161.

Panofsky, H.A., Brier, G.W., 1958. Some Applications of Statistics to Meteorology. Pennsylvania State University, University Park, PA.

Penman, H.L., 1948. Natural evaporation from open water, bare soil and grass. Proc. Roy. Soc. London A193, 120–146.

Rao, G.A., 1986. Drought probability maps. Cog-VIII rapportuer on drought probability maps WMO. WCP (2470). World Meteorological Organization, Geneva.

Şahin, A.D., Sırdaş, S., 2002. A new graphical approach between solar irradiation variables and drought. In: EPMR-2002 International Conference Environmental Problems of The Mediterranean Region, 12–15 April 2002. Near East University, Northern Cyprus.

Şen, Z., 2005. Kuraklık Afet ve Modern Hesaplama Yöntemleri (Drought Hazard and Modern Calculation Methods). Su Vakfı (Water Foundation), Istanbul, Turkey, 248 pp.

Şen, Z., 2008. Wadi Hydrology. Taylor and Francis, CRC Press, New York, 347 pp.

Şen, Z., 2010. Fuzzy Logic and Hydrological Modeling. Taylor and Francis, CRC Press, Boca Raton, Florida, 340 p.

Şen, Z., 2012. Innovative trend analysis methodology. J. Hydrol. Eng. 17 (9), 1042–1046.

Şen, Z., 2014a. Applied and Practical Hydrogeology. Elsevier, Amsterdam, 406 p.

Şen, Z., 2014b. Trend identification simulation and application. J. Hydrol. Eng. 19 (3), 635–642.

Şen, Z., Al-Jadid, A.G., 1999. Automated average areal rainfall calculation in Libya. Water Resour. Manag. 14, 405–416.

Shafer, B.A., Dezman, L.E., 1982. Development of a surface water supply index (SWSI) to assess the severity of drought conditions in snow pack runoff areas. In: Proceedings of the Western Snow Conference, pp. 164–175.

Sırdaş, S., 2002. Meteorolojik kuraklik modellemesi ve Türkiye uygulamasi (Meteorological Drought Modelling and Application to Turkey, in Turkish) (Ph.D. thesis). Istanbul Technical University, Istanbul, Turkey.

Sirdas, S., Şen, Z., 2003. Spatio-temporal drought analysis in the Trakya region, Turkey. J. Hydrol. Sci. 48, 809–819.

Smith, D.I., Hutchinson, M.F., McArthur, R.J., 1993. Australian climatic and agricultural drought: payments and policy. Drought Network News 5 (3), 11–12.

Soulé, P.T., 1992. Spatial patterns of frequency and duration for persistent near-normal climatic events in the contiguous United States. Clim. Res. 2, 81–89.

Thom, H.C.S., 1958. A note on the gamma distribution. Mon. Weather Rev. 86 (4), 117–122.

Thom, H.C.S., 1966. Some methods of climatological analysis. WMO technical note 81. Secretariat of the WMO, Geneva, Switzerland, pp. 1–53.

Thornthwaite, C.W., 1948. An approach toward a rational classification of climate. Geogr. Rev. 38 (1), 55–94.

Thornthwaite, C.W., 1963. Drought. Encycl. Britannica 7, 699–701.

van Rooy, M.P., 1965. A rainfall anomaly index independent of time and space. Notos 14, 43.

Weghorst, K.M., 1996. The Reclamation Drought Index: Guidelines and Practical Applications. Bureau of Reclamation, Denver, CO, 6 pp. (Available from Bureau of Reclamation, D-8530, Box 25007, Lakewood, CO 80226.).

Wilhite, D.A., Glantz, M.H., 1985. Understanding the drought phenomenon: the role of definitions. Water Int. 10, 111–120.

Wilhite, D.A., Pulwarty, R.S., 2005. Drought and water crises: lessons learned and the road ahead. In: Wilhite, D.A. (Ed.), Drought and Water Crises: Science, Technology, and Management Issues. CRC Press, Boca Raton, FL, pp. 389–398.

Willeke, G., Hosking, J.R.M., Wallis, J.R., Guttman, N.B., 1994. The national drought atlas. Institute for water resources report 94-NDS-4, U.S. Army Corps of Engineers.

WMO, 1975a. Drought: special environmental report no. 5. WMO No. 403. World Meteorological Organization, Geneva.

WMO, 1975b. Drought and agriculture. WMO technical note 138. World Meteorological Organization, Geneva.

WMO, 1983. Catalogue of meteorological training publications and audiovisual aids, third ed. Education and training programme report no. 4, WMO/TD-No. 124. World Meteorological Organization, Geneva.

Zadeh, L.A., 1968. Fuzzy algorithms. Inf. Control. 12 (2), 94–102.

Chapter | THREE

Temporal Drought Analysis and Modeling

CHAPTER OUTLINE

Applied Drought Modeling, Prediction, and Mitigation

3.1 GENERAL

In practice, most often droughts are examined from the temporal variations point of view, because anyone living at a given location is concerned first with the temporal variation without initial consideration of the regional or areal effects. If temporal drought variations at a set of single points or stations (meteorology, groundwater wells, etc.) are known, then one can draw equal drought lines in the form of contour maps that reflect the regional drought variation pattern ("see chapter: Regional Drought Analysis and Modeling"). It is also possible to sequence different temporal maps so as to appreciate the spatiotemporal drought pattern tracing ("see chapter: Spatio-Temporal Drought Analysis and Modeling").

The objective of this chapter is to provide simple probabilistic, statistical, stochastic analytical, empirical, and simulation solutions to various drought problems of wet and dry periods in meteorological and hydrologic time series. Such solutions are helpful for planners and decision makers for water resource systems design and management. The hydrometeorological time series are random in character and, therefore, their internal and structural variables (wet and dry spells) also have random behaviors. One such variable is drought, which is very significant from a water resources supply and demand point of view, because once its duration and magnitude are found objectively, it is then possible to plan for water resources management, especially during dry periods. Among the management strategies are possible water transportation alternatives in known quantities to a drought-stricken area, if needed, either from other water resource alternatives (mainly from nearby groundwater storages, ie, aquifers) or from water impoundments in reservoirs during wet periods ("see chapter: Introduction"). Droughts are representative of dry periods in a time series during which successive precipitation, streamflow, soil moisture, and other values are less than a given threshold level ("see chapter: Basic Drought Indicators"). For example, when a drought control scheme is considered, then the problem is to supply the necessary demand during dry periods. In this case, the truncation level may be considered as equal to demand level. The threshold level does not need to be constant at all times, but in this chapter most often it is assumed as constant, in addition to monthly and seasonal drought predictions.

The initiation of each drought is due to precipitation at less than the expected level. Snow- and ice-affected regions experience their lowest flows during the winter due to snow accumulation and streamflow reduction as a result of freezing. It is, therefore, necessary to differentiate between the two events of summer and winter droughts in order to make a consistent analysis (Tallaksen and Hisdal, 1997). A detailed description of how this can be done is given by Hisdal et al. (2001).

Preliminary studies concerning wet and dry periods have made use of the assumption that successive observations are independent, but later serially dependent hydrometeorological variables are also treated by the use of various stochastic processes. Drought analysis for serially correlated time series has been undertaken by Saldarriaga and Yevjevich (1970). They have approximated the empirical probability distribution function (PDF) and, in turn, various statistical properties of wet and dry periods are identified and formulated by employing a finite tetrachoric series expansion as presented by Kendall (1948). For small serial correlation coefficients, tetrachoric series converge quite rapidly, but in case of large serial correlation coefficients the convergence is rather slow; therefore, one has to be satisfied with practically approximate results. However, in this chapter, explicit and exact analytical formulations are presented concerning the statistical behaviors of wet and dry periods by making use of various stochastic models such as Markov processes, which simplify the problem, leading to exact analytical results that are very useful in practical applications.

The majority of drought analyses have concentrated on temporal assessments. The first classical statistical analysis approach was about the evaluation of the instantaneously smallest value in a measured sequence of basic variable such as rainfall, runoff, or soil moisture recorded at a single site (Gumbel, 1963). This method gives information on the maximum drought duration magnitude within a prescribed period of time such as 2-, 10-, 25-, 50-, or 100-year return periods ("see chapter: Climate Change, Droughts and Water Resources"). Yevjevich (1967) presented the first objective definition of temporal droughts, which led to applications by various researchers (Downer et al., 1967; Llamas and Siddiqui, 1969; Millan and Yevjevich, 1971; Guerrero-Salazar and Yevjevich, 1975; Şen, 1976, 1977, 1980a). Due to the analytical difficulties, the study of regional droughts has been carried out to a relatively smaller extent. The first study along this line was done by Tase (1976), who performed many computer simulations to explore various drought properties. Different analytical solutions of drought occurrences have been proposed by Şen (1980b) through the random field concept. However, these studies are limited in the sense that they investigate regional drought patterns without temporal considerations ("see chapter: Regional Drought Analysis and Modeling").

In the context of temporal hydrological droughts, Yevjevich (1967) and Şen (1977) defined severity as the cumulative shortage or deficit sum (total deficit) with reference to a truncation level; therefore, severity has the unit of mm (rainfall) or cubic meters per time (runoff). After a threshold selection, a drought severity analysis can be conducted based on drought duration as a function of the selected threshold level. The severity of drought is a function of the drought duration and PDF of the hydrometeorological variable and its autocorrelation structure. The frequency analysis of critical droughts is helpful in choosing design criteria in many water resources projects (hydrological drought), and the selection of a cropping system or pattern (agricultural drought). The longest duration and the greatest severity can be taken as the statistics for critical drought (Şen, 1980c; Sharma, 2000). The severity is crucial for hydrological droughts while critical duration, even with less severity, is important for agricultural droughts.

Among the purposes of this chapter is the presentation of a systematic approach for the calculation of temporal drought occurrences by simple probability procedures and then their numerical solutions, which are confirmed by extensive Monte Carlo simulation studies. Improvements in statistical methods have even tended to place a new emphasis on rainfall studies, particularly with respect to a better understanding of persistence effects (Şen, 1989, 1990). In fact, persistence is used to estimate possible runs of dry periods. Quantitative drought analysis techniques are presented based on objective methodologies, which can provide a common basis for drought analysis provided that there are enough data. Drought is a partial time event over the whole period, but it may reappear rather frequently depending on the meteorological, climatological, hydrological, and environmental conditions.

This chapter also provides the statistical objective definitions of droughts through mathematical formulations and later the average properties such as mean, standard deviation, skewness, kurtosis, and other quantities will be explained with practical applications. Herein, the probability, statistical, and stochastic methodologies play a fundamental role in deriving objective and reliable drought descriptors. Although the forecasting aspects are very significant for drought preparedness and early warning studies, unfortunately, many proposed methodologies in the literature pose some difficulties in their adaptations for practical problem solutions. In this chapter, after a detailed probabilistic, statistical, and stochastical drought modeling and development of basic formulations, their practical uses are presented through simple examples and empirical formulations.

3.2 NUMERICAL DEFINITION OF DROUGHTS

Human relationships with water resources and supply depend temporally and spatially on the available water amount, which can be categorized linguistically (qualitatively by words) as "deficit," "surplus," or "insufficient" and "inadequate" and "stress" or "shortage." For instance, according to precipitation amounts any period (days, weeks, months, and years, multiyear) can be classified into two categories as "rainy" and "nonrainy" and, similarly, runoff as "dry" and "wet," which is also the case for soil moisture. It is well appreciated in practice that each word implies a fuzzy set (Şen, 2010). Simply any day may be described as "rainy" provided that there is rainfall, snow, hail, or dew. If a time series record is available, then a convenient amount, for instance, arithmetic average, can be taken as a threshold value for the identification of distinctive drought categories, which can be expressed subjectively by words and objectively by numerical values. Even in the case of verbal expressions there are corresponding sets of numbers that may vary temporally and spatially. This is very obvious from many classification tables presented in "chapter: Basic Drought Indicators".

An additional point in an objective definition of drought is the basic time interval (hour, day, week, month, season, or year). If the basic time interval is day, then nonrainy day is a representative of "dry" spell. For instance, in Fig. 3.1 there are 12 time intervals with drought durations, D_d, of 1, 8, and 3 time intervals.

FIG. 3.1 Temporal dry units and drought lengths.

After the determination of basic time interval (day, week, month, or year), there are some principles that are necessary for the determination of "wet" and "dry" spell characteristics. For instance, in the case of a hydrometeorological time series, X_i $(i = 1, 2, ..., n)$, its truncation at a threshold level, X_0, gives rise to the following drought features (Fig. 3.2).

The succession of "dry" intervals overtime preceded and succeeded by a "wet" state is the "dry" (drought), D, length, which is also referred to as the negative run-length in mathematical statistics (Feller, 1967).

The use of a daily time resolution introduces two special problems as dependency among droughts and the presence of minor droughts. During a prolonged dry period, it may be observed that the flow exceeds the threshold level for a very short period of time; thereby, a large drought is divided into a number of minor droughts that are mutually dependent (Fig. 3.3). To avoid these problems that could distort an extreme value modeling, a consistent definition of drought events should include some kind of pooling in order to define an independent sequence of droughts (Tallaksen, 2000).

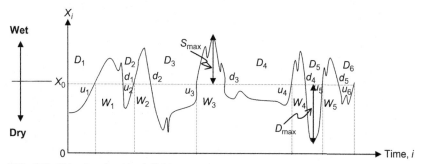

FIG. 3.2 Objective drought definitions.

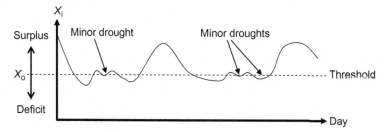

FIG. 3.3 Minor droughts.

By consideration of drought spells in Fig. 3.2 one can define basic and objective quantities concerning various drought characteristics as follows:

1. A "wet" state $(X_i > X_0)$, occurs at any time interval, i, with its "surplus" amount as $(X_i - X_0) > 0$.
2. Otherwise a "dry" state $(X_i < X_0)$, exists with its corresponding "deficit" amount as $(X_0 - X_i) < 0$.
3. A "wet spell" is defined as the delimitation of consecutive uninterrupted wet spells by at least one "wet" state. If the dry state locations at the beginning and the end of a "wet spell" are $(X_i < X_0)$ and $(X_j < X_0)$, respectively, then the duration of this wet spell is defined quantitatively as $(j–i)$.
4. Similarly, a "dry spell" is defined as an uninterrupted sequence of "dry" states that are delimited at the two ends by at least one "wet" state. If the "wet" state locations at the beginning and the end of a "dry spell" are "wet" states as $(X_k > X_0)$ and $(X_l > X_0)$, then the "dry spell" duration is equal to $(l–k)$.
5. If there is a transition from a "dry spell" to a subsequent "wet spell," then such a transition point is defined by two successive states as "dry" $(X_i < X_0)$ and "wet" $(X_{i+1} > X_0)$. This means that at ith location there is an upcrossing point, u_i.
6. Similarly, any transition between a "wet spell" and the following "dry spell" implies a downcrossing, d_i, point defined by a "wet" state $(X_i > X_0)$ and the next "dry" state $(X_{i+1} < X_0)$.
7. Among the "dry spell" durations at a given truncation level there is one, which is the maximum (minimum) and this duration is referred to as the "critical dry duration" ("critical wet duration"). Such critical periods play a significant role in any capacity design or water storage system, and especially, in water resources operations and management.
8. The summation of surpluses (deficits) along with any wet duration (dry duration) is referred to as the "total deficit", D_T ("total surplus", S_T). These are also known as the magnitudes in practical studies. It is also possible to consider the "gross deficit", D_G ("gross surplus", S_G) by considering the totals over the whole record period. Comparison of these two gross values leads to the following three alternatives:
   a. If $D_G = S_G$, then at this location, temporarily, the summation of deficits is in balance with surpluses during the whole record or planned period considered. It is possible to balance the whole deficit occurrence during this duration by surplus amounts, provided that the surpluses are stored in convenient storage. For instance, if the water demand level of a settlement area is considered as the truncation level, then the total surplus around this level can meet the whole deficit (water shortage) occurrences without any water shortage provided that the proper storage units are constructed. Such a situation corresponds to ideal reservoir design, which is referred to as the Ripple diagram in hydrology (Chow, 1964; Linsley, 1982).
   b. If $D_G > S_G$, then at this location one expects a drought effect, and in order to offset such a case, it is necessary to transfer water from other locations.

   c. If $D_G < S_G$, there is a wet spell expectation at this location with extra water amounts and it may be possible to transfer the extra water to nearby water shortage or drought-stricken areas.

9. The duration of each "wet spell" ("dry spell") is referred to as "wet duration, D_w ("dry duration", D_d). At a given truncation level, either $D_W = D_L$ or the difference ($D_W - D_L$ or $D_L - D_W$) is equal to 1.

10. The maximum deficit (surplus) during a dry (wet) period is the maximum dry amount, D_{max}, which can be calculated as

$$D_{max} = \max\left(X_o - X_i\right) \quad (i = 1, 2, ..., n_d) \tag{3.1}$$

In a similar manner, the maximum surplus value, S_{max}, can be calculated as

$$S_{max} = \max\left(X_i - X_o\right) \quad (i = 1, 2, ..., n_w) \tag{3.2}$$

11. The magnitude of each "wet spell," M_w ("dry spell," M_d) is defined as the summation of surpluses (deficits) along each duration. This is equivalent to the earlier definition of total deficit (surplus). If there are m "wet" ("dry") states during any "wet spell" ("dry spell") then the magnitude is defined objectively as

$$M_w = \sum_{j=1}^{m} \left(X_j - X_o\right) \tag{3.3}$$

where j is a counter for the number, m, of dry states in the "dry spell" duration. Similarly,

$$M_d = \sum_{j=1}^{m} \left(X_o - X_j\right) \tag{3.4}$$

12. The intensity of "wet spell," I_w ("dry spell," I_d) is defined as the ratio of magnitude to the duration of the same spell. These are given for "wet spell" and "dry spell" durations as

$$I_w = \frac{M_w}{D_w} = \frac{\sum_{j=1}^{m}\left(X_o - X_j\right)}{D_w} \tag{3.5}$$

and

$$I_d = \frac{M_d}{D_d} = \frac{\sum_{j=1}^{m}\left(X_j - X_o\right)}{D_d} \tag{3.6}$$

respectively.

Each one of these definitions can be calculated if there is a reliable sequence of data at hand with objectively defined threshold level. It is also possible to make future predictions of these quantities for different return periods (design durations) such as 5-, 10-, 25-, 50-, or 100-year, which will be dealt with later in this chapter and also in "chapter: Climate Change, Droughts and Water Resources". These were studied analytically by Şen (1976, 1977) through mathematical statistical, probability, and stochastic methodologies.

Example 3.1 Drought Factor Calculations

In a given time series, if there are n records out of which n_d and $n_w = n - n_d$ are the dry and wet spell numbers, respectively, then the percentage (probability, risk) of a drought can be calculated simply as

$$P_d = n_d/n \tag{3.7}$$

and the complementary percentage of wet spells is

$$P_w = 1 - P_d \tag{3.8}$$

This last expression is equivalent to the exceedance probability of the water demand level (X_o) over the record span. It is obvious that as the water demand level changes the probability of dry (wet) spell also changes. Decrease in the water demand level causes increase in the wet spell probability, but just the opposite is valid for dry spell probability. Hence, it is possible to show these variations in a graph, which is referred to as a dry or wet period percentage curve as in Fig. 3.4.

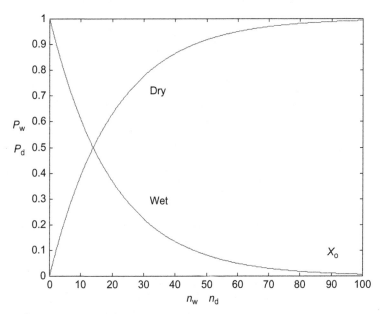

FIG. 3.4 Dry and wet spell time percentage curves.

Şen (1989) provided the average, D_a, and standard deviation, D_s, values for the number of dry spells, as follows:

$$D_a = nP_d(1 - P_d) \tag{3.9}$$

and

$$D_s = nP_d(1 - P_d)(1 - 3P_d + 3P_d^2) \tag{3.10}$$

Example 3.2 Dry and Wet Spell Calculations

It is observed that during a 35-year annual runoff (m^3/s) record there are 7 years with below the water demand level. Calculate the wet and dry period probability of occurrences in addition to the mean and standard deviation. With these parameters, assume that the annual records abide with the normal (gaussian) PDF, and hence, predict the possibility of drought severity expectation within the next 25-year period with probability of occurrence equal to 20%.

The empirical probabilities of dry and wet spell on the basis of given information can be calculated simply from Eqs. (3.7) and (3.8), respectively, which yield $P_d = 7/35 = 0.20$, and hence, its complementary probability of wet spell is $P_w = 1.00 - 0.20 = 0.80$.

The statistical parameters, namely, the mean and standard deviation for the prediction from a normal PDF, can be calculated from Eqs. (3.9) and (3.10) as

$$D_a = 35 \times 0.20 \times 0.80 = 5.6 \text{ m}^3/\text{s}$$

and

$$D_s = 35 \times 0.20 \times 0.80 \times (1.00 - 3 \times 0.20 + 3 \times 0.2^2) = 2.9 \text{ m}^3/\text{s}$$

As the nonexceedance probability of a single drought occurrence expectation within the next 25-year period is $1/25 = 0.04$ (exceedance probability of 0.96), then one can calculate from the normal PDF tables or from any convenient software such as MATLAB$^®$ by the statement **norminv(0.96,5.6,2.9)**, which yields 10.677 m^3/s.

3.3 THE THRESHOLD LEVEL METHOD

The threshold level can be any value between the maximum and minimum records within a given time series. Although the deviations from the mean value are not stable even during the 30-year "normal" data duration ("see chapter: Introduction"), but for comparison purposes this duration is accepted internationally as the "normal period"; therefore, even in the cases of longer data availability, most often the deviations are calculated on the basis of the same normal period. If the drought properties vary locally, then the deviations from the mean value at different places indicate distinctive patterns due to wide scatter of deviations even for wet or dry spells. In general, the use of mean value as the threshold implies 50% percentile, which may not be the same at different locations because each location record may have different PDFs. It is possible to compare the drought features on the basis of a convenient percentile such as the 10% of the cumulative PDF of the precipitation distribution, which has been applied in Australia by Gibbs and Maher (1967).

The best that can be done is to compare the drought features at different truncation levels. This is due to the fact that there may be differences in drought properties for "low," "medium," and "high" precipitation, streamflow, or soil moisture data. In general, "low" values are considered as the causes of droughts but as they are a part of the whole record, one may derive useful information from the drought features at different threshold levels. Hydrological drought analyses, in terms of streamflow deficits, are studied over a season or longer time periods and in a regional context. However, short-term (less than a season) streamflow deficits might also be defined as droughts and treated at a fixed point in space (Zelenhasic and Salvai, 1987). Unfortunately, there is no distinctive understanding between the drought indices and their connection to drought event definition ("see chapter: Basic Drought Indicators").

The threshold level method depends on runs (positive or negative) below or above a given threshold; in statistics, this is known as the run or crossing theory (Feller, 1967). The sequence of less than the threshold values constitutes the run-length (drought duration, D_d), run-sum (total surplus, S_T, or total deficit, D_T), which are among the features of a wet (dry) spell. The crossing theory was first suggested by Rice (1945) and later extended and summarized by Cramer and Leadbetter (1967). The first applications in the hydrology context are due to the studies by Yevjevich (1967), where the method is based on the statistical theory of runs for analyzing a time series. Statistical and PDF properties of water deficits such as the duration and total deficit are recommended for at-site drought definitions. It is also possible to define the maximum surplus, S_{max} (maximum deficit, D_{max}) and its time of occurrence (see Fig. 3.2). The minimum value in the sequence of measurements is also another simple indicator of a drought event. In Fig. 3.2, the start, t_s, and the end time, t_e, instances of a drought correspond to successive upcrossing and downcrossing instances, respectively. The selection of threshold level may dependent on one of the following points:

1. For the probabilistic and statistical analysis of drought features, the threshold value may be selected as any level between the minimum and maximum values. This helps to identify the internal drought structure of any given hydrometeorological time series.
2. The threshold level may correspond to some physical quantity in water resources planning, operation, management, and maintenance stages. For instance, in the case of a water impoundment reservoir, the threshold level may correspond to safe yield of the reservoir storage.
3. In some practical applications, such as the "low" flow maintenance for ecological life sustenance, the threshold value can be taken as some percentage

of the mean value of the event ("see chapter: Basic Drought Indicators, Section 2.19").

4. Again, to sustain certain ecosystem equilibrium, the threshold value can be adopted as the average of a certain subsequent time period such as 5 or 10 days, and so on.

5. A duration curve represents the relationship between the magnitude and frequency of daily, monthly, or some other time interval ("see chapter: Basic Drought Indicators, Section 2.19"). The duration curve has the form as in Fig. 2.32, where the horizontal axis represents the event values versus percentage occurrences on the vertical axis. Such a curve is equivalent with the exceedance PDF of the variable concerned.

The threshold level may change from one season to another depending on the water resources activities and requirements. A variable threshold can thus be used to define periods of precipitation or streamflow deficiencies as departures or anomalies from the "normal" seasonal "low" flow volumes. A daily varying threshold level can also be defined as an exceedance probability of duration curves. In order to increase the sampling range and smooth the threshold cycle, the short duration such as daily exceedance can be calculated within an n-day window. For example, in applying a monthly window, the exceedance on the first day of the month should be calculated from all discharges recorded between 15th of the previous month and 15th of the current month.

Example 3.3 Threshold Calculations as Windows

Drought analysis is necessary for future predictions of some features from the past records, which provide bases for the probabilistic and statistical models. Once the PDFs of the drought descriptors are available, it is then possible to make predictions based on a certain risk level. The necessary data for drought analysis can be derived from the whole record either by considering the lowest values during a certain time period such as month or year, which is the sequence of annual maximum or minimum series (AMS) or partial duration series (PDS) where all the low values are considered below a certain threshold level. In case of low-threshold levels, the occurrence of zero-drought years may significantly reduce the information content of the AMS.

The definition of the AMS is more straightforward provided that the hydrometric year can be properly defined (Tallaksen and Hisdal, 1997). The PDS model has a more consistent definition of the extreme value assessment by considering several extreme values. In the PDS, however, minor droughts (Fig. 3.3) may significantly distort the extreme value modeling, and a procedure for exclusion of minor droughts should be imposed. Whether to use the AMS or PDS model depends on the available data and the type of analysis to be carried out.

For instance, 7-day window discharge for 2-year (7Q2) and 10-year (7Q10) annual low-flow statistics are based on an AMS of the smallest values of mean discharge computed over any 7 consecutive (10 consecutive) days during the annual period. It is possible to fit a PDF to the annual series of 7-day minima, and then to calculate 7Q2 statistic, with a 2-year recurrence interval (nonexceedance probability of 50%). For example, the Sep. 7Q2 and 7Q10 flow statistics are calculated by fitting a PDF to the annual series of 7-day minimums computed just from daily mean flows in Sep. of every year.

The AMS for 23 years at a surface water measurement station are given in the second column of Table 3.1.

The third column includes ranked flows and the fourth one has the empirical probabilities, P_m, calculated according to Eq. (2.24), where m is the rank of the

TABLE 3.1 Low Flow Data

Data No.	Annual Low Flow (m^3/s)	Ranked Flow (m^3/s)	Empirical Probability
1	41.604	17.394	0.042
2	118.576	20.765	0.083
3	463.524	39.904	0.125
4	125.510	41.604	0.167
5	17.394	45.568	0.208
6	20.765	46.181	0.250
7	54.078	54.078	0.292
8	72.666	56.481	0.333
9	69.395	57.290	0.375
10	179.000	69.395	0.417
11	95.581	72.666	0.458
12	56.481	89.721	0.500
13	45.568	90.091	0.542
14	120.165	95.581	0.583
15	39.904	112.784	0.625
16	188.483	118.576	0.667
17	46.181	120.165	0.708
18	90.091	125.510	0.750
19	89.721	155.037	0.792
20	57.290	179.000	0.833
21	155.037	188.483	0.875
22	112.784	224.529	0.917
23	224.529	463.524	0.958

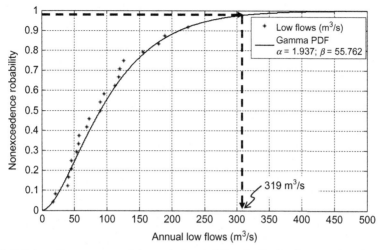

FIG. 3.5 Low flow scatter and theoretical probability distribution function (PDF).

data in ascending order and n is the number of data. The scatter of the ranked flows against their corresponding empirical probabilities is given in Fig. 3.5.

In Fig. 3.5, the annual low flows are fitted with the theoretical Gamma PDF, which has shape and scale parameters as $\alpha = 1.937$ and $\beta = 55.762$, respectively. One can benefit from the given Gamma PDF parameters and Fig. 3.5 for calculation of any low-flow occurrence prediction within the next, say, 50-year planning duration. In this case, the probability of low-flow occurrence during this duration is $1/50 = 0.02$, which gives the nonexceedance probability as $1 - 0.02 = 0.98$. The corresponding value can be found from the MATLAB software by entering these probability and parameter values into the statement of inverse cumulative distribution function (CDF) for Gamma PDF as **gaminv (0.98,1.937,55.762)** and the result appears as 319 m³/s ("see chapter: Basic Drought Indicators, Section 2.14").

3.4 DROUGHT FORECASTING

In terms of long-term drought forecasting, some studies are underway in which the time series of Palmer drought severity index (PDSI), precipitation, temperature, and streamflow are correlated with the El Niño Southern Oscillation (ENSO)-related variables, such as the southern oscillation index (SOI) or sea surface temperature (SST) (Piechota and Dracup, 1996). The other useful correlative variable that has been identified is the geopotential height, which augurs well for forecasting drought periods in association with the SOI (Cordery and McCall, 2000). On a short time scale, such as a month or a season, it may be possible to indicate the probable timings of inception and termination of drought (Beran and Rodier, 1985). Chang and Kleopa (1991) indicated that drought monitoring offers a means of providing some clues on short-term forecasting, which essentially employ one or more combinations of the following procedures (Panu and Sharma, 2002):

1. *Linear regression models:* These involve weather variables, such as air pressure, air and SSTs, wind velocities and directions, and the recent records of precipitation data. Among models for forecasting agricultural droughts are regression approaches involving such variables as crop yield, antecedent precipitation indices, number of wet days, and so forth. (Kumar and Panu, 1997).

2. *Time series models:* These are forecasting algorithms that have been used for short-term forecasting by employing the notions of Markov and autoregressive integrated moving average (ARIMA) stochastic models, nonhomogeneous Poisson processes coupled with conditional probabilities (Lohani and Lognathan, 1997).

3. *Empirical models:* Recession rates of streamflow hydrographs, stage graphs of other water bodies, and indices based on the status of soil moisture and vegetation in the region under consideration have been used for short-term forecasting (Zelenhasic and Salvai, 1987).

4. *Teleconnections:* These are links between SST and inland weather, wind in East Africa, and monsoon in India, location of the intertropical convergence zone, jet streams, ENSO, and others. These form the basis of medium- to long-term drought forecasting (Cordery and McCall, 2000).

The best and most reliable way of drought prediction is to derive the drought behavior from the available past records and then to develop a suitable model to extrapolate these features into the future. For reliable and robust drought descriptions, not only is the historical data is insufficient but also one should try to obtain additional information numerically and linguistically about the hydroclimatic features of the study area, vegetation cover, soil type, and ecological and environmental factors. Even though the mathematical models for drought prediction may be well developed and effective, it is not sufficient to rely only on the mathematical model results. Droughts are embedded in the climatological behavior of the area concerned and they are extreme events such as floods. Their extremes appear if necessary precautions and mitigation policies are not considered prior to their silent start. Droughts, in climatic systems, have occurred many times at any place and they are expected to occur again in the future. Apart from climatic features, droughts are expected to impact society more frequently in the future due to growing water needs, imbalance between water supply and demand, extravagant style of life with water consumption increase, global warming, and climate change. It is recommended that drought impacts should be handled by using the risk management approach ("see chapter: Climate Change, Droughts and Water Resources") rather than the crisis-based approach, as it is the most practiced way today in many societies and countries.

The most important task in any drought study is to derive the local and temporal drought features from the available data and to predict its future behavior through reliable and efficient methodologies. Unfortunately, drought prediction has not enjoyed the wide spectrum of use that flood prediction has; therefore, there is much room for development, assessment, and validation of simple drought predictor methodologies. The prediction must depend on cause and effect relationships, which may also include verbal (fuzzy) information that is very important in the prediction model construction (Şen, 2010). As causative agents, droughts begin due to persisting high pressure that results in dryness because of air subsidence, more sunshine irradiation and evaporation, and the deflection of precipitation bearing storms. It is, therefore, necessary to have sufficient information concerning the local or distant influences that might generate such atmospheric blocking patterns ("see chapter: Introduction").

Without reliable, effective, and validated drought prediction methodologies, it is not possible to plan and manage drought consequences. Drought prediction has utmost importance if risk management is to be used for drought combat, mitigation, and adaptation tasks ("see chapter: Climate Change, Droughts and Water Resources"). In the literature, drought variable behaviors are examined through reasonably well-established frequency domain (empirical PDF) based on the historical measurement sequence of precipitation, streamflow, or soil moisture.

Monitoring and forecasting the onset of droughts needs a range of methodologies and technologies. Water availability models are dependent on rainfall, runoff, and reservoir storage and groundwater level measurements. A series of such measurements provides the opportunity for the identification of departures from the mean, especially total deficits in river flows (run-sums) and precipitation anomalies in terms of long-term percentages and averages. On the other hand, as supportive facilities, the satellite monitoring of vegetation, soil moisture, and other parameters also assist to assess the regional progression pattern of a drought.

To check water balance, ground-based observations and measurements should be monitored including soil moisture, surface and groundwater abstraction rates on the basis of an effective well inventory, and water quality indicators. Apart from this numerical information, local verbal knowledge and information, especially from public enquiries, complaints, and suggestions (although fuzzy in content), are very useful at the onset of drought modeling and prediction studies.

In the preliminary drought prediction process, precipitation and temperature data and their forecasts are widely used as in climate change studies. They provide basic information about the medium- and long-range forecasts about meteorological parameters and drought indices such as the standardized precipitation index (SPI) as explained in "chapter: Basic Drought Indicators". Numerical weather prediction software of different types and origin provide reliable predictions only for lead times of 3–4 days, after which one can use completely probabilistic prediction methodologies for long-range applications.

The employment of regressions and other types of statistical methods are helpful to relate future rainfall to indicators such as SST and surface pressure differences (Troccoli and Palmer, 2007). It is already mentioned in "chapter:

Introduction" that droughts are associated with a number of regional and global-scale features of the atmospheric-oceanic circulation. Global consideration of these features paves the way to more reliable and accurate drought prediction background information. However, whatever the global information sources, local influences should be considered especially during a drought.

Example 3.4 Drought Features

In the southern part of Turkey near the Tigris River is the city of Diyarbakir, and its meteorology station has monthly total rainfall records from 1960 to 2008. Monthly total rainfall amounts are presented in Fig. 3.6 for each water year with the monthly averages, which starts from Sep. onward. It is obvious that the rainfall fluctuations (standard deviation) are smaller during the Jun.–Sep. period.

The monthly averages are taken as the truncation level and then the software (Appendix 3.1) is developed and applied for the calculation of drought features. Fig. 3.7 shows few features in the monthly drought time series.

Out of $49 \times 12 = 588$ months there are 125 drought periods, which means that on average there are $588/125 \cong 5$-year drought duration expectations with maximum drought duration of 7 months (Fig. 3.7A). Table 3.2 presents the statistical properties of these drought quantities.

The maximum deficit, total deficit, and intensity values appear in Fig. 3.7 as negative values, but they are entered as absolute values in this table.

Prior to theoretical modeling and prediction of drought, this example provides practical drought quantity prediction, by simple methodology of extreme values. For a complete description of this prediction procedure, the following steps are necessary.

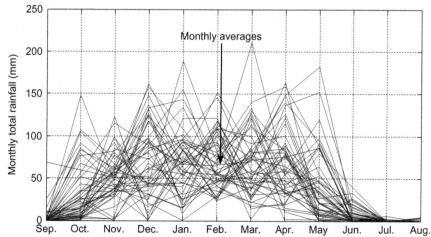

FIG. 3.6 Diyarbakir meteorology station monthly total rainfall (mm).

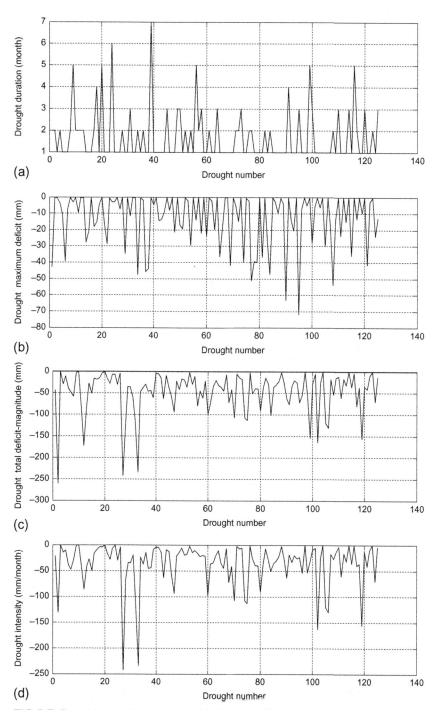

FIG. 3.7 Drought quantities time series: (A) duration, (B) maximum deficit, (C) total deficit (magnitude), (D) intensity.

TABLE 3.2 Drought Quantity Statistics

Drought Quantity	Minimum	Average	Maximum	Standard Deviation
Duration (month)	1	2	7	1.18
Maximum deficit (mm)	0.04	12.87	71.79	16.23
Total deficit (mm)	0.20	49.14	281.09	53.10
Intensity (mm/month)	0.07	37.04	242.07	42.79

1. Practical steps:
   a. Identify the drought features as for the crossing points (upcrossing and downcrossing), drought duration, drought total deficit (magnitude), and drought intensity quantities by use of the software in Appendix 3.1.
   b. Rank each drought quantity in ascending order by assuming that they are completely independent of each other.
   c. Calculate the exceedance probabilities for each ranked drought quantity according to Eq. (2.24).
   d. Plot the scatter diagram of ranked drought quantities versus their probabilities.
2. Theoretical steps:
   a. Identify the best PDF for the drought quantity at hand by checking a set of PDFs. Herein, the reader is recommended to try with the following PDF in sequence:
      i. Logarithmic-normal PDF
      ii. Two-parameter Gamma PDF
      iii. Exponential PDF
      iv. Weibull PDF
      v. Truncated normal PDF
   b. Calculate adapted PDF parameters.
   c. Calculate the CDF values with the parameters found in the previous step.
   d. Check the suitability of the CDF with the scatter diagram in part A by one of the goodness-of-fit tests such as chi-square, Kolmogorov–Smirnov, or similar tests.
   e. Calculate the drought quantity future predictions for the next 5-, 10-, 25-, 50-, 100-, and 250-year return periods.

Fig. 3.8 presents the results of applying the above steps to different drought quantities.

It is obvious that each drought quantity is matched with the two-parameter Gamma CDF and the parameters are given on the same graphs. Even visual inspections indicate that the CDFs fit well the measurements. In Fig. 3.8A there are many drought durations with the same length, and the smaller the drought length the more frequent their appearance. This means that as the drought quantity increases, the frequency of occurrence decreases.

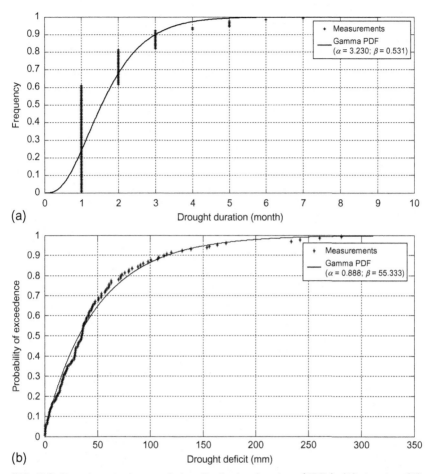

FIG. 3.8 Drought quantity cumulative distribution functions (CDFs): (A) duration, (B) maximum deficit,

(Continued)

By knowing the CDF of each drought duration, it is possible to make predictions for a given risk level as shown in Table 3.3. This can be achieved by a MATLAB statement provided that the type and parameters of the fitted CDF are known. For instance, future drought period predictions for a set of given risk levels can be obtained by the following MATLAB statement

$$\textit{gaminv}\left(\left[0.50\ 0.80\ 0.90\ 0.96\ 0.98\ 0.99\right], 3.230, 0.531\right)$$

where **gaminv** implies inverse calculations from the two-parameter Gamma CDF ("see chapter: Basic Drought Indicators"). It should be noted that the MATLAB statement is entered with reliability values, because they correspond to nonexceedance probabilities as in Fig. 3.8. Other drought quantity predictions

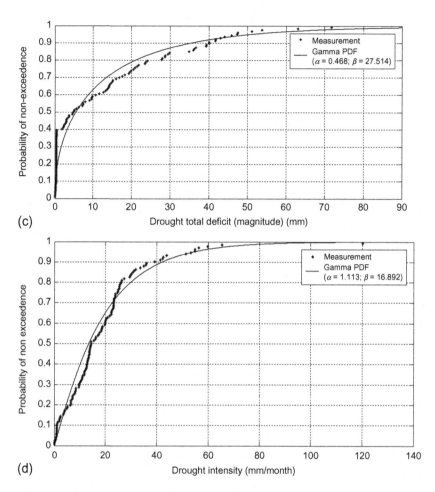

(c)

(d)

FIG. 3.8—CONT'D (C) total deficit (magnitude), (D) intensity.

TABLE 3.3 Drought Quantity Predictions

Return Period (month)	Risk (%)	Reliability (%)	Duration (month)	Maximum Deficit (mm)	Total Deficit (mm)	Intensity (mm/month)
2	0.50	0.50	2	32.39	5.50	13.57
5	0.20	0.80	2	79.80	21.06	29.99
10	0.10	0.90	3	116.51	35.29	42.17
25	0.04	0.96	4	165.55	55.71	58.13
50	0.02	0.98	4	202.89	71.92	70.13
100	0.01	0.99	5	240.38	88.91	82.09

can be calculated accordingly by importing the corresponding two-parameter Gamma CDF parameter values.

Risk and reliability concepts applications will be presented in "chapter: Climate Change, Droughts and Water Resources" in detail for various drought predictions.

3.4.1 Drought Models

One of the characteristics of many precipitation climates throughout the world is that wet or dry spells may occur in a variety of durations. Daily rainfall may be modeled simply using the rainy or nonrainy day criteria ("see chapter: Regional Drought Analysis and Modeling"), which utilize critical magnitudes of daily rainfall and any fluctuation or runs of wet and dry spells may be identified (Summer, 1988).

For different purposes, full information about the number of wet and dry spells becomes more pressing in many fields, such as in water management projects, agricultural planning, flood studies, and in various industrial activities. Thus, the geographical distribution of these meteorological parameters for any area is helpful in cases of new water resources development and planning projects, especially related to agricultural activities and environmental engineering (Theoharatos and Tselepidaki, 1990). For instance, in reservoir design, if the total amount of water is in excess of the demand, then during wet periods the excess water is stored for consumption during dry spells, if necessary, either from the previously stored water in the reservoir and/or, if possible, from alternative water resources. Thus the accumulated water excess plays an important role in reservoir design. In addition, some other important factors that require the prediction of wet and dry spells are water pollution, sedimentation, problems of erosion, and size of pipes (Aboammoh, 1991). In general, one can say that any period is wet or dry naturally due to a relationship between water supply and demand; that is, if in any period of time water supply is smaller than demand, then this period is considered as a dry period, otherwise, it is a wet period.

In practice, most of the drought models are concentrated on the available precipitation (meteorological drought) and streamflow (hydrologic drought) historic records. The early models are developed on the basis of a single location; therefore, preliminary drought prediction methodologies consider temporal variations only. However, their validation in a small area around the main point is also considered as a simple and local evaluation. Early models are based on the probabilistic, statistical, and stochastic modeling approaches, predominantly on the frequency analysis methodologies. In the application and development of the necessary formulations, there are basic assumptions such as the structural independence, stationarity, Bernoulli trials, Markov processes, chains, an so on. In the literature, most frequently, drought studies are based on annual data; therefore, a year is the minimum time scale in their predictions. However, in this chapter monthly drought prediction models are also presented.

Models are well established and developed for drought durations but drought severity modeling studies are less developed in the literature. No doubt the duration aspect is of importance in terms of frequency characterization and forecasting, yet the severity aspect is no less important. For instance, severity is the crucial parameter in sizing storage reservoirs toward combating droughts (Şen, 1980a; Sharma, 2000). There is also a need to standardize and to accept a unique definition of severity that could be used in all drought scenarios universally. The same statement also applies to total deficit (surplus) volume (ie, magnitude of a drought). There is still ambiguity in the usage of the above terms in the vocabulary of drought studies (Panu and Sharma, 2002).

Concerns such as the probabilities of a dry day following another dry or wet day or the probability for a succession of dry or wet days are appropriately addressed by different authors. Several authors have pointed out that a first-order Markov chain model may provide good results (Gabriel and Neumann, 1957, 1962; Green, 1964; Wienser, 1965; Feyerherm and Bark, 1967, 1973). Many other models have been mentioned, and the intricate problem attracted the attention of various authors (Dyer and Tyson, 1977; Şen, 1976, 1977, 1980b, 1989, 1990; Eljadid and Şen, 1997a,b). All cannot be included here, but some interesting approaches may be singled out. Essenwanger (1986) maintained a binomial model, which may be helpful but there are significant discrepancies between observed and model data. One handicap of the binomial model is the fact that the total series of runs must be known a priori before parameters can be computed. Theoharatos and Tselepidaki (1990) determined the number of rainy days by using the Polya statistical method and probabilities. They obtained results showing that the Polya PDF satisfactorily fitted the real daily rainfall values. To achieve maximum efficiency, it is necessary that during wet periods rainfall should be conserved for exploitation in subsequent dry periods (Eljadid, 1997).

Although meteorological and hydrologic droughts can be studied at a specific site, agricultural drought studies need regional model developments ("see chapters: Regional Drought Analysis and Modeling; Spatio-Temporal Drought Analysis and Modeling"). They affect the production of basic necessities, such as food, and thus are more likely to affect the economy of a region and, along its continuation, socioeconomic drought effects start to inflict society on a large scale. Annual periods are not sufficient for agricultural drought modeling and prediction; monthly or seasonal periods must be considered and, accordingly, annual model outputs must be modified to suit the agricultural droughts. As mentioned in the previous chapter, PDSI is developed for agricultural drought descriptions. In the literature, there is not a complete agreement for the effective use of this indicator in different climate conditions. A suggested by Panu and Sharma (2002) the weekly moisture adequacy index based on the food and agriculture organization (FAO) method of water balancing can be examined for its feasibility as a parameter in statistical forecasting of agricultural droughts. Also, further progress in the pattern recognition techniques is needed before they can be incorporated in agricultural drought monitoring and early warning systems. As for the agricultural drought effects, the most important issue is to obtain a reliable functional relationship between the soil moisture conditions and crop growth and, especially, yield. In order to achieve such a goal, pattern recognition techniques must enter the agricultural drought assessment and prediction studies.

In this chapter, systematic modeling and prediction approaches are presented for the calculation of temporal drought occurrences by simple but effective probability procedures. Recent improvements in statistical methods have even tended to place a new emphasis on rainfall studies, particularly with respect to a better understanding of persistence (continuity of dry spells) effects (Şen, 1989, 1990).

There is a need to develop methods and techniques to forecast the initiation and/or termination points of droughts. The ARIMA models, pattern recognition techniques, physically based techniques using PDSI, Palmer hydrological drought index (PHDI), and SPI ("see chapter: Basic Drought Indicators") a moisture adequacy index involving Markov chains, or the notion of conditional probability, seem to offer potential to develop reliable and robust forecasts toward this goal. Such research efforts would be of considerable importance in mitigating the impacts of short-term hydrological droughts and/or agricultural droughts (Panu and Sharma, 2002).

3.5 DROUGHT FEATURES

The three most significant characteristics of temporal droughts are duration, total deficit (magnitude), and intensity (see Example 3.4). As has already been defined in "chapter: Basic Drought Indicators", quantitatively that intensity refers to the degree of the precipitation shortfall, which is measured by the departure of some climatic parameter (precipitation), indicator (reservoir levels), or index (SPI) from normal and is closely linked to duration in the determination of impact. Drought duration is the time difference between the initiation and termination times, which is rather fuzzy due to the creeping property of drought. The establishment of drought at a noticeable level usually requires about 2–3 months and then continues for many consecutive months, years, and even decades. The drought impacts are closely related to the timing of the onset of the precipitation shortage, its intensity, and the duration.

In general, the creeping nature of drought makes it particularly challenging to quantify the impact, and may make it more challenging to provide disaster relief than for other natural hazards. These drought characteristics have hindered development of accurate, reliable, and timely estimates of severity and impacts (ie, drought early warning systems) and, ultimately, the formulation of drought preparedness plans.

In the context of uncertainty techniques, it is possible to calculate almost all objective drought quantities probabilistically and statistically provided that a historical record is available. Among the statistical parameters are average, standard deviation, correlation coefficient, and skewness values in addition to grouped data evaluation in terms of histograms and empirical PDFs. On the other hand, if the interest lies in the drought frequency of occurrence, then theoretical PDF statements may also be calculated from the same record (Şen, 1985).

3.6 TEMPORAL DROUGHT MODELING METHODOLOGIES

Statistical theory of runs provides a common basis for the objective definition and modeling of drought events given a time series. Simply, a constant truncation level divides the whole time series into two complementary parts, those greater than the truncation level, which is referred to as the positive-run in statistics, wet spell in hydrology, and accordingly, a negative run or dry spell (Fig. 3.2). Feller (1967) gave a definition of runs based on recurrence theory and Bernoulli trials. A sequence of n events, S (success) in statistics, or wet spell in water science and F (failure) or dry spell, contains as many S-runs of length r as there are nonoverlapping uninterrupted blocks containing exactly r events S each. This definition is not convenient practically, because it does not say anything about the start and the end of the run. On the other hand, a definition of runs given by Mood (1940) seems to be most revealing for the analysis of various run properties (drought features) because a run is defined as a succession of similar events preceded and succeeded by different events with the number of similar events in the run referred to as its length (see Fig. 3.2).

3.6.1 First Order Markov Process Drought Properties

Hydrometeorological time series simulation is achieved most frequently by the first-order Markov process. The only difference from the Bernoulli trials is that each state, rather than having $+1$ or -1 values only, there are actual values as surpluses or deficits, which are random in character. In general, a stationary first-order Markov process is used for annual values and it relates the current hydrometeorological value, X_i, to the previous one, X_{i-1}, as follows:

$$X_i = \mu + \rho(X_{i-1} - \mu) + \sigma\sqrt{1 - \rho^2}\varepsilon_i \tag{3.11}$$

where μ, σ, ρ, and ε_i are the long-term arithmetic average, standard deviation, serial correlation coefficient, and completely independent normal random variable with zero mean and unit standard deviation. The application of this process requires that X_is as a time series constitute a normal (gaussian) PDF with drought properties as shown in Fig. 3.2.

In general, one can define the surplus, p, and deficit, q, complementary probabilities at any instant, i, as follows:

$$p_i = P(X_i > X_0) = \int_{X_0}^{+\infty} f_i(X)\,\mathrm{d}X \tag{3.12}$$

and

$$q_i = P(X_i \leq X_0) = \int_{-\infty}^{X_0} f_i(X)\,\mathrm{d}X \tag{3.13}$$

respectively, where $f_i(X)$ is the PDF of X and $p_i + q_i = 1$.

If the multidimensional PDF of a hydrometeorological variable is given as $f(X_1, X_2, \ldots, X_{k+j})$, then it is possible to write the joint probability of k deficit and j surplus occurrences as

$$P(k^-, j^+) = \int_{-\infty}^{X_0} \cdots (k\,\mathrm{times}) \cdots \int_{-\infty}^{X_0} \int_{X_0}^{+\infty} \cdots (j\,\mathrm{times}) \cdots$$
$$\int_{X_0}^{+\infty} f(X_1, X_1, \ldots, X_{k+j})\,\mathrm{d}X_1\,\mathrm{d}X_2\cdots\mathrm{d}X_{k+j} \tag{3.14}$$

On the other hand, in case of identical and independent normal PDF, multivariate probabilities can be written as multiplications,

$$P(k^-, j^+) = \prod_{i=1}^{k} \int_{-\infty}^{X_0} f(X_i) dX_i \prod_{i=k+1}^{j+k} \int_{X_0}^{+\infty} f(X_i) dX_i \qquad (3.15)$$

or expression of integrals in terms of basic state probabilities (Eqs. 3.12 and 3.13) and one can obtain,

$$P(k^-, j^+) = \prod_{i=1}^{k} P(X_i \le X_0) \prod_{i=k+1}^{j+k} \int_{X_0}^{+\infty} P(X_i > X_0) \qquad (3.16)$$

On the other hand, for the first-order Markov process, the multidimensional probability function can be factorized down to a multiplicative sequence of first order (lag-one) conditional probabilities as follows:

$$f(X_1, X_1, \ldots, X_{k+j}) = f(X_1) f(X_2/X_1) f(X_3/X_2) \cdots f(X_{k+j}/X_{k+j-1}) \qquad (3.17)$$

Substitution of this expression into Eq. (3.14) leads to

$$P(k^-, j^+) = \int_{-\infty}^{X_0} f(X_1) \prod_{i=2}^{k} \int_{-\infty}^{X_0} f(X_i/X_{i-1}) dX_i \prod_{i=k+1}^{j+k} \int_{X_0}^{+\infty} f(X_i/X_{i-1}) dX_i \quad (3.18)$$

In terms of basic probabilities this expression becomes

$$P(k^-, j^+) = P(X_1 \le X_0) \prod_{i=2}^{k-1} P(X_i < X_0/X_{i-1} < X_0) P(X_{k+1} > X_0/X_k < X_0)$$
$$\prod_{i=k+2}^{j+k} P(X_i > X_0/X_{i-1} > X_0) \qquad (3.19)$$

Hence, multiple integration is rendered into simple conditional probability, which depends on two successive states in terms of joint probability as follows:

$$P(X_i > X_0/X_{i-1} > X_0) = \frac{P(X_i > X_0, X_{i-1} > X_0)}{p_i} \qquad (3.20)$$

If the hydrometeorological sequence of events abide with the first-order normal Markov process, then the joint probability on the right-hand side can be written as (Benjamin and Cornell, 1970),

$$P(x_i > x_0/x_{i-1} > x_0) = \int_{x_0}^{+\infty} \int_{x_0}^{+\infty} \frac{1}{2\pi\sqrt{1-\rho^2}}$$
$$\exp\left\{ -\frac{1}{2(1-\rho^2)} \left[x_i^2 - 2\rho x_i x_{i-1} + x_{i-1}^2 \right] \right\} dx_i dx_{i-1} \qquad (3.21)$$

where ρ is the first order autocorrelation parameter. In this expression, the standard hydrometeorological variable is employed according to Eq. (3.21). By considering this standard variate with truncation level, x_o, the following four elementary conditional (transitional) probability statements can be defined (Şen, 1976):

$$
\begin{aligned}
P(x_i > x_o, x_{i-1} > x_o) + \frac{p}{q} P(x_i > x_o, x_{i-1} > x_o) &= p \\
P(x_i > x_o, x_{i-1} > x_o) + P(x_i < x_o, x_{i-1} > x_o) &= p \\
P(x_i < x_o, x_{i-1} > x_o) + \frac{q}{p} P(x_i < x_o, x_{i-1} < x_o) &= q \\
P(x_i < x_o, x_{i-1} < x_o) + P(x_i > x_o, x_{i-1} < x_o) &= q
\end{aligned}
\tag{3.22}
$$

For the case of $p=q=1/2$, Sheppard (1918) has suggested the following practical expression:

$$
P(x_i > 0, x_{i-1} > 0) = \frac{1}{4} + \frac{1}{2\pi} \arcsin \rho
\tag{3.23}
$$

The analytical solution of Eq. (3.20) leads to

$$
P(x_i > x_o / x_{i-1} > x_o) = \frac{1}{2} + \frac{1}{\pi} \arcsin \rho
\tag{3.24}
$$

It is possible to benefit from the definition of autorun coefficient, r, by Şen (1976, 1978, 1980a).

$$
r = P(x_i > x_o / x_{i-1} > x_o)
\tag{3.25}
$$

Consideration of this equation as a basis other conditional probabilities in Eq. (3.22) can be calculated as follows:

$$
\begin{aligned}
P(x_i > x_o / x_{i-1} < x_o) &= \frac{p}{q} r \\
P(x_i < x_o / x_{i-1} < x_o) &= 1 - \frac{p}{q}(1 - r) \\
P(x_i < x_o / x_{i-1} > x_o) &= (1 - r)
\end{aligned}
\tag{3.26}
$$

With the use of these conditional probability statements and making the necessary mathematical treatment, Eq. (3.19) takes the following form:

$$
P(k^-, j^+) = p(1 - r) \left[1 - \frac{p}{q}(1 - r) \right]^{k-1} r^{j-1}
\tag{3.27}
$$

or for $k=0$ this expression becomes

$$
P(j^+) = p r^{j-1}
\tag{3.28}
$$

If interest lies in finding the negative run-length, L, probability to be greater than or equal to a given period, j, then the following expression given by Feller (1967) can be used:

$$P(L \geq j) = P(j^+) + \sum_{i=1}^{\infty} P(k^-, j^+) \qquad (3.29)$$

The substitution of Eqs. (3.26) and (3.27) into this last expression leads after some algebra to

$$P(L \geq j) = r^{j-1} \qquad (3.30)$$

For $\rho = 0$, $r = p$, one can obtain that

$$P(L = j) = (1 - r)r^{j-1} \qquad (3.31)$$

Double integral in Eq. (3.21) can be calculated with difficulty for $x_o \neq 0$, but Cramer and Leadbetter (1967) provided the following expression for this purpose:

$$P(x_i > x_0, x_{i-1} > x_0) = \frac{1}{2\pi} \int_0^p \exp\left\{ -\frac{x_o^2}{2(1+z)} (1 - z^2)^{-1/2} \right\} dz \qquad (3.32)$$

The numerical solutions of his equation for a set of ρ, r, and x_o values are given in Table 3.4.

TABLE 3.4 r Values for Different Truncation Levels

	Standard Truncation Level, x_o				
ρ	**-0.524**	**-0.253**	**0.0**	**0.253**	**0.524**
0.0	0.700	0.600	0.500	0.400	0.300
0.1	0.718	0.625	0.532	0.437	0.340
0.2	0.736	0.650	0.564	0.474	0.382
0.3	0.754	0.676	0.597	0.514	0.426
0.4	0.774	0.703	0.631	0.555	0.472
0.5	0.795	0.732	0.666	0.598	0.520
0.6	0.818	0.762	0.704	0.43	0.574
0.7	0.844	0.796	0.747	0.694	0.632
0.8	0.873	0.835	0.795	0.752	0.701
0.9	0.911	0.884	0.856	0.826	0.788

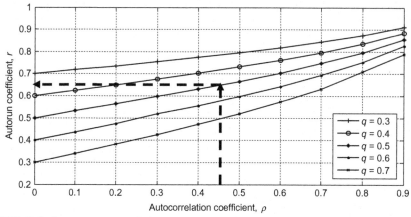

FIG. 3.9 Autorun–autocorrelation parameter relationships.

The relationship between the autocorrelation and autorun coefficients (conditional probability) at different truncation probability levels is shown in Fig. 3.9, which provides a simple chart to convert the first-order serial correlation coefficient to the autorun coefficient that is the main key in the drought calculations.

Example 3.5 Drought Duration Probability Prediction

In a given annual rainfall series, the percentages of dry spells is 0.31 at 500 mm truncation level, and the records comply by the normal (gaussian) PDF with average and standard deviation values as 420 and 180 mm, respectively. The first-order serial correlation coefficient is computed as 0.3. What is the expected drought duration in the future?

It is first necessary to calculate the standard truncation level for the records. This is possible by consideration of the standardization formulation given in Eq. (2.1) which yields

$$x_0 = \frac{500 - 420}{180} = 0.444$$

This value is not available exactly in Table 3.4; therefore, interpolation is necessary between 0.253 and 0.524 and, similarly, for finding the value of r, which leads to 0.426. Hence, the substitution of this r value into Eq. (3.30) gives

$$P(L=j) = (1 - 0.426) \times 0.426^{j-1} = 0.574 \times 0.426^{j-1}$$

It is now possible to calculate the probability of any drought duration probability. For this purpose the change of the drought length probability with drought length is given in Fig. 3.10.

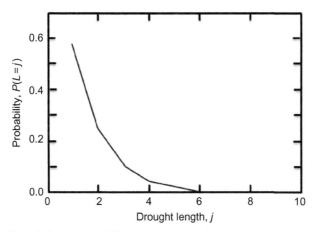

FIG. 3.10 Drought length probability.

3.6.1.1 DROUGHT DURATION STATISTICAL PROPERTIES

From the above probability values one can calculate "wet" and "dry" spell statistical parameters. For instance, truncation of an independent normal process at its mean value gives rise to wet and dry period durations equal to 2. Additionally, if the variance is calculated, then the statistical properties of droughts can be described easily. Şen (1976) gave "wet" and "dry" spell statistical parameters as the expectation of dry duration, which can be calculated in general as

$$E(L) = \sum_{i=1}^{\infty} jp(L=j) \tag{3.33}$$

or

$$E(L) = \sum_{i=1}^{\infty} P(L \geq j) \tag{3.34}$$

Substitution of Eq. (3.29) into this expression gives

$$E(L) = \frac{1}{1-r} \tag{3.35}$$

Furthermore, substitution of values in Table 3.1 into this last expression yields dry period statistical characteristics for the first-order Markov process. For example, if the truncation level is equal to the average of the first-order Markov process, where $p=q$, then the substitution of Eqs. (3.24) and (3.25) by considering Eq. (3.35) yields the drought duration expectation (long-term arithmetic average) as

$$E(L) = \frac{2\pi}{\pi - 2\arcsin\rho} \tag{3.36}$$

In general, the dry period duration average on a given truncation level, q, is equal to the average wet period on truncation level $p = 1 - q$. Hence, it is possible to derive the following equation for wet, L_w, and dry, L_d, period expectations on symmetrical truncation levels around the average value as

$$E(L_w) = E(L_d) \tag{3.37}$$

On the other hand, at a given truncation level the average of the dry and subsequent wet period duration summations, L_T, is equal to the separate summations of dry and wet period averages,

$$E(L_T) = E(L_d) + E(L_w) \tag{3.38}$$

The necessary second-order statistical moment for finding the variance of dry and wet spells is given in the form of expectation as (Feller, 1967)

$$E(L^2) = \sum_{i=1}^{\infty} j^2 P(L=j)$$

Combining this expression with Eq. (3.30) gives

$$E(L^2) = \frac{1+r}{(1-r)^2} \tag{3.39}$$

As the variance definition is given, in general, as

$$V(L) = E(L^2) - E^2(L) \tag{3.40}$$

the substitution of Eqs. (3.35) and (3.39) into this last expression leads to the desired result as

$$V(L) = \frac{r}{(1-r)^2} \tag{3.41}$$

In practical applications most often standard deviation is used, which is the square root of this last expression. If the truncation level is equal to the arithmetic average of the first-order Markov process, then consideration of Eqs. (3.24) and (3.25) leads to

$$V(L) = \frac{2\pi(\pi + 2\arcsin\rho)}{(\pi - \arcsin\rho)^2} \tag{3.42}$$

Even though the basic first-order gaussian Markov process values have zero skewness, the dry and wet period skewnesses are not equal to zero. Skewness of dry and wet periods is dependent on the third statistical moment. Its analytical derivation has been given by Şen (1976) as follows:

$$S(L) = \frac{r(1-r)}{r^{3/2}} \tag{3.43}$$

This expression takes the following form when the truncation level is equal to the arithmetic average of the first-order Markov process:

$$S(L) = \frac{3\pi + 2\arcsin\rho}{[2\pi(\pi + 2\arcsin\rho)]^{1/2}} \tag{3.44}$$

After all these analytical derivations, one can calculate the coefficient of variation for dry and wet periods as

$$C(L) = \sqrt{r} \tag{3.45}$$

In case of truncation level equal to the arithmetic average, one can obtain that

$$C(L) = \left(-\frac{1}{2} - \frac{\arcsin\rho}{\pi}\right) \tag{3.46}$$

In Table 3.5, dry and wet period statistical parameters are given for a set of truncation levels and different correlation coefficients.

There are two benefits in the application of stationary processes. One of them is to identify whether a given process is serially dependent or independent. For instance, if a given time series is truncated at $p = q = 0.5$ level and the average dry and wet periods are found as equal to 2, then this process has an independent internal structure; otherwise serial dependence is valid and the degree of dependence can be determined according to the expected (average) drought duration, $E(L)$, from Table 3.5. The second benefit lies in the prediction of drought lengths at any given truncation level.

Example 3.6 Drought Property Calculation

By considering the Çine River quantities in Turkey as in Table 3.6, determine different drought properties of this catchment at $q = 0.5$ truncation level, which corresponds to arithmetic average level.

TABLE 3.5 Dry and Wet Spell Statistical Quantities

ρ	r	E(L)	V(L)	S(L)	D(L)
q = 0.30 truncation level					
0.0	0.700	3.333	7.777	2.032	0.837
0.1	0.718	3.543	9.028	2.027	0.847
0.2	0.734	3.788	10.560	2.023	0.857
0.3	0.754	0.065	12.459	2.019	0.868
0.4	0.774	4.425	15.154	2.016	0.879
0.5	0.795	4.878	18.917	2.013	0.891
0.6	0.818	5.498	24.695	2.010	0.901
0.7	0.844	6.410	34.681	2.007	0.919
0.8	0.873	7.874	54.126	2.004	0.934
0.9	0.911	11.235	115.010	2.002	0.954
q = 0.40 truncation level					
0.0	0.600	2.500	3.750	2.065	0.774
0.1	0.625	2.667	4.444	2.055	0.790
0.2	0.650	2.853	5.306	2.046	0.806
0.3	0.676	3.006	6.439	2.038	0.821
0.4	0.703	3.367	7.969	2.031	0.838
0.5	0.732	3.741	10.191	2.024	0.855
0.6	0.762	4.201	13.452	2.018	0.873
0.7	0.796	4.902	19.127	2.013	0.892
0.8	0.835	6.060	30.670	2.008	0.914
0.9	0.884	8.620	65.695	2.003	0.940
q = 0.50 truncation level					
0.0	0.500	2.000	2.000	2.121	0.707
0.1	0.532	2.136	2.136	2.100	0.730
0.2	0.564	2.293	2.967	2.082	0.751
0.3	0.597	2.481	3.675	2.066	0.772
0.4	0.631	2.710	4.634	2.053	0.794
0.5	0.666	3.378	5.970	2.041	0.816
0.6	0.704	3.952	8.035	2.030	0.839
0.7	0.747	4.878	11.670	2.021	0.864

0.8	0.795	6.944	18.917	2.013	0.891
0.9	0.856		42.280	2.006	0.925
q = 0.60 truncation level					
0.0	0.400	1.667	1.111	2.213	0.632
0.1	0.437	1.776	1.379	2.173	0.661
0.2	0.475	1.904	1.624	2.140	0.689
0.3	0.519	2.057	2.176	2.111	0.717
0.4	0.555	2.247	2.803	2.087	0.745
0.5	0.598	2.487	3.700	2.066	0.773
0.6	0.643	2.801	5.045	2.048	0.802
0.7	0.694	3.268	7.441	2.033	0.833
0.8	0.752	4.032	12.226	2.020	0.867
0.9	0.826	5.747	27.282	2.009	0.909
q = 0.70 truncation level					
0.0	0.300	1.428	0.621	2.373	0.548
0.1	0.340	1.515	0.780	2.298	0.583
0.2	0.382	1.618	0.999	2.236	0.618
0.3	0.426	1.742	1.292	2.185	0.653
0.4	0.472	1.894	1.639	2.143	0.687
0.5	0.520	2.083	2.257	2.107	0.721
0.6	0.574	2.347	3.163	2.077	0.757
0.7	0.632	2.717	4.667	2.052	0.795
0.8	0.710	3.344	7.841	2.032	0.837
0.9	0.788	4.717	17.532	2.014	0.887

TABLE 3.6 River Flow Discharge Characteristics

Catchment Area	River	Station Number	Observation Number (year)	Correlation Coefficient, ρ
Çine	Çine stream	701	32	0.44

For the drought duration statistical parameter calculations, it is necessary, first, to find the corresponding key parameter that is the autorun coefficient from Fig. 3.9. The entrance of the autocorrelation value of 0.44 on the horizontal axis in the same figure yields the autorun coefficient as 0.66, based on the truncation level percentage (probability) at $q=0.5$. The procedure is shown in Fig. 3.9 by the two broken arrows. With the autorun coefficient at hand, one is then able to calculate various statistical parameters of the drought duration as follows:

1. The average value of expected drought duration in the future can be estimated from Eq. (3.35) as $E(L)=1.0/(1-0.66)=2.94$ year, which can be taken as 3 years in practical studies. Because $E(L)$ is greater than 2, the time series has a dependent structure. Accordingly, the first-order serial correlation coefficient, 0.45, from Table 3.5 is almost equal to the one given in Table 3.6.

2. The standard deviation of the drought duration can be calculated by taking the square root of Eq. (3.41). First, the variance of the drought duration is $V(L)=0.66/(1-0.66)^2=5.71$; hence, the standard deviation is equal to ±2.39, which means that around the average drought duration, as calculated in the previous step, there might be expected ±2.39 deviations. To be on the safe side in any drought planning study, it is recommended to augment the expected drought duration by the standard deviation. Finally, recommended risky drought duration expectation has the theoretical value equal to $2.94 +2.39=5.33$ year. In practice, one can recommend drought preparation plans for a 6 year period in the future at this location.

3. In order to appreciate whether the PDF of the drought occurrences at this location has symmetrical (gaussian) distribution, it is helpful to look at the skewness value, again with the autorun coefficient, from Eq. (3.43), which yields $0.66 \times (1-0.66)/0.66^{3/2}=0.415$. This indicates that the drought PDF has significant skewness.

In hydrological studies, one of the most frequently encountered questions is what is the probability of expecting drought during the next, say, 5 years? For the drought occurrence probability calculation within the next 5 years, Eq. (3.31) helps and the final answer is $P(L=5)=(1-0.66)(0.66)^{5-1}=0.0646$. Its return period is $1/0.0646 \cong 16$ years.

3.6.1.2 CRITICAL DROUGHT DURATION STATISTICAL PROPERTIES

Şen (1980c) developed an analytical approach for the calculation of the critical drought duration in a first-order Markov process. The critical drought has already been defined as the longest among all the drought durations at a given truncation level (Fig. 3.11). The top figure indicates the drought durations at a truncation level, X_o; the middle one corresponds to drought lengths at the first appearance times; and the bottom figure indicates the total deficit (drought magnitude), which is equal to the summation of deficits during a dry spell.

In the previous section, the probability of a drought duration to be greater than any given j duration has been expressed by Eq. (3.30), where the autorun

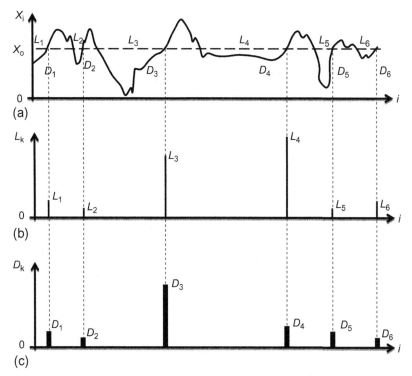

FIG. 3.11 (A) Drought durations on a given truncation level, (B) drought durations, (C) total deficits.

coefficient has the main role. In an independent process the same probability takes the following form:

$$P(L \geq j) = p^{j-1} = (1-q)^{j-1} \tag{3.47}$$

On the other hand, close inspection of Fig. 3.11 reveals the following significant modeling information:

1. At a given truncation level, there is a random number of droughts and let this number be k.
2. Magnitude and intensity of a drought period are also random.

These two points indicate that the drought durations occur as a random number from random variables. In the same figure, drought period durations are shown at upcrossing points ($X_i \leq X_o$, $X_{i-1} \leq X_o$), but likewise downcrossing points can also be adapted. In either way, there is no significant difference in the calculations. The drought lengths, L_j, and total deficit (drought magnitudes), D_j, appear completely independently from each other at any given truncation level. It is also assumed that in a given sample length, n, the number of upcrossing points, N_u, at a given truncation level abides by the Poisson process, which has the expectation (average) value as

$$E(N_u) = nq(1-r) \tag{3.48}$$

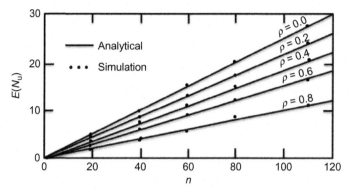

FIG. 3.12 Average of upcrossing numbers versus sample length at truncation level $q = 0.5$.

Accordingly, the Poisson process (PDF) can be given as follows (Parzen, 1962):

$$P(N_u = i) = \exp\left[-nq(1-r)\frac{\left[nq(1-r)^i\right]}{i!}\right] \tag{3.49}$$

Fig. 3.12 includes the number of upcrossings (downcrossings) according to this theoretical Poisson process and the simulation studies for sample length, n. Computer simulation results correspond to the above referred theory with rather minor differences.

After all the aforementioned explanations, it is now clear that the critical drought duration has the longest drought duration among the random number of random drought variables at a given truncation level. The PDF of the maximum of the random number of random variables is presented as a theory by Todorovich and Woolhiser (1976). The longest drought duration, L, and the maximum total deficit, D_n (magnitude), of drought variations in a sample are given as mathematical definitions

$$L_n = \max_{0 \leq j \leq k} (L_j) \tag{3.50}$$

and

$$D_n = \max_{0 \leq j \leq k} (D_j) \tag{3.51}$$

It is clear that these variables never decrease but they are accumulatively increasing processes, which means that $L_n \leq L_{n+1}$ and $D_n \leq D_{n+1}$. In practice, drought durations are integers, whereas drought magnitude has decimal numbers. If at the ith interval, the upcrossing number is shown by N_i then for $j > i$, it is essential that $N_i \leq N_j$.

Generally, it is possible to write the probability of a critical drought duration to be equal to or smaller than j by considering the maximum of the random number of random variables from Todorovich (1970) theory as

$$P(L_n \leq j) = P(N_n = 0) + \sum_{i=1}^{\infty} [P(L \leq j)]^i P(N_n = i) \tag{3.52}$$

The completion of the necessary algebraic manipulations gives

$$P(L_n \leq j) = \exp\{-nq(1-r)[1 - P(L \leq j)] \tag{3.53}$$

or

$$P(L_n = j) = \exp[-nq(1-r)P(L > j+1)] - \exp[-nq(1-r)P(L > j)] \tag{3.54}$$

Consideration of Eq. (3.30) with this last expression provides the following new form as

$$P(L_n = j) = \exp\left[-nq(1-r)r^{j-1}\right]\left\{\exp\left[nq(1-r)^2 r^{j-1}\right] - 1\right\} \tag{3.55}$$

More specifically, if the data is truncated at the median level ($p=q=\frac{1}{2}$) this expression may be simplified to

$$P(L_n = j) = \exp\left(-\frac{n}{2^{j+1}}\right)\left[\exp\left(\frac{n}{2^{j+2}}\right) - 1\right] \tag{3.56}$$

On the other hand, maximum total deficit to be equal to or smaller than s can be written as follows:

$$P(D_s \leq s) = \exp\{-nq(1-r)[1 - P(D \leq s)] \tag{3.57}$$

or

$$P(D_s \leq s) = \exp[-nq(1-r)P(D > s)] \tag{3.58}$$

Example 3.7 Critical Drought Duration

Numerical solutions of Eq. (3.55) are presented in Table 3.7 for a set of sample lengths and serial correlation coefficients. Also, a computer simulation technique is used for the first-order Markov process to obtain the average critical drought durations, which are given in the columns under the heading of Sim. (for simulation).

In this table, the relative errors between the analytical and simulation results are calculated as $100 \times$ (Ana. $-$ Sim.)/Sim, and the error percentages in the Sim. columns are all well less than $\pm 5\%$, which is practically acceptable. Furthermore, as already shown by Millan and Yevjevich (1971) the differences between the analytical solutions and the simulation results are within the acceptable error limits and they are rather small.

TABLE 3.7 Critical Drought Duration Expectation Values

	Truncation Level, q = 0.5											
	ρ = 0.0			ρ = 0.1			ρ = 0.2			ρ = 0.3		
Sample Length	Ana.	Sim.	Dif. (%)	Ana.	Sim.	Dif. (%)	Ana.	Sim.	Dif. (%)	Ana.	Sim.	Dif. (%)
25	3.97	3.97	0.00	4.22	4.23	−0.23	4.47	4.51	−0.89	4.75	4.86	−2.26
50	4.98	5.00	−0.40	5.32	5.35	−0.56	5.68	5.75	−1.21	6.10	6.20	−1.61
100	5.98	6.06	−1.32	6.42	6.48	−0.92	6.89	7.00	−1.57	7.45	7.60	−1.97
200	6.98	7.03	−0.71	7.50	7.34	2.18	8.10	8.10	0.00	8.80	8.94	−1.57
500	8.29	8.25	0.48	8.97	9.05	−0.88	9.71	9.82	−0.11	10.56	10.81	−2.23

Ana., analytical; Sim., simulation; Dif., difference.

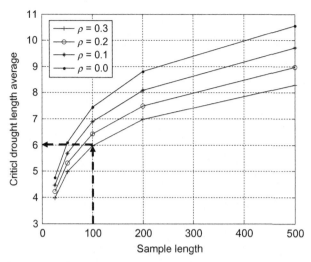

FIG. 3.13 Critical drought length average change by sample length.

Fig. 3.13 provides the relationship between the critical drought duration averages and the sample (record) length.

Example 3.8 Critical Drought Duration Length

At a meteorology station annual precipitation records during a 35-year period abide with the gaussian PDF and the first-order Markov process generating mechanism. The serial correlation coefficient is calculated as $\rho = 0.3$. Hence, estimate the possible critical drought duration within the next 100-year duration.

The solution can be read off from Fig. 3.13, where the horizontal axis is entered at 100 years with a broken arrow and its continuation as a horizontal broken arrow with reflection at $\rho = 0.3$ curve appears on the vertical axis as the critical drought duration equal to 6 years.

As for the critical drought duration probability to be equal to or smaller than a given duration for $\rho = 0$ at $q = 0.5$ level is given in Fig. 3.14 for different sample sizes. It is obvious that the differences between the analytical and simulation results are well within the practically acceptable limits.

As the sample size increases the probability of the critical drought duration to be less than a given duration also increases. On the other hand, similar probability variations are shown in Fig. 3.15 for independent and a set of dependent process correlation coefficients. Again, the analytical results match well with the simulation results.

A similar probability graph is given in Fig. 3.16 for the critical drought duration in an independent process at a set of truncation level percentages, q.

It is seen from this figure that an increase in the truncation level causes an increase in the probability of the critical drought duration. A similar situation is

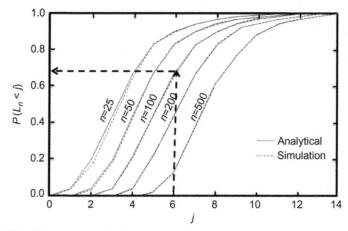

FIG. 3.14 Critical drought duration probabilities for a set on sample lengths.

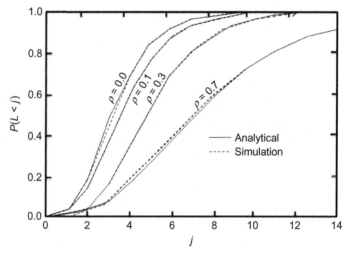

FIG. 3.15 Critical drought duration probabilities for a set of correlation coeffieients.

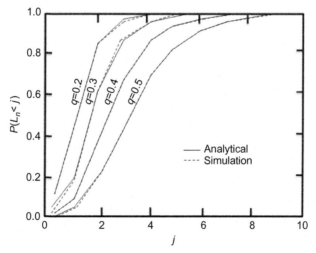

FIG. 3.16 Critical drought duration probabilities at different truncation levels.

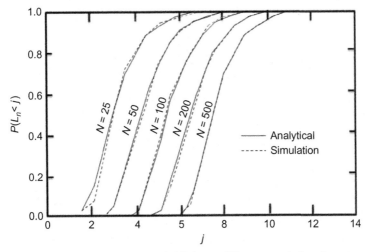

FIG. 3.17 Critical drought duration probabilities at different sample lengths.

valid in Fig. 3.17 for a dependent process critical drought duration probabilities at autocorrelation coefficient $\rho = 0.3$ level.

Provided that the truncation level and the sample length remain the same, an increase in the serial correlation coefficient causes an increase in the critical drought duration. This point indicates the significance of the serial correlation coefficient (hence, related autorun coefficient) consideration in the critical drought duration calculations.

Example 3.9 Critical Drought Duration

Calculate the critical drought duration probability for the same process given in Example 3.7.

It is already calculated in the previous example that the critical drought duration is 6 years. Hence, the entrance of this value in Fig. 3.14 on the horizontal axis with a broken arrow and its reflection at the sample size of $n = 100$ yields the probability value as 0.68 on the vertical axis.

3.6.1.3 MAXIMUM DEFICIT SUMMATION

It is rather difficult to derive an analytical expression for the deficit summation, D_j, along a drought duration of length j. According to central limit theorem in statistics, it is possible to make an approximation by assuming a normal PDF for the hydrometeorological variable. Hence, it is necessary to find the mean and the standard deviation of the total deficit along a given drought duration. Şen (1980b) gave the mean, μ_D, and variance, σ_D^2, of a single deficit analytical expressions as follows, based on the truncated gaussian PDF:

$$\mu_D = \frac{\exp\left(-\dfrac{x_o^2}{2}\right)}{p}\sqrt{2\pi} + x_o \tag{3.59}$$

and

$$\sigma_D^2 = 1 + x_o \frac{\exp\left(-\dfrac{x_o^2}{2}\right)}{p}\sqrt{2\pi} - \frac{\exp\left(x_o^2\right)}{2}\pi p^2 \tag{3.60}$$

Hence, total deficit mean, μ_s, and variance, σ_s^2, can be expressed as

$$\mu_s = i\mu_D \tag{3.61}$$

or approximately,

$$\sigma_s^2 = \sigma_D^2\left[i + 2\rho\frac{i(1-\rho) - \left(1 - \rho'\right)}{\left(1 - \rho\right)^2}\right] \tag{3.62}$$

Consequently, it is possible to write the normal PDF of the total deficit as

$$P(D \leq s) = \frac{1}{\sqrt{2\pi}\sigma_s}\int_0^\infty \exp\left[-\frac{1}{2}\left(\frac{D - \mu_s}{\sigma_s}\right)^2\right] \tag{3.63}$$

To find the maximum total deficit PDF, it is necessary to substitute this last expression into Eq. (3.57). Here, the numerical solutions are presented for s variations from 0 to 20 with $\Delta s = 0.50$ increments. Table 3.8 presents the comparison between the analytical derivation and simulation results.

In this table the relative errors are calculated as in Table 3.7. Due to the assumptions, the relative error percentages in the case of maximum total deficit are more than the critical drought duration estimation, but still they are well below the $\pm 5\%$ error limit and, therefore, practically acceptable. On the other hand, Fig. 3.18 provides the relationship between the maximum total deficit and the sample length.

Example 3.10 Maximum Total Deficit Calculation

Again, for the process as given in Example 3.7, one can search for the expected average maximum deficit amount during the next, say, 150-year duration.

For this purpose, the vertical broken arrow is entered from the horizontal axis in Fig. 3.18 until its reflection at $\rho = 0.3$, and then onward a horizontal broken arrow reaches the vertical axis at the value of maximum total deficit as equal to 5.9 mm.

TABLE 3.8 Maximum Total Deficit Expectation Values

| Sample Length | Truncation Level, q = 0.5 | | | | | | | | | | | | | | |
| | $\rho = 0.0$ | | | $\rho = 0.1$ | | | $\rho = 0.2$ | | | $\rho = 0.3$ | | |
	Ana.	Sim.	Dif. (%)	Ana.	Sim.	Dif. (%)	Ana.	Sim.	Dif. (%)	Ana.	Sim.	Dif. (%)
25	3.61	3.60	0.28	3.81	3.90	−2.30	4.17	4.22	−1.18	4.50	4.60	−2.17
50	4.58	4.55	0.66	4.87	5.01	−2.79	5.42	5.50	−1.45	5.90	6.08	−2.96
100	5.50	5.55	−0.90	5.91	6.15	−3.90	6.70	6.85	−2.19	7.50	7.70	−2.59
200	6.42	6.41	0.15	6.95	7.03	−1.14	7.95	7.95	0.00	6.80	6.99	−2.71
500	7.65	7.56	1.19	8.41	8.64	−2.66	9.35	9.73	−3.90	10.80	11.00	−1.82

Ana., analytical; Sim., simulation; Dif., difference.

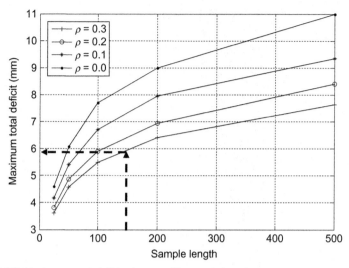

FIG. 3.18 Maximum total deficit change with sample length.

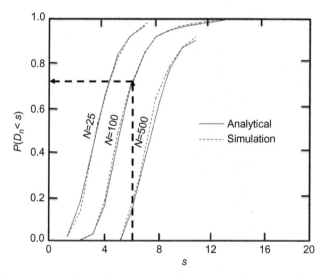

FIG. 3.19 Maximum total deficits PDF in independent process for different sample lengths.

On the other hand, the PDFs of the maximum total deficit are presented in Figs. 3.19–3.21 for independent ($\rho=0.0$) and dependent ($\rho=0.3$) processes at $q=0.5$ truncation level, respectively. Again, in all cases the relative error percentage between the analytical and simulation solutions is always less than $\pm 5\%$, which indicates the practical reliability of the derived analytical functions. Comparison of these figures implies that for a given sample length and

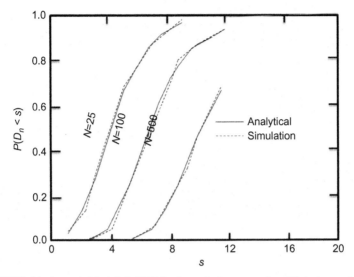

FIG. 3.20 Maximum total deficit PDF in dependent process for different sample sizes ($\rho = 0.3$).

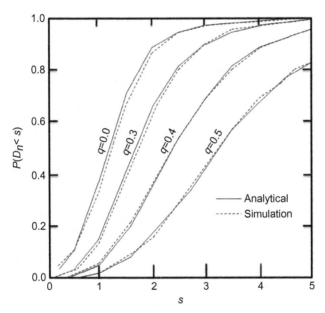

FIG. 3.21 Maximum total deficit PDF in independent process for different truncation levels.

truncation level the probability of maximum total deficit in dependent processes is more than independent processes.

3.6.1.4 DEPENDENT PROCESS MAXIMUM TOTAL DEFICIT
Fig. 3.21 presents maximum total deficit PDF's for sample length 5 at different truncation levels.

Example 3.11 Maximum Total Deficit Probability

Figs. 3.19 and 3.20 help to calculate the maximum total drought probability for any desired maximum total deficit amount. Hence, one can use the maximum deficit amount in the previous example, which was 5.9 mm, for the probability calculation. For this purpose, Fig. 3.19 is entered from the horizontal axis with a broken arrow until it reaches the, say, 100-year future period, and then the horizontal broken arrow shows on the vertical axis that the desired probability is equal to $P(D_{100} < 5.9) = 0.73$.

Example 3.12 Critical Drought Duration Probability

The methodology developed above is applied to various river flows, the characteristics of which are presented in Table 3.9.

Fig. 3.22 indicates, similar to Fig. 3.12, by consideration of Eq. (3.48), the observation (empirical) calculation and the theoretical upcrossing number

TABLE 3.9 Annual Flow Characteristics

River	Location	Sample Length (year)	Average Flow (ft³/s)	ρ
Rhine	Basle, Switzerland	150	36,253	0.077
Danube	Orshove, Romania	120	189,455	0.096
Mississippi	St. Louis, ABD	96	175,119	0.294
Missouri	Montana, ABD	65	7635	0.593

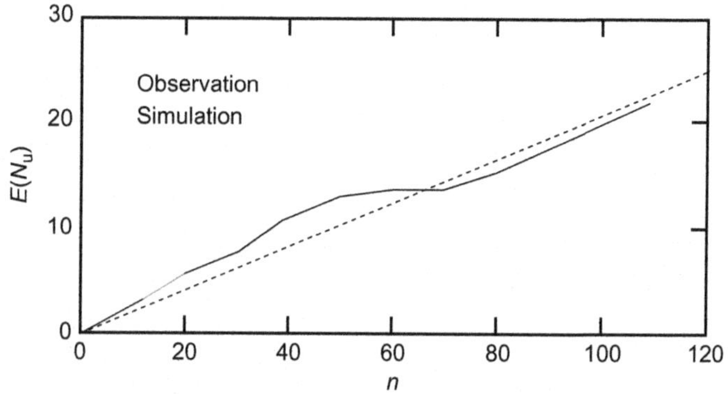

FIG. 3.22 Rhine River upcrossing numbers.

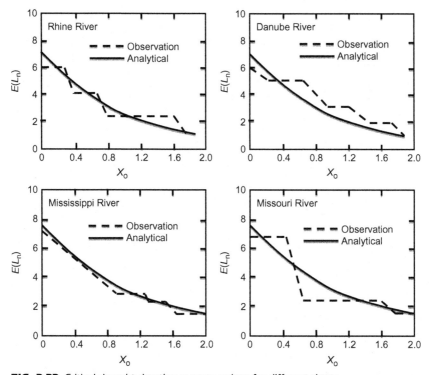

FIG. 3.23 Critical drought duration average values for different rivers.

changes and it is observed how close the upcrossing numbers are at large sample lengths for the Rhine River.

Fig. 3.23 exposes observed and calculated critical drought durations at truncation levels, x_0, between 0 and 2.

In general, observed and analytical critical drought durations yield results very close to each other. The difference between the two approaches is due to the following reasons:

1. In finite sample lengths, there are sampling errors in the calculation of critical drought durations. Here, there is only one hydrometeorology time series and hence only one critical drought duration; therefore, this causes a sampling error, where the analytical solution represents the combined result of ensemble solutions.

2. Critical drought duration calculation from the given example is an integer number, whereas analytical solutions are in decimal forms.

Fig. 3.24 presents observed and analytical results of the maximum total deficit for the Rhine and Danube rivers, respectively, at different truncation levels. The Rhine River has a serial correlation coefficient, $\rho = 0.077$, and consequently, observed maximum total deficit is always greater than the analytical counterpart. In other words, a first-order Markov process underestimates the maximum total deficit value.

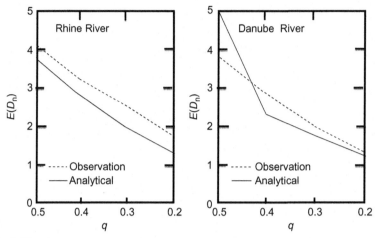

FIG. 3.24 Maximum total deficit variation with truncation level for Rhine and Danube rivers.

3.7 CRITICAL DROUGHT DURATION

In the literature, some studies exist concerning the longest (critical) negative run-length. For instance, the expression of the PDF of the longest critical negative run-length in finite but independent series has been given as an approximation by Feller (1967). On the other hand, Millan (1972) gave an exact formula for the longest negative run-length distribution for an irreducible Markov chain of two ergodic stages on the basis of combinatorial analysis. These formulae are tedious to evaluate and an approximation is necessary to critical drought duration features by stochastic processes.

3.7.1 Enumeration Model of Independent Bernoulli Trials

To discuss the enumeration method explicitly, first independent processes are considered (Şen, 1980d). Truncation of such a process at a constant level yields two elementary and mutually distinct events; namely, a surplus $(X_i > X_o)$ and a deficit $(X_i < X_o)$ with probabilities $p = P(X_i > X_o)$ and $q = P(X_i < X_o) = 1 - p$, respectively. If the probability of the longest positive run-length L in a sample size i equal to j is denoted by $P_i(L = j)$, then for sample size $n = 1$ one can simply write

$$P_1(L = 0) = q$$
$$P_1(L = 1) = p \tag{3.64}$$

In the case of $i = 2$, there are 2^2 combinations of surplus and deficit events, out of which the one with a sequential surplus–surplus event constitutes the longest run-length equal to zero $(L = 0)$. On the other hand, the combination of surplus–deficit or deficit–surplus events give rise to the longest run-length equal to one $(L = 1)$, whereas a deficit–deficit event results in the longest run-length equal to two $(L = 2)$, and they are shown in Fig. 3.25.

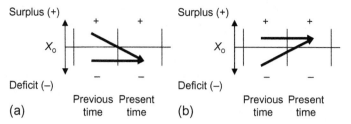

FIG. 3.25 Probability calculating structure.

Because at this stage the elementary events are assumed to be independent and mutually exclusive, the probabilities of various combinations can be deduced as

$$P_2(L=0) = P_1(L=0)q$$
$$P_2(L=1) = P_1(L=1)q + P_1(L=0)p \qquad (3.65)$$
$$P_2(L=2) = P_1(L=1)p$$

In the same way, for $i=3$, there are $2^3 = 8$ combinations, which include different mutually exclusive and collectively exhaustive alternatives as deficit–deficit–deficit ($L=0$), surplus–deficit–deficit ($L=1$), deficit–surplus–deficit ($L=1$), deficit–deficit–surplus ($L=1$), deficit–surplus–surplus ($L=2$), surplus–deficit–surplus ($L=2$), surplus–surplus–deficit ($L=2$), and surplus–surplus–surplus ($L=3$). After a close inspection of these combinations, the probabilities of the longest run-length can be found in terms of the previous step probabilities.

$$P_3(L=0) = P_2(L=0)q$$
$$P_3(L=1) = P_2(L=1)q + P_2(L=0)p + P_1(L=1)qp$$
$$P_3(L=1) = P_2(L=2)q + [P_2(L=1)p - P_1(L=1)q]p \qquad (3.66)$$
$$P_3(L=3) = P_2(L=2)p$$

Similar types of recurrence relationships can be obtained for $i=4$, and there are $2^4 = 16$ cases out of which the following probability pattern emerges:

$$P_4(L=0) = P_3(L=0)q$$
$$P_4(L=1) = P_3(L=1)q + P_3(L=0)p + P_2(L=1)qp$$
$$P_4(L=2) = P_3(L=2)q + [P_3(L=1)p - P_2(L=1)q]p + P_2(L=2)qp \quad (3.67)$$
$$P_4(L=3) = P_3(L=3)q + [P_3(L=2)p - P_2(L=2)q]p$$
$$P_4(L=4) = P_3(L=3)p$$

It is possible to continue writing the probabilities of the longest run-length for sample sizes, n, more than 4 ($n>4$). However, a close inspection of Eq. (3.67) and previous expressions reveals the general probability pattern of relationships in sample size, n, for zero longest positive run-length as

$$P_n(L=0) = P_{n-1}(L=0)q \qquad (3.68a)$$

Furthermore, it is also straightforward to write a recurrence relationship for the longest positive run-length to be equal to the sample size considered by the comparison of the last lines in Eqs. (3.64)–(3.67) leading to

$$P_n(L=n) = P_{n-1}(L=n-1)q \tag{3.68b}$$

However, the general probability expression for the longest run-length of size j where $0 < j < n$ becomes as (Şen, 1980d)

$$P_n(L=j) = P_{n-1}(L=j)q + \left[P_{n-1}(L=j-1) - \sum_{i=1}^{k_1} P_{n-i-1}(L=j-1)qp^{i-1} \right]$$
$$p + \sum_{i=1}^{k_2} P_{n-i-1}(L=j)qp^i$$

$$\tag{3.68c}$$

where $k_1 = \min(n-j, j-1)$ and $k_2 = \min(n-j-1, j)$. Eqs. (3.68a)–(3.68c) represent completely the exact probability of the longest positive run-length. By interchanging p and q, the probability of longest negative run-length, N, that is, critical drought duration, can be obtained as

$$P_n(N=0) = P_{n-1}(N=0)p$$
$$P_n(N=j) = P_{n-1}(N=j)p + \left[P_{n-1}(N=j-1) - \sum_{i=1}^{k_1} P_{n-i-1}(N=j-1)pq^{i-1} \right]$$
$$q + \sum_{i=1}^{k_2} P_{n-i-1}(N=j)pq^i$$
$$P_n(N=n) = P_{n-1}(N=n)q$$

$$\tag{3.69}$$

The validity of these probability expressions can be checked numerically such that the summation of all possible longest positive (negative) run-lengths should be equal to one.

$$\sum_{i=1}^{n} P_n(L=j) = \sum_{i=0}^{n} P_n(N=i) = 1 \tag{3.70}$$

A set of numerical solutions for Eq. (3.74) is given in Fig. 3.26 for different truncation level probability, p. These figures are obtained by the MATLAB code written software, which has been given in Appendix 3.2.

The theoretical mean (expectation), $E(.)$, and the variance, $V(.)$, of the longest run-length can be calculated similar to expressions Eqs. (3.33) and (3.40) as

$$E(L) = \sum_{j=1}^{n} jP(L=j) \tag{3.71}$$

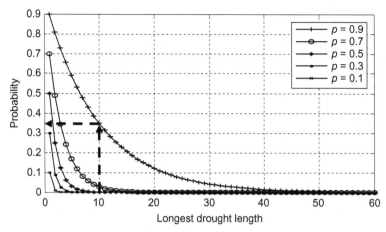

FIG. 3.26 Different PFDs for independent process.

and

$$V(L) = \sum_{j=1}^{n} j^2 P(L=j) - E^2(L) \tag{3.72}$$

respectively.

Example 3.13 Critical Drought Duration Probability in Independent Time Series

It has been observed that at a location the annual flow series does not have serial correlation and the structure is entirely independent. In such a case, what will be the expected probability of having the longest drought length equal to 10 at $p=0.9$ level?

The solution can be read off from Fig. 3.26 directly by entering the value of 10 on the horizontal axis; hence, the vertical axis yields the probability as 0.34.

3.7.2 Dependent Bernoulli Trials

The exact distribution of the longest positive (negative) run-length in n trials for the case of dependent events has been given by Millan (1972) through the application of combinatorial analysis. The result of each trial is dependent only on the outcome of the previous trial. For derivation of the longest run-length, two probability information sources are necessary.

1. Truncation level surplus probability, similar to the independence case in the previous subsection.
2. Information about the transitions (conditional state probabilities) between two successive states.

There exist four possibilities of transitions between successive trials as in Fig. 3.27. These four probabilities are as follows:

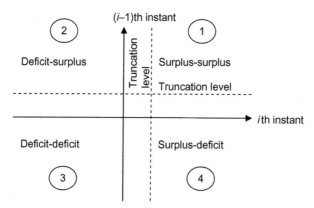

FIG. 3.27 Four events at the same truncation level.

1. Transition from a surplus to a surplus, $P(+/+)$.
2. Transition from a surplus to a deficit, $P(-/+)$.
3. Transition from a deficit to a surplus, $P(+/-)$.
4. Transition from a deficit to a deficit, $P(-/-)$.

Prior to the probability calculations of these events, Fig. 3.28 shows the joint probability notations for each subdomain.

Because homogeneous (steady state) Bernoulli trials are considered, the transition as well as the state probabilities are independent from absolute time. Hence, the state probabilities on the second step in time can be written in terms of transition probabilities as

$$p = P(+/+)p + P(+/-)q$$
$$q = P(-/+)p + P(-/-)q \qquad (3.73)$$

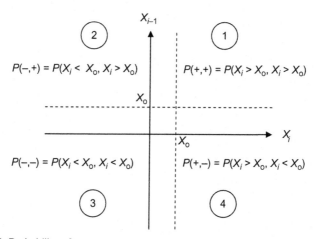

FIG. 3.28 Probability of events.

where $p+q=1$; $P(+/+)+P(-/+)=1$; and $P(-/+)+P(-/-)=1$. After all these definitions, the enumeration method can be applied similar to the case of independent process. Probabilities of the longest positive run-length, L, being equal to an integer j value in a sample of size i with a surplus state at the final stage is denoted by $P_i^+(L=j)$. At the first stage, there exists no transition and one can simply write similar to Eq. (3.64) with new notations as

$$P_1^-(L=0)=P(-)=q$$
$$P_1^+(L=1)=P(+)=p$$

(3.74)

The reader should notice that in the dependent case, the final stage plays a significant role in the run-length probability calculations. For $i=2$, there are $2^2=4$ transitions and they are dependent on the previous time interval probabilities as follows:

$$P_2^-(L=0)=P_1^-(L=0)P(-/-)$$
$$P_2^+(L=1)=P_1^-(L=0)P(+/-)$$
$$P_2^-(L=1)=P_1^-(L=1)P(-/+)$$
$$P_2^+(L=2)=P_1^-(L=1)P(+/+)$$

(3.75)

Continuation of the enumeration in the same way leads to the general formulations as (Şen, 1980d)

$$P_i^-(L=0)=P_{i-1}^-(L=0)P(-/-/)$$
$$P_i^+(L=j)=\sum_{m=0}^{k_1}P_{i-j}^-(L=m)P(+/-)P^{j-1}(+/+)\sum_{m=1}^{k_2}P_{i-m}^-(L=m)P^{m-1}(+/+)$$
$$P_i^-(L=i)=P_{i-1}^+(L=j)P(-/+)+P_{i-1}^-(L=j)P(-/-) \quad \text{if } i-1=j \text{ then } P(-/-)=0$$
$$P_i^+(L=i)=P_{i-1}^+(L=i-1)P(+/+)$$

(3.76)

where $k_1=\min(i-j-1,j)$ and $k_2=\min(i-j-1,j-1)$. The probabilities on the longest drought duration can be evaluated from Eq. (3.76) provided that $P(+/+)$, $P(+/-)$, $P(-/+)$ and $P(-/-)$ are interchanged by $P(-/-)$, $P(-/+)$, $P(+/-)$ and $P(+/+)$, respectively. Of course, in Eq. (3.76) the probability of the longest run-length can be obtained without dependence on the final stage as

$$P_i(L=j)=P_i^+(L=j)+P_i^-(L=j)$$

(3.77)

The numerical solution of the probability equations set is possible through computer software, and such a program has been developed in MATLAB given in Appendix 3.3. This software also calculates the statistical parameters of the longest positive (negative) run-length.

The analytical PDFs given in Eq. (3.76) are solved at $q=0.4$ truncation level, for sample size $n=25$ and serial correlation coefficients $\rho=0.1$, 0.3, and 0.5. The range of correlation coefficients covers the situations encountered in practice. The transition probabilities for these correlation coefficients are presented in Table 3.10.

The numerical solutions of Eq. (3.76) are given with the values in Fig. 3.29 together with the Monte Carlo simulation results. An inspection of the figure shows that the developed analytical methodology results are in good agreement with the simulations within ±1% relative error limits.

The agreement between the analytical and simulation approaches is satisfactory even for big correlations such as $\rho = 0.7$. The same figure indicates that for a given probability of exceedance (risk), the more the correlation coefficient, the longer the drought length. Therefore, persistent annual flows require larger reservoir capacities than independent processes. The longest run-length is quite sensitive to the correlation coefficient (hence, autorun coefficient), which must be estimated from a given observation sequence with great care and accuracy.

The expected (average) longest run-lengths are given in Fig. 3.30, which shows that the increase in the longest run-lengths is rather slow in large sample sizes. However, in small sample sizes, the increase is quite rapid.

Fig. 3.31 represents the change of variance of the longest drought duration with the sample size, where the same arguments are valid as for Fig. 3.30.

Although the above formulations are exact for independent processes, it is a very good approximation for Markov chains and yields practically satisfactory results. A prerequisite for the application of the formulations in an actual design situation is the calculation of the transition probabilities from a given data sequence.

TABLE 3.10 Transition Probabilities

ρ	P(+/+)	P(+/−)	P(−/+)	P(−/−)
0.1	0.44	0.38	0.56	0.62
0.3	0.52	0.32	0.48	0.68
0.4	0.69	0.20	0.31	0.80

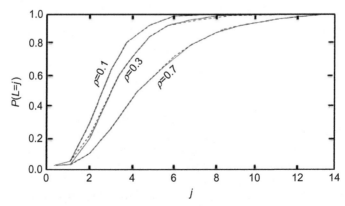

FIG. 3.29 Numerical solution result of the maximum run-length in a dependent process.

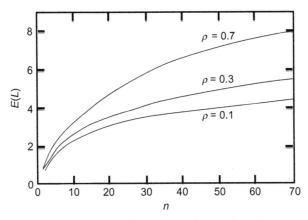

FIG. 3.30 Average of the longest positive run-length variation with sample size.

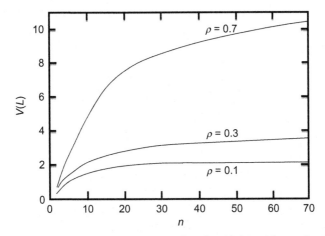

FIG. 3.31 Variance of the longest positive run-length variation with sample size.

3.7.3 Second-Order Markov Process Drought Properties

In this process, present time hydrometeorological variable, X_i, is a linear function of the two previous time instant variables, X_{i-1} and X_{i-2}, with the addition of a random component, ε_i. It is assumed that the hydrometeorological variables have a normal (gaussian) PDF,

$$X_i = aX_{i-1} + bX_{i-2} + \varepsilon_i \qquad (3.78)$$

where a and b are the process coefficients that are functions of the first- and the second-order serial coefficients ρ_1 and ρ_2 and ε_i shows ith error, which has a completely independent structure that abides with the normal independent PDF with zero mean. Şen (1990) has presented the dry and wet duration properties of this process. The same rational steps as in the first-order Markov process are considered but instead of two successive time intervals, herein three consecutive time intervals are needed for basic treatments. In the second-order Markov process case, there are $2^3 = 8$ different subdomains; again, each with specific probability statement as in Fig. 3.32.

The second-order Markov process requires three-interval basic transitional probabilities in addition to two-interval probabilities as in Fig. 3.25A and B for which the surplus and deficit probabilities remain as they were in the previous subsection. The new set of transitional probabilities can be defined as eight ($i = 2^3 = 8$) alternatives, which are mutually exclusive and collectively exhaustive. For example, one of the eight alternatives is

$$P(+/+-) = P(X_i > X_0, X_{i-1} > X_0, X_{i-2} < X_0)$$

Herein, $P(+/+-)$ refers to the probability of surplus at the current interval given that two previous successive intervals have deficit and surplus states, respectively. On the other hand, mutual exclusiveness implies that $P(+/++) + P(-/++) = 1$; $P(-/-+)$ $+ P(-/+-) = 1$; $P(+/-+) + P(-/-+) = 1$; and $P(+/--) + P(-/--) = 1$.

According to the notation in Fig. 3.32, each one of the transition (conditional) probabilities can be conditioned on the previous two stages as follows:

$$
\begin{aligned}
P(+/++) &= P(X_i > X_0/X_{i-1} > X_0, X_{i-2} > X_0) \\
P(-/++) &= P(X_i < X_0/X_{i-1} > X_0, X_{i-2} > X_0) \\
P(+/+-) &= P(X_i > X_0/X_{i-1} > X_0, X_{i-} < X_0) \\
P(-/+-) &= P(X_i < X_0/X_{i-1} > X_0, X_{i-2} < X_0) \\
P(+/-+) &= P(X_i > X_0/X_{i-1} < X_0, X_{i-2} > X_0) \\
P(-/-+) &= P(X_i < X_0/X_{i-1} < X_0, X_{i-2} > X_0) \\
P(+/--) &= P(X_i > X_0/X_{i-1} < X_0, X_{i-2} < X_0) \\
P(-/--) &= P(X_i < X_0/X_{i-1} < X_0, X_{i-2} < X_0)
\end{aligned}
\tag{3.79}
$$

The first probability statement implies a surplus at the present time interval on the condition that two previous time intervals are also at surplus states.

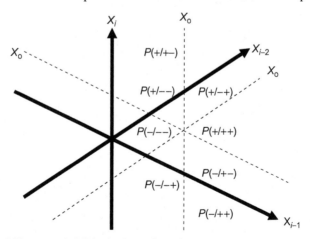

FIG. 3.32 Different probabilities in three-dimensional space.

The reader can interpret the other probability statements similarly. On the basis of the conditional probability statements, mutually exclusiveness and exhaustiveness lead to the following equalities:

$$P(+/++) + P(-/++) = 1$$
$$P(-/-+) + P(-/+-) = 1$$
$$P(+/-+) + P(-/-+) = 1 \tag{3.80}$$
$$P(+/--) + P(-/--) = 1$$

The second-order Markov process has the same critical drought duration probabilities as in the first-order Markov process during the initial (Eq. 3.64) and the next time (Eq. 3.75) intervals. With the sample length $i = 3$, the $2^3 = 8$ mutually exclusive alternatives can be combined as follows:

$$P_3(L=0/++) = P_2(L=0/+)P(+/++)$$
$$P_3(L=1/-+) = P_2(L=0/+)P(-/++) + P_2(L=1/-)P(-/+-)$$
$$P_3(L=1/++) = P_2(L=1/-)P(+/+-)$$
$$P_3(L=1/+-) = P_2(L=1/+)P(+/-+) \tag{3.81}$$
$$P_3(L=2/--) = P_2(L=1/+)P(-/-+)$$
$$P_3(L=2/+-) = P_2(L=2/-)P(-/+-)$$
$$P_3(L=3/--) = P_2(L=2/-)P(+/--)$$

For instance, $P_3(L=1/++)$ indicates the maximum (critical) drought duration length of one ($L=1$), provided that the two previous time intervals have surplus states. For the subsequent sample lengths, the probabilities are functions of the probability statements in the two previous sample lengths and, accordingly, it is possible to write down general probabilities for sample length n as (Şen, 1990)

$$P_n(L=0/++) = P_{n-1}(L=0/++)P(+/++)$$
$$P_n(L=1/-+) = P_{n-1}(L=0/++)P(-/++) + P_{n-1}(L=1/+-)P(-/+-)$$
$$\quad + P_{n-1}(L=1/+-)P(-/+-)$$
$$P_n(L=1/-+) = P_{n-1}(L=1/+-)P(+/+-) + P_{n-1}(L=1/+-)P(-/+-)$$
$$P_n(L=1/+-) = P_{n-1}(L=0/+-)P(+/+-)$$
$$P_n(L=j/--) = P_{n-1}(L=j-1/-+)P(-/-+) + P_{n-1}(L=j/-+)P(-/-+)$$
$$P_n(L=j/-+) = P_{n-1}(L=j/++)P(-/++) + P_{n-1}(L=j/+-)P(-/+-)$$
$$P_n(L=j/++) = P_{n-1}(L=j/++)P(+/++) + P_{n-1}(L=j/+-)P(+/+-)$$
$$P_n(L=j/+-) = P_{n-1}(L=j/-+)P(-/-+) + P_{n-1}(L=j/++)P(+/++)$$
$$\text{where } 2 \le j \le (n-2)$$
$$P_n(L=n-2/--) = P_{n-1}(L=n-3/-+)P(-/-+)$$
$$P_n(L=n-2/-+) = P_{n-1}(L=n-3/--)P(-/+-)$$
$$P_n(L=n-2/++) = P_{n-1}(L=n-3/+-)P(+/+-)$$
$$P_n(L=n-2/+-) = P_{n-1}(L-n-3/--)P(-/--)$$
$$P_n(L=n-1/+-) = P_{n-1}(L=n-1/--)P(+/--)$$
$$P_n(L=n/--) = P_{n-1}(L=n-1/--)P(-/--)$$
$$\tag{3.82}$$

In general, the PDFs of the critical drought duration at truncation level, $q = 0.5$, indicates that all the critical drought PDFs have a positive skewness, which increases with increasing sample length; there is an important effect of the positive dependence term, $P(+/+ +)$, on the critical drought duration. Increase in its value reduces the frequency of the critical drought duration occurrence and the effect of other transitional probabilities is not as effective as for $P(+/+ +)$.

Expectation of the critical drought duration is calculated from Eq. (3.71) and the results are presented in Fig. 3.33 for different transition probabilities.

Comparison of these figures shows that the effect of the positive dependence probability statement $P(+/+ +)$ on the expectation values is more sensitive for transition probabilities greater than 0.6.

Example 3.14 Critical Drought Duration Probability in Second-Order Dependent Time Series

The above methodology is applied to three different river basins, the characteristics of which are presented in Table 3.11. The statistical properties of these flows have already been presented by Yevjevich (1963) and Şen (1990). Table 3.12 includes critical drought durations according to different truncation probability levels. Substitution of these probability values into Eq. (3.82) gives the critical drought probabilities, which are then substituted into Eq. (3.71) for the expectation value calculations.

In Fig. 3.34 the correspondence between the observed and predicted values is very good, at least for small sample sizes. In the three cases prediction values are greater than observation counterparts for $L > 20$, and the differences are due to sampling errors.

3.7.4 Seasonal Bernoulli Trials

The analytical derivation of the longest run-length for seasonal series, say monthly, hydrometeorological variables is quite difficult but under the light of the two previous sections, one can simplify the analytical solutions (Şen, 1980c). The transitional probabilities between two successive time intervals (from $i - 1$ to i) of a seasonal hydrometeorological variable can be written in a matrix form as follows:

$$P_i = \begin{array}{c} \\ + \\ - \end{array} \begin{array}{c} + \qquad\quad - \\ \begin{bmatrix} P_i(+/+) & P_i(-/+) \\ P_i(+/-) & P_i(-/-) \end{bmatrix} \end{array} \qquad (3.83)$$

Similar to all the previous notational convenience, here again $+$ and $-$ correspond to surplus and deficit states, respectively. Each one of the elements in this matrix shows conditional probability, where the summation of row elements is equal to 1 due to the mutual exclusiveness of the probability principle. If there are k time intervals (seasons) in a process then, similar to Eq. (3.83), there are a set of transitional probabilities that can be denoted by

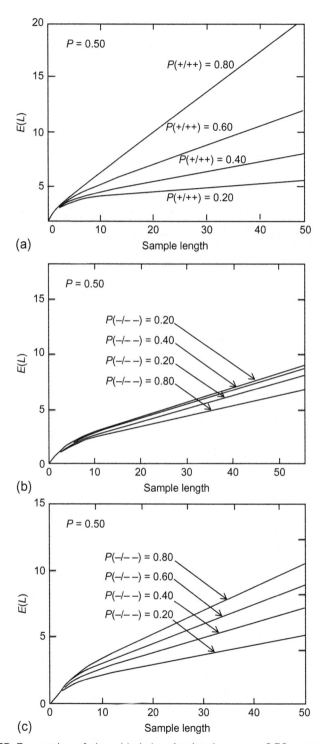

FIG. 3.33 Expectation of the critical drought duration at $p=0.50$ truncation level: (A) $P(+/+ +)$, (B) $P(-/- -)$, (C) $P(-/- -)$.

TABLE 3.11 River Flow Characteristics

River	Location	Sample Length (year)	Average Discharge (cfs)
Rhine	Basle	150	36,253
Tuna	Romania	120	189,455
Mississippi	St. Louis	96	175,119

TABLE 3.12 Observed Transition Probabilities (s for surplus; d for deficit).

	Present Year, i		
i − 2	i − 1	s	d
Rhine			
	s	0.49	0.51
s	d	0.25	0.24
d	s	0.23	0.28
s	s	0.10	0.13
d	d	0.14	0.11
	d	0.12	0.12
		0.12	0.17
Tuna			
	s	0.48	0.52
s	d	0.21	0.27
d	s	0.25	0.27
s	s	0.09	0.14
d	d	0.11	0.15
	d	0.13	0.12
		0.12	0.14
Mississippi			
s	s	0.45	0.55
d	d	0.25	0.21
s	s	0.20	0.34
d	s	0.16	0.10
	d	0.10	0.11
	d	0.06	0.15
		0.14	0.19

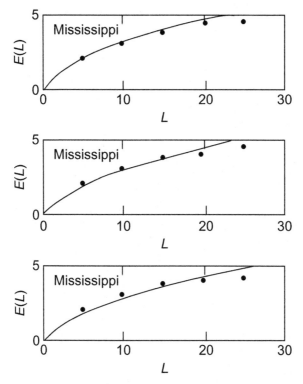

FIG. 3.34 Observed and predicted critical drought duration expectations for three rivers.

$P_2, P_3, \ldots, P_k$. These can be calculated either from a given theoretical joint PDF of two subsequent variables or from a given observation sequence by calculating transitional probabilities. For the development of a convenient model, one should keep in mind that for any given hydrometeorological sequence, the critical drought duration in sample length $(i+1)$ is a function of the critical drought duration in the previous sample size i and the transitional probabilities between the two time intervals. Let $P^+(L=j)$ denote the probability of the critical drought duration length equal to j with the end state at surplus (+). The probabilities for the first stage are exactly the same as for dependent Bernoulli trials in Section 3.7.2.

In the second time interval, there are $2^2 = 4$ complementary alternatives, which can be written according to dependent Bernoulli trials case with time indicative subscripts as the number of sample size.

$$
\begin{aligned}
P_2^-(L=0) &= P_1^-(L=0)P_2(-/-) \\
P_2^+(L=1) &= P_1^-(L=0)P_2(+/-) \\
P_2^-(L=1) &= P_1^+(L=1)P_2(-/+) \\
P_2^+(L=2) &= P_1^+(L=1)P_2(+/+)
\end{aligned}
\tag{3.84}
$$

Similarly, during the transition from sample size $i=2$ to $i=3$, the probabilities take the following forms:

$$
\begin{aligned}
P_3^-(L=0) &= P_2^-(L=0)P_3(-/-) \\
P_3^+(L=1) &= P_2^-(L=0)P_3(+/-) + P_2^-(L=1)P_3(+/-) \\
P_3^-(L=1) &= P_2^+(L=1)P_3(-/+) + P_2^-(L=1)P_3(-/-) \\
P_3^+(L=2) &= P_2^+(L=2)P_3(+/+) \\
P_3^-(L=2) &= P_2^+(L=2)P_3(-/+) \\
P_3^+(L=3) &= P_2^+(L=2)P_3(+/+)
\end{aligned}
\tag{3.85}
$$

Finally, for $i=4$ with $2^4 = 16$ alternatives, it is possible to write with the same thoughts the probabilities as

$$
\begin{aligned}
P_4^-(L=0) &= P_3^-(L=0)P_4(-/-) \\
P_4^+(L=1) &= P_3^-(L=0)P_4(+/-) + P_3^-(L=1)P_4(+/+) \\
P_4^-(L=1) &= P_3^+(L=1)P_4(-/+) + P_3^-(L=1)P_4(-/-) \\
P_4^+(L=2) &= P_3^+(L=1)P_4(+/+) + P_3^-(L=2)P_4(+/-) \\
P_4^-(L=2) &= P_3^+(L=2)P_4(-/+) + P_3^-(L=2)P_4(-/-) \\
P_4^+(L=3) &= P_3^+(L=2)P_4(+/+) \\
P_4^-(L=3) &- P_3^+(L-3)P_4(-/+) \\
P_4^+(L=4) &= P_3^+(L=3)P_4(+/+)
\end{aligned}
\tag{3.86}
$$

A close inspection of these equation systems for critical drought duration to be equal to zero in a sample of length n leads after a rational deduction as

$$
P_n^-(L=0) = P_{n-1}^-(L=0)P_n(-/-)
\tag{3.87}
$$

On the other hand, for the last sample length, it is possible to decide that

$$
P_n^+(L=n) = P_{n-1}^+(L=n-1)P_n(+/+)
\tag{3.88}
$$

However, for sample lengths in between $(1 \le j \le n-1)$, the probability statements with + or − endings have the following general formulations:

$$
P_n(L=j) = P_n^+(L=j) + P_n^-(L=j)
\tag{3.89}
$$

$$
\begin{aligned}
P_n^+(L=j) ={}& \sum_{m=0}^{k_1} P_{n-j}^-(L=m)P_{n-j+1}(+/-)\prod_{k=1}^{j-1}P_{n-j+k+1}(+/+) \\
&+ \prod_{m-1}^{k_1} P_{n-m}^-(L=j)P_{n-m+1}(+/-)\prod_{k-1}^{m-1}P_{n-m+k+1}(+/+)
\end{aligned}
\tag{3.90}
$$

and

$$P_n^-(L=j)=P_{n-1}^+(L=j)P_n(-/+)+P_{n-1}^-(L=J)P_n(-/-) \quad (\text{if } n-1=j$$

$$\text{then } P(-/-)=0) \tag{3.91}$$

Herein, $k_1=\min(n-j-1, j)$ and $k_2=\min(n-j-1, j-1)$. Eqs. (3.87)–(3.91) give in detail the probability of critical drought duration. The most critical drought duration (longest negative run-length) can be obtained by sign alteration in the above expressions as

$$P_n^+(N=0)=P_{n-1}^+(N=0)P_n(+/+)$$
$$P_n^-(N=j)=\sum_{m=0}^{k_1}P_{n-j}^+(N=m)P_{n-j+1}(-/+)\prod_{k=1}^{j-1}P_{n-j+k+1}(-/-)$$
$$+\sum_{m=1}^{k_2}P_{n-m}^+(N=j)P_{n-m+1}(-/+)\prod_{k=1}^{m-1}P_{n-m+k+1}(-/-)$$
$$P_n^+(N=j)=P_{n-1}^-(N=j)P_n(+/-)+P_{n-1}^+(N=j)P(+/+) \quad \text{if } n-1=j$$
$$\text{then } P(+/+)=0$$
$$P_n^-(N=n)=P_{n-1}^-(N=n-1)P(-/-)$$

$$\tag{3.92}$$

It is possible to check the validity of this set of probability equations whether their summation is equal to 1.

In the case of monthly flows, in addition to the two initial time interval state probabilities, there are $2 \times 12 = 24$ transition probabilities that reflect the internal dependence structure of the given hydrometeorology time series. The probability of "high" flows to follow "high" flows, $P_i(+/+)$, and "low" flows to follow "low" flows, $P_i(-/-)$, is greater that the probability of a "high" flow to follow "low", $P_i(+/-)$ and "low" flow to follow "high" flow, $P_i(-/+)$.

Example 3.15 Critical Drought Duration Probability in Seasonal Time Series-1

Herein, application is performed for the first four different synthetic scenario transition probability cases given in Table 3.13.

In case I, the transition probability from the surplus to the next surplus spell, $P_i(+/+)$, first increases from 0.1 to 0.9 and then likewise decreases to 0.1; in case II, the opposite situation is valid; in case III, there are alternate increases and decreases; and in case IV, there is a sudden decrease from 0.9 to 0.1. In the scenario and later on real cases the truncation level is assumed to pass through the median value of the PDF, which means that $p=q=0.5$. In case of standard variable with zero mean and unit standard deviation the transitional probabilities are related to the serial correlation coefficient, ρ, as (Şen, 1980d)

$$\rho=\sum_{i=1}^{12}[P_i(+/+)P_i(+)-P_i(-/+)P_i(+)-P_i(+/-)P_i(-)+P_i(-/-)P_i(-)]$$

$$\tag{3.93}$$

TABLE 3.13 Scenario Probabilities

(i)	Month	Case I			Case II			Case III			Case IV		
		$P_i(+)$	$P_i(+/+)$	$P_i(-/-)$	$P_i(+)$	$P_i(+/+)$	$P_i(-/-)$	$P_i(+)$	$P_i(+/+)$	$P_i(-/-)$	$P_i(+)$	$P_i(+/+)$	$P_i(-/-)$
1	Jan.	0.50	0.10	0.90	0.50	0.90	0.10	0.50	0.10	0.90	0.50	0.90	0.10
2	Feb.	0.50	0.20	0.80	0.50	0.80	0.20	0.50	0.90	0.10	0.50	0.90	0.10
3	Mar.	0.50	0.40	0.60	0.50	0.60	0.40	0.50	0.10	0.90	0.50	0.90	0.10
4	Apr.	0.50	0.60	0.40	0.50	0.40	0.60	0.50	0.90	0.10	0.50	0.90	0.10
5	May	0.50	0.80	0.20	0.50	0.20	0.80	0.50	0.10	0.90	0.50	0.90	0.10
6	Jun.	0.50	0.90	0.10	0.50	0.10	0.90	0.50	0.90	0.10	0.50	0.90	0.10
7	Jul.	0.50	0.80	0.20	0.50	0.20	0.80	0.50	0.10	0.90	0.50	0.10	0.90
8	Aug.	0.50	0.60	0.40	0.50	0.40	0.60	0.50	0.90	0.10	0.50	0.10	0.90
9	Sep.	0.50	0.40	0.60	0.50	0.60	0.40	0.50	0.10	0.90	0.50	0.10	0.90
10	Oct.	0.50	0.20	0.80	0.50	0.80	0.20	0.50	0.90	0.10	0.50	0.10	0.90
11	Nov.	0.50	0.10	0.90	0.50	0.90	0.10	0.50	0.10	0.90	0.50	0.10	0.90
12	Dec.	0.50	0.10	0.90	0.50	0.90	0.10	0.50	0.90	0.10	0.50	0.10	0.90
	Average	0.50	0.43	0.57	0.50	0.57	0.43	0.50	0.50	0.50	0.50	0.50	0.50

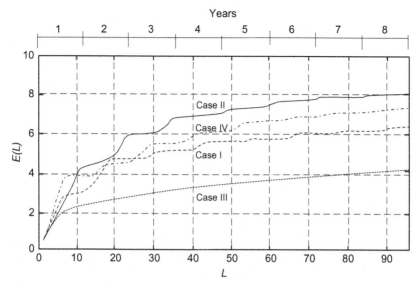

FIG. 3.35 The change of critical drought expectation by transition probability sets and the sample length.

For each case in Table 3.1 the suitable probability equations are applied and the final results are presented in Fig. 3.35.

It is possible to conclude that the following points are among the most significant features:

1. If there is seasonality in the transition probabilities, then critical drought duration expectation variation of the same process has the reflections of the seasonality in the expected drought duration, $E(L)$, variation with the desired drought length, L.

2. Increase in the sample length gives rise to decrease in the seasonality effect on the expectation values.

3. The order of the transition probabilities is more important than their values. For instance, in Table 3.13 in cases III and IV, there are six probability numbers with values 0.1 and 0.9, but in case of IV their sequence gives rise to more pronounced critical drought duration averages.

Example 3.16 Critical Drought Duration Probability in Seasonal Time Series-2

The application is made also with factual data given for the rivers; namely, Fırat (Euphrates), Kızılırmak, Manavgat, and B. Menderes in Turkey (see Fig. 3.36). The necessary state and transitional probabilities are calculated from the given annual flow sequences. The characteristics of these river flows are presented in Table 3.14.

Execution of the following steps is necessary to calculate the estimations of various probabilities, $\hat{P}_i(+)$ and $\hat{P}_i(+/+)$.

FIG. 3.36 Location of four seasonal flow measurement stations.

TABLE 3.14 Flow Characteristics

River Name	Station Name	Basin Area (km²)	Record Duration (year)
Büyük Menderes	Çine	948	33
Manavgat	Manavgat	928	31
Kızılırmak	Kızılırmak	15,582	32
Fırat	Keban	63,874	34

1. Each year's ith and $(i+1)$th monthly flows are truncated at appropriate monthly demand levels.
2. Joint surplus (at subsequent two states) number $(n_{ss})_i$ is counted for the whole duration. On the other hand, for each month surplus number $(n_s)_i$ is also found,
3. If there are n observations, then one can calculate the following probability estimations:

$$\hat{P}_i(+) = \frac{(n_s)_i}{n} \qquad (3.94)$$

and

$$\hat{P}_i(+/+) = \frac{(n_{ss})_i}{n} \qquad (3.95)$$

4. Finally, transitional probabilities can be calculated from classical probability frequency definition as

$$\hat{P}_i(+/+) = \frac{\hat{P}_i(+,+)}{P_i(+)} \qquad (3.96)$$

Table 3.15 provides the result of complete transition probability values that are the preliminary requirements for the critical drought duration expectations (means), variances, and any other statistical parameter calculations.

TABLE 3.15 Actual Probabilities

(I)	Month	Büyük Menderes			Manavgat			River Names Kızılırmak			Fırat (Euphrates)		
		$P_i(+)$	$P_i(+/+)$	$P_i(-/-)$	$P_i(+)$	$P_i(+/+)$	$P_i(-/-)$	$P_i(+)$	$P_i(+/+)$	$P_i(-/-)$	$P_i(+)$	$P_i(+/+)$	$P_i(-/-)$
1	Jan.	0.363	0.833	0.762	0.415	0.855	0.606	0.375	0.833	0.750	0.441	0.933	0.947
2	Feb.	0.455	0.466	0.722	0.484	0.667	0.813	0.406	0.769	0.840	0.323	0.818	0.739
3	Mar.	0.845	0.687	0.765	0.613	0.737	0.916	0.406	0.846	0.842	0.353	0.750	0.909
4	Apr.	0.485	0.500	0.529	0.484	0.733	0.500	0.312	0.800	0.773	0.323	0.636	0.782
5	May	0.545	0.555	0.600	0.484	0.733	0.750	0.468	0.533	0.882	0.323	0.636	0.826
6	Jun.	0.576	0.542	0.857	0.516	0.750	0.800	0.500	0.500	0.562	0.529	0.33	0.687
7	Jul.	0.576	0.894	0.857	0.451	0.929	0.823	0.500	0.750	0.750	0.470	0.875	0.777
8	Aug.	0.515	1.000	0.857	0.516	0.875	1.000	0.469	0.600	0.588	0.441	0.867	0.842
9	Sep.	0.545	0.889	0.933	0.548	0.882	0.928	0.406	0.847	0.789	0.471	0.812	0.889
10	Oct.	0.515	1.000	0.937	0.548	0.941	0.928	0.406	0.769	0.842	0.471	1.000	1.000
11	Nov.	0.545	0.944	1.000	0.548	1.000	1.000	0.437	0.857	0.944	0.500	0.882	0.941
12	Dec.	0.455	0.933	0.777	0.581	0.777	0.766	0.469	0.600	0.705	0.441	0.533	0.526
	Average	0.535	0.795	0.801	0.515	0.823	0.819	0.429	0.725	0.772	0.421	0.756	0.822

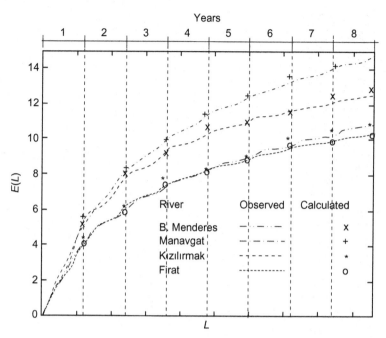

FIG. 3.37 Variation of t critical drought duration with sample length.

The substitutions of the probabilities in Table 3.15 into Eq. (3.92) provide the variation of the critical drought duration with sample length as in Fig. 3.37.

3.8 ANALYTICAL DERIVATION OF LONGEST RUN-LENGTH

The probability of the longest drought spell length in an independent process was derived first by De Moivre (1738) based on mathematical combinatorial analysis. In an independent process of length n, the PDF, $f_n(L)$, of the longest drought duration, L, has been given as

$$f_n(L) = \frac{(y-1)(1-qy)}{(L+1)-Lyp} \frac{1}{y^{n+1}}$$ (3.97)

where

$$y = 1 + pq^L + (L+1)(pq^L)^2 + (L+1)^2(pq^L)^3 + \cdots$$ (3.98)

Eq. (3.69) provides the critical drought duration length, L, probabilities in a sample size of n. By definition the summation of all these probabilities, $P_n(L=i)$, for $i = 1, 2, \ldots, n$ is equal to 1. Table 3.16 includes expected values (averages) according to Eqs. (3.97) and (3.69) with relative errors, which are very small. This indicates that the derivation of Eq. (3.69) is reliable.

TABLE 3.16 Longest (Critical) Drought Duration Expected Values

n	Eq. (3.97)	Eq. (3.69)	Relative Error (%)
25	3.99	4.04	1.23
50	4.99	5.01	0.40
100	5.99	5.98	0.14

Eq. (3.75) helps to find the probability of no critical drought of length L in a given sample size of n as follows:

$$P_n(j < L) = \sum_{i=1}^{L-1} P_n(L = i) \tag{3.99}$$

Numerical solutions of Eqs. (3.96) and (3.69) are presented in Table 3.17 for very small n values.

It is obvious that the relative error is less than 3%. After all these calculations it is possible to state that Eq. (3.96) results are approximate but the results from Eq. (3.69) are exact. Numerical solution of Eq. (3.69) at different truncation levels as $p = 0.1$, 0.3, and 0.5 show the variation of longest run-length (critical drought duration) with the CDF at a set of sample lengths, n, and truncation level surplus probabilities as $p = 0.1$, 0.3, and 0.5 (Fig. 3.38).

It is possible to deduce the following significant points after visual inspection of Fig. 3.38:

1. At a given truncation level, any increase in the sample length implies longer dry duration occurrences.
2. Whatever the truncation level, the CDFs become closer to each other for large sample lengths.
3. Increase in the truncation level, p, causes increase in the dry spell duration. Fig. 3.39 shows the expected value of the longest drought duration in various sample lengths at different truncation levels.

TABLE 3.17 Probability of No Critical Drought of Length $L = 2$

n	Eq. (3.97)	Eq. (3.69)	Relative Error (%)
2	0.76631	0.75000	2.12
3	0.62996	0.62500	0.78
4	0.50156	0.50000	0.31
5	0.40677	0.40625	0.13

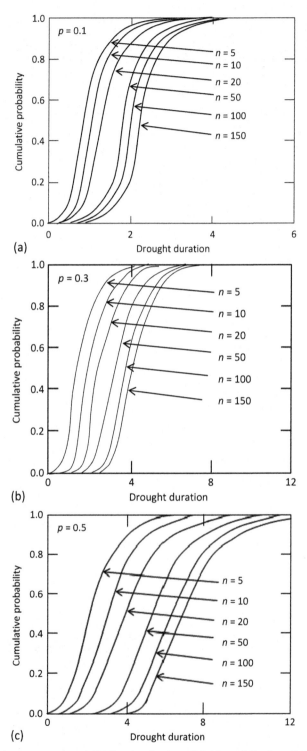

FIG. 3.38 Longest run-length PDF at levels a) $p = 0.1$, b) $p = 0.3$, c) $p = 0.5$.

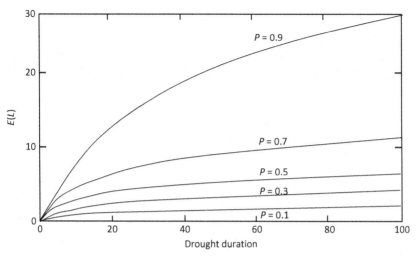

FIG. 3.39 Longest run-length expectation variation.

Up to now, all the derivations of the longest drought duration are achieved according to probability equations by Feller (1967). On the other hand, Mood (1940) gave another solution for the longest run-length (critical drought duration) probability by considering that at the beginning and the end of a drought spell, there are different state types. For instance, in the above derivations, even for $i = 1$, the probability of L being equal to 1 or 0 are written. For sample length, n, it is assumed that there is a drought duration, which is not possible in Mood's approach. On the other hand, in Mood's approach, definition of droughts start from sample size $n = 3$, because this is the first case where there are different states at the beginning and the end of the only drought duration of 1. In a sample of size, n, drought duration may have at the maximum $n - 2$ size. If $P'_n(L = j)$ is considered as Mood's probability definition of drought duration equal to j in sample size n, then this probability can be written in terms of Feller probabilities as follows:

$$P'_n(L=j) = P_n(L=j) - 2p^j q \sum_{i=1}^{k_3} P_{n-j-1}(L=i-1) + I_n \qquad (3.100)$$

Here, for $k_3 = \min(n-j, j)$

$$I_n = \begin{cases} 0 & \text{for } j < m \\ p^j P_{n-2j}(L=0)p^j & \text{for } j > m \end{cases} \qquad (3.101)$$

and hence,

$$n = \begin{cases} \dfrac{n}{2} & \text{for even } n \\ \dfrac{n}{2} - \dfrac{1}{2} & \text{for odd } n \end{cases} \qquad (3.102)$$

TABLE 3.18 Expectation and Variance

| n | Expectation | | Variance | |
	Feller	Mood	Feller	Mood
2	1.000	0.000	0.500	0.000
3	1.375	0.125	0.734	0.109
4	1.688	0.375	0.965	0.359
5	1.938	0.625	1.184	0.672
6	2.156	0.981	1.382	1.035
7	2.344	1.148	1.554	1.392
8	2.512	1.387	1.703	1.729
9	2.662	1.605	1.829	2.032
10	2.799	1.807	1.938	2.299
20	3.729	3.104	2.521	3.835
25	4.036	3.502	2.667	3.966
30	4.289	3.821	2.773	4.090
40	4.692	4.313	2.914	4.134
50	5.007	4.686	3.006	4.146

Table 3.18 shows expectation and variance of Feller and Mood drought durations definition for different sample sizes at $p=q=0.5$ truncation level.

It is obvious that drought durations are comparatively longest in the case of a Feller definition rather than Mood's solution. The difference between the two approaches decreases as the sample length increases. On the contrary, the variances of drought duration for a Feller definition are smaller than Mood's definition for $n > 8$.

Example 3.17 Critical Drought Duration Probability with Analytical Formulations

For the application of the previous methodologies, data from three rivers from different parts of the world are presented in Table 3.19.

TABLE 3.19 Flow Characteristics

River	Location	Sample Length, n (year)	Serial Correlation	Average Discharge (cfs)
Cherry	Ca, ABD	45	0.013	368
Tuna	Romania	120	0.096	189,455
Rhine	Switzerland	150	0.077	36,253

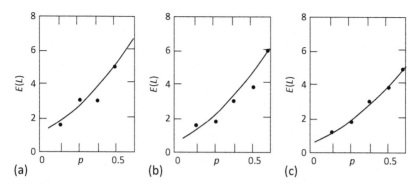

FIG. 3.40 Observed and theoretical critical drought duration: (A) Cherry, (B) Danube, and (C) Rhine rivers.

The longest drought duration expectations at different truncation levels are presented in Fig. 3.40. It is obvious that the analytical (continuous line) model results in these figures are very close to empirical scatter points.

3.9 CROSSING PROBABILITIES

In Fig. 3.2 crossing points as upcrossing (downcrossing) are shown as transitions from dry (wet) spells to wet (dry) spells. The length between an upcrossing (downcrossing) and the following downcrossing (upcrossing) points is referred to as a wet (dry) period. The more the crossing points the more frequent the wetness and dryness; hence, such location is subjected to water shortages, stresses, droughts, and floods. It is, therefore, possible to benefit from the number of these crossings for the description of water problems in an area.

3.9.1 Identically and Independently Distributed Variables

Crossing points in any hydrometeorology process at a given truncation level provide useful information about the drought properties of the underlying generation mechanism. In each time interval, there are two alternatives (surplus and deficit), which constitute a Bernoulli sequence of events (Section 3.7.1). For a given sample size of n, the possibility of k dry periods is identical with the number of upcrossings and their probabilities at a certain truncation level (see Fig. 3.11). Şen (1990) derived the complete description of upcrossing number, N_u, probabilities through the enumeration method starting from a sample size of $n=2$, where the probabilities of none or one upcrossing are given as

$$P_2(N_u = 0) = p^2 + pq + q^2$$
$$P_2(N_u = 1) = qp$$

$$(3.103)$$

Likewise, in the case of sample length $n=3$, $2^3=8$ crossing probability combinations can be written as

$$P_3(N_u = 0) = P_2(N_u = 0)q + p^3$$
$$P_3(N_u = 1) = (pq + q^2)pqpq + qp^2 \qquad (3.104)$$

In sample size $n=4$, the possible number of upcrossings increases to three crossing probabilities, which can be derived as

$$P_4(N_u = 0) = P_3(N_u = 0)q + p^4$$
$$P_4(N_u = 1) = P_2(N_u = 0)qp + P_3(N_u = 1)q + [P_3(N_u = 1) - P_2(N_u = 1)q]p$$
$$P_4(N_u = 2) = qpqp = q^2p^2$$

$$(3.105)$$

In the same manner of enumeration, the probabilities for $n=5$ with $2^5=32$ alternatives become

$$P_5(N_u = 0) = P_4(N_u = 0)q + p^5$$
$$P_5(N_u = 1) = P_3(N_u = 0)qp + P_4(N_u = 1)q + [P_4(N_u = 1) - P_3(N_u = 1)q]p$$
$$P_5(N_u = 2) = P_3(N_u = 0)qp + qpqpq + qpqq^2$$

$$(3.106)$$

For $n=6$ with $2^6=64$ alternatives can be grouped into four probability statements as follows:

$$P_6(N_u = 0) = P_5(N_u = 0)q + p^6$$
$$P_6(N_u = 1) = P_4(N_u = 0)qp + P_5(N_u = 1)q + [P_5(N_u = 1) - P_4(N_u = 1)q]p$$
$$P_6(N_u = 2) = P_4(N_u = 0)qp + P_5(N_u = 2)q + P_5(N_u = 1)qp^2 + qpqq^3$$
$$P_6(N_u = 3) = qpqpqp = q^3p^3$$

$$(3.107)$$

For $n=7$ there are $2^5=128$ mutually exclusive alternatives and they can be grouped into four categories as

$$P_7(N_u = 0) = P_6(N_u = 0)q + p^7$$
$$P_7(N_u = 1) = P_5(N_u = 0)qp + P_6(N_u = 1)q + [P_6(N_u = 1) - P_5(N_u = 1)q]p$$
$$P_7(N_u = 2) = P_5(N_u = 0)qp + P_6(N_u = 2)q + [P_6(N_u = 2) - P_5(N_u = 2)q]p$$
$$P_7(N_u = 3) = P_5(N_u = 2)qp + P_6(N_u = 3)q + P_6(N_u = 3)p$$

$$(3.108)$$

and, similarly, for $n=8$ one can write

$$P_8(N_u = 0) = P_7(N_u = 0)q + p^8$$
$$P_8(N_u = 1) = P_6(N_u = 0)qp + P_7(N_u = 1)q + [P_7(N_u = 1) - P_6(N_u = 1)q]p$$
$$P_8(N_u = 2) = P_6(N_u = 1)qp + P_7(N_u = 2)q + [P_7(N_u = 2) - P_6(N_u = 2)q]p$$
$$P_8(N_u = 3) = P_6(N_u = 2)qp + P_7(N_u = 3)q + [P_7(N_u = 3) - P_6(N_u = 3)q]p$$
$$P_8(N_u = 4) = qpqpqpqp = q^4p^4$$

$$(3.109)$$

The above sequence of formulations provides the general pattern between successive sample sizes; hence, it is possible to deduce the general formulation for sample size n as follows:

$$
\begin{aligned}
&P_n(N_u=0) = P_{n-1}(N_u=0)q + p^n \\
&P_n(N_u=j) = P_{n-2}(N_u=j-1)qp + P_{n-1}(N_u=j)q \\
&\quad + [P_{n-1}(N_u=j) - P_{n-2}(N_u=j)q]p \quad \text{where } 1<j<n/2 \\
&P_n(N_u=n/2) = q^{n/2}p^{n/2} \quad \text{for even } n \\
&P_n(N_u=n/2) = P_{n-2}(N_u=n/2-1)qp \\
&\quad + P_{n-1}(N_u=n/2)q + P_{n-1}(N_u=n/2)p \quad \text{for odd } n
\end{aligned}
\tag{3.110}
$$

A close inspection of this last equation system reveals that it is a function of the two previous upcrossing probability formulations. It is significant to notice that for each sample length, the summation of probabilities $P_n(N_u=0), P_n(N_u=1), \ldots,$ and $P_n(N_u=n/2)$ is equal to 1. The numerical solution of this last equation system for a given set of parameters n and p $(q=1-p)$ provides the probability values of upcrossing numbers. For instance, in Fig. 3.41 the upcrossing probabilities are

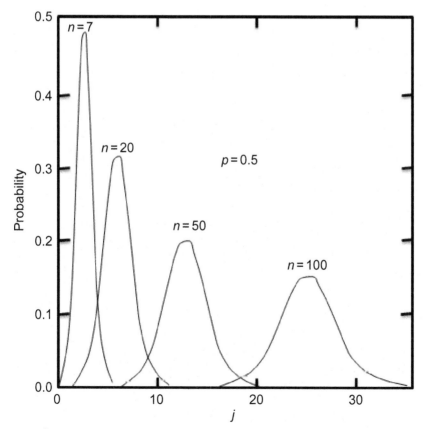

FIG. 3.41 Uncrossing probabilities for various sample lengths (p = 0.50).

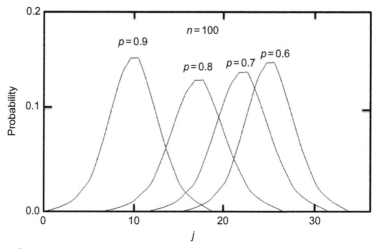

FIG. 3.42 Upcrossing probabilities for various exceedance probabilities ($n=100$).

given for a set of sample lengths at $p=0.6$ truncation level. The MATLAB software for this purpose is given in Appendix 3.4.

It is possible to conclude from this figure that each PDF of upcrossings is symmetrical and they can be expressed by a normal (gaussian) PDF.

Fig. 3.42 indicates various upcrossing probabilities for sample length $n=100$ at various truncation levels. These have more or less the same shapes again as a normal PDF.

The total number of crossings can be calculated from Eq. (3.48) for independent process, where $r=p$ as follows:

$$E(N_u) = np(1-p) \tag{3.111}$$

Fig. 3.43 gives the expected number of uncrossings at various truncation levels, p, and sample lengths.

The average number of upcrossings increases with the sample length but decreases as the truncation level increases. In relatively small sample sizes, expected value calculations are more sensitive. The maximum upcrossing (downcrossing) number occurs at 0.5 truncation level. In general, such a truncation level corresponds to average = mode = median value in symmetrical distributions but to median value in unsymmetrical PDFs.

Approximate formulation that expresses the variance of upcrossing numbers can be given as follows:

$$V(N_u) = np(1-p)(1-3p+3p^2) \tag{3.112}$$

Comparison of these two last equations leads to

$$V(N_u) = E(N_u)(1-3p+3p^2) \tag{3.113}$$

Upcrossing number standard deviation variation by a set of sample lengths and truncation levels are given in Fig. 3.44.

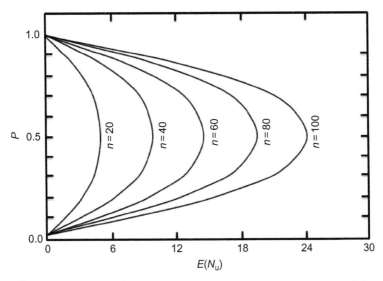

FIG. 3.43 Average upcrossing number for different truncation levels and sample lengths.

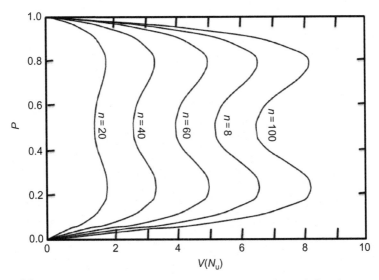

FIG. 3.44 Upcrossing number standard deviation for a set of sample lengths.

All the PDFs in Figs. 3.41 and 3.42 can be represented by a normal PDF with mean value as in Eq. (3.111) and standard deviation from Eq. (3.113) as its square root value.

Upcrossing number arithmetic mean and variance values for various p values are given in Figs. 3.45 and 3.46, respectively.

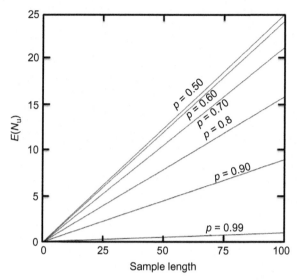

FIG. 3.45 Upcrossing number arithmetic mean variation by sample length and exceedance probability.

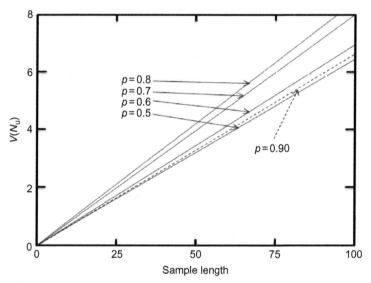

FIG. 3.46 Upcrossing number variance variation by sample length and exceedance probability.

3.9.2 Unidentically and Independently Distributed Variables

In the case of unidentical PDF of hydrometeorological variable by time, the upcrossing probabilities can be derived by considering Eqs. (3.103)–(3.110) for independent processes. In each time interval, there are two-state Bernoulli events with different exceedance and nonexceedance probabilities as p_i and $q_i = 1 - p_i$, respectively. Consideration of Eq. (3.103) for sample size $n = 2$ gives

$$
\begin{aligned}
P_2(N_u = 0) &= p_1 p_2 + p_1 q_2 + q_1 q_2 \\
P_2(N_u = 1) &= q_1 p_2
\end{aligned}
\tag{3.114}
$$

For sample size $n = 3$ from Eq. (3.104), one can obtain

$$
\begin{aligned}
P_3(N_u = 0) &= P_2(N_u = 0)q_3 + p_1 p_2 p_3 \\
P_3(N_u = 1) &= (p_1 q_2 + q_1 q_2)p_3 + q_1 p_2 q_3 + q_1 p_2 p_3
\end{aligned}
\tag{3.115}
$$

In general, for sample size n similar to Eq. (3.110), it is possible to write the following system of equations:

$$
\begin{aligned}
P_n(N_u = 0) &= P_{n-1}(N_u = 0)q_n + \prod_{i=1}^{n} p_i \\
P_n(N_u = j) &= P_{n-2}(N_u = j - 1)q_{n-1}p_n + P_{n-1}(N_u = j)q_n \\
&\quad + [P_{n-1}(N_u = j) - P_{n-2}(N_u = j)q_{n-1}]p_n \quad \text{where } 1 < j < n/2 \\
P_n(N_u = n/2) &= \prod_{k=1,3,5\ldots}^{n/2} q_k \prod_{k=2,4,6\ldots}^{n/2} p_k \quad \text{for even } n \\
P_n(N_u = n/2) &= P_{n-2}(N_u = n/2 - 1)q_{n-1}p_n + P_{n-1}(N_u = n/2)q_n \\
&\quad + P_{n-1}(N_u = n/2)q_n + P_{n-1}(N_u = n/2)p_n \quad \text{for odd } n
\end{aligned}
\tag{3.116}
$$

The validation of this last set of probability equations can be checked as to its summation, which should be equal to 1 and the statistical parameters are also calculated through the software given in Appendix 3.5. By this software, similar graphs to the independent case can be obtained provided that the heterogeneous probabilities, $p_1, p_2, p_3, \ldots$ are known.

3.10 ANNUAL FLOW TOTALS

One of the most important variables that should to be taken into consideration in any water resources system planning is the total deficit (magnitude) during a drought duration. An empirical solution has already been presented in Example 3.4 where from a given time series by use of the software in Appendix 3.1, the total drought duration time series (Fig. 3.7C) and also CDF (Fig. 3.8C) are presented with the Gamma PDF. This section concentrates on the analytical derivation of the total deficit or surplus amounts based on what has been explained in the previous sections and chapters. Here, the derivations are presented for total deficits, but in the case of symmetrical PDFs total surpluses formulations are similar to the total deficit, D, say, at $p = 0.7$ truncation level is equivalent to the total surplus at the $p = 0.3$ truncation level.

The total deficit, D_n, over a drought duration, n, can be expressed similar to Eq. (3.4) as

$$D_n = \sum_{i=1}^{n} (x_0 - x_i) \tag{3.117}$$

If each water deficit is redefined as $d_i = x_0 - x_i$, then this expression can be rewritten as follows:

$$D = \sum_{i=1}^{n} d_i \tag{3.118}$$

Herein, deficit amounts and drought duration have random characters, which mean that the total deficit is equal to random summation of random variables. According to probability theory, the joint probability, $P(D_n, n)$, of two random variables is equal to the multiplication of one of the random variable conditional probabilities, $P(D_n/n)$, given the other random variable marginal probability, $P(n)$, as

$$P(D_n, n) = P(D_n/n)P(n) \tag{3.119}$$

For the marginal PDF of the total deficit, it is necessary to sum up the joint probability from $n=0$ to ∞, which yields that

$$P(D) = \sum_{n=0}^{\infty} P(D_n/n)P(n) \tag{3.120}$$

where D is the total deficit. The arithmetic average value (expectation in population domain) can be calculated as

$$E(D) = \int_0^{\infty} DP(D)dD \tag{3.121}$$

The substitution of Eq. (3.120) into this last expression gives

$$E(D) = \sum_{n=0}^{\infty} P(n) \int_0^{\infty} D_n P(D_n/n)dD_n \tag{3.122}$$

The integration term represents conditional expectation as $E(D/n)$, and hence, one can write succinctly,

$$E(D) = \sum_{n=0}^{\infty} E(D_n/n)P(n) \tag{3.123}$$

The conditional expectation can be expressed for a given n value as $E(D/n) = nE(d_i)$; hence,

$$E(D) = E(d) \sum_{n=0}^{\infty} nP(n) \qquad (3.124)$$

The summation on the right-hand side expresses the expectation of drought duration average, n_{d}, which helps to simplify the previous expression to

$$E(D) = E(d)E(n_{\mathrm{d}}) \qquad (3.125)$$

It is possible to benefit from Eq. (3.118) for the second-order moment calculation by taking first the square and then application of the expectation operator on both sides leads to

$$E(D^2/n) = \sum_{i=1}^{n} E(d_i^2) + \sum_{i=1}^{n} \sum_{\substack{j=1 \\ i \neq j}}^{n} E(d_i d_j)$$

or by consideration of the variance definition in terms of expectation (Feller, 1967), it is possible to write this expression in a more explicit form as

$$E(D_n^2/n) = \sum_{i=1}^{n} [V(d) + 2E(d)] + \sum_{i=1}^{n} \sum_{\substack{j=1 \\ i \neq j}}^{n} [\mathrm{Cov}(d_i, d_j) + E^2(d)]$$

or as

$$E(D_n^2/n) = nV(d) + 2nE(d) + \sum_{i=1}^{n} \sum_{\substack{j=1 \\ i \neq j}}^{n} \mathrm{Cov}(d_i, d_j) \qquad (3.126)$$

If the correlation coefficient definition, $\rho_{i,j} = \mathrm{Cov}(y_i, y_j)/V(y_i)$, is taken into consideration, due to the stationary property of the hydrological process, $\rho_{i,j} = \rho_{|i-j|}$, then after some algebraic manipulations one can obtain (Şen, 1974)

$$E(D_n^2/n) = nV(d) + 2nE(d) + 2V(d) \sum_{i=1}^{n-1} (n-i)\rho_i \qquad (3.127)$$

This conditional expectation can be converted to unconditional second-order expectation by considering the probabilities of the durations, which leads to after algebraic manipulation

$$E(D^2) = V(y_i)E(n_{\mathrm{d}}) + 2E(d)E(n_{\mathrm{d}}) + 2V(d) \sum_{n=0}^{\infty} \sum_{i=1}^{n-1} (n-i)\rho_i P(n) \qquad (3.128)$$

Consideration of Eq. (3.125) leads to the following water deficit summation variance value as

$$V(D) = E(D^2) - E^2(D)$$

$$= V(d)E(n_d) + E^2(d)V(n_d) + 2V(d)\sum_{n=0}^{\infty}\sum_{i=1}^{n-1}(n-i)\rho_i P(n) \qquad (3.129)$$

For the simplest independent hydrometeorological processes $\rho_i = 0$; consequently, variance formulation becomes

$$V(S) = E(S^2) - E^2(S) = V(yd)E(n_p) + E^2(yd)V(n_p) \qquad (3.130)$$

or

$$V(D) = \frac{qV(d) + pE^2(d)}{q^2} \qquad (3.131)$$

In this manner, both the arithmetic average (Eq. 3.125) and the variance (Eq. 3.131) of the water deficit summation are derived separately.

Example 3.18 Deficit PDF Calculations

Example 3.4 has given the PDF of deficits as two-Gamma PDF with shape and scale parameters $\alpha = 0.888$ and $\beta = 55.333$, respectively. The mean, $E(d)$, and variance, $V(d)$, of the two-parameter Gamma PDF are given by Feller (1967) as

$$E(d) = \alpha\beta \qquad (3.132)$$

and

$$V(d) = \alpha\beta^2 \qquad (3.133)$$

respectively. Accordingly, the substitution of these expressions into Eqs. (3.125) and (3.131) leads to the two-parameter Gamma PDF expectation and variance of total deficit as

$$E(D) = \alpha\beta E(n_d) \qquad (3.134)$$

and

$$V(D) = \frac{q\alpha\beta^2 + p\alpha^2\beta^2}{q^2} = \frac{\alpha\beta^2(q\beta + p\alpha)}{q^2} \qquad (3.135)$$

respectively. With the parameters at hand, their numerical values are $E(d) = 0.888 \times 55.33 = 49.13$ mm and $V(y) = 0.888 \times 55.33^2 = 2718.551$ mm^2, but the standard deviation is 52.14 mm.

After the substitution of drought duration expectation from Eq. (3.34) into Eq. (3.134) gives the total deficit expectation as

$$E(D) = \frac{\alpha\beta}{1-r} \qquad (3.136)$$

The correlation coefficient of the Diyarbakır monthly rainfall total time series is 0.41, which corresponds to the autorun coefficients, at $q=0.5$ truncation percentage approximately, as $r=0.631$ from Table 3.5. With all the numerical values at hand, the variance and expectation of the total deficit for this location can be calculated from Eqs. (3.135) and (3.136) as $V(D)=0.888 \times 55.33^2 \times (0.5 \times 55.33 +0.5 \times 0.888)/0.5^2 = 354.59 \times 10^3$ mm^2, but the standard deviation as the square root is equivalent to 595.48 mm. The expectation value from Eq. (3.135) becomes $E(D)=0.888 \times 55.33/(1-0.631) \cong 134$ month.

The reader should notice that the expectation (average) and the standard deviation calculated in this section are different than the ones given empirically in Table 3.2. The reason is that herein the monthly averages are not adapted as the truncation level but a constant level is taken into consideration. It is well appreciated that in this case, deviations are greater than the periodic truncation levels.

3.10.1 Independent Processes

Although various moments of total deficit in the case of normal independent process have been analytically given by Downer et al. (1967), the same results are driven on the basis of the methodology developed in this section. They have employed the data generation method in order to derive the run properties of the skewed variables. In general, the deficits, d, are distributed according to the truncation PDF, $f_t(d)$, with the truncation level at x_o. The truncated distribution can be obtained from the overall PDF, $f(x)$, as

$$f_t(d) = \frac{1}{N}f(x) \qquad (3.137)$$

where N is the normalizing factor, which makes the area under $f_t(x)$ equal to unity. Thus,

$$N = \int_{x_o}^{\infty} f(x)\mathrm{d}x$$

or

$$N = 1 - q = p$$

Hence, Eq. (3.137) can be written as

$$f_t(d) = \frac{1}{p}f(x) \qquad (3.138)$$

3.10.1.1 NORMALLY DISTRIBUTED PROCESSES

If the original random variables are normally distributed with mean, μ, and standard deviation, σ, then the standardized deficits below the standardized truncation level, x_o, are distributed according to a standard truncated normal distribution,

$$f_t(X) = \frac{1}{p(2\pi)^{1/2}} \exp\left(-\frac{x^2}{2}\right) \quad X_o \leq x < -\infty \tag{3.139}$$

The first two moments about x_o of Eq. (3.137) can be found after some algebra as

$$E(d) = \frac{1}{p(2\pi)^{1/2}} \int_{x_o}^{\infty} x \exp\left(-\frac{1}{2}x^2\right) dx = \frac{\exp\left(-\frac{1}{2}x_o^2\right)}{p(2\pi)^{1/2}} \tag{3.140}$$

and

$$E(d^2) = \frac{1}{p(2\pi)^{1/2}} \int_{x_o}^{\infty} x^2 \exp\left(-\frac{1}{2}x^2\right) dx = 1 + \frac{x_o \exp\left(-\frac{1}{2}x_o^2\right)}{p(2\pi)^{1/2}} \tag{3.141}$$

or, the moments about the origin become

$$E(d) = x_o + \frac{\exp\left(-\frac{1}{2}x_o^2\right)}{p(2\pi)^{1/2}} \tag{3.142}$$

and

$$V(d) = 1 + \frac{x_o \exp\left(-\frac{1}{2}x_o^2\right)}{p(2\pi)^{1/2}} - \frac{\exp\left(-x_o^2\right)}{2\pi p^2} \tag{3.143}$$

Now, it is possible to find various run-sum properties by substituting Eq. (3.131) into Eq. (3.125) with consideration that $E(n_d) = 1/p$, which yields

$$E(D) = \frac{\exp\left(-\frac{1}{2}x_o^2\right)}{pq(2\pi)^{1/2}} + \frac{x_o}{q} \tag{3.144}$$

The variance of the run-sum can be calculated from Eq. (3.131).

Example 3.19 Truncated Normal PDF Drought Magnitude Calculations

Danube River annual flow series are considered for the truncated normal PDF application in predicting the drought deficit calculations. In Fig. 3.47A natural flow histogram and matching logarithmic-normal probability distribution frequency function are given. The annual flow series have slightly positive skewness coefficient equal to 0.416. In the same figure, logarithmic-normal PDF parameters (A and B) numerical values are given in addition to the mean and the variance values after the logarithmic transformation. In order to avoid the skewness in the annual flows, a logarithmic transformation is applied and the resultant histograms with the normal probability frequency function are presented in Fig. 3.47B.

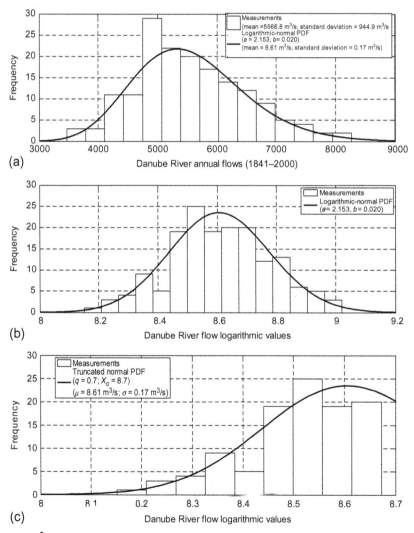

(a)

(b)

(c)

FIG. 3.47 Danube River annual flow histogram and theoretical frequency functions for a) natural data, b) logarithmic transformation, c) truncated data.

The measured and theoretical values match each other closely with relative error less than 5%. Now the drought quantities can be calculated theoretically by considering a constant truncation level, which is adapted as $q=0.7$, and this corresponds to logarithmic annual value of 8.7 (see Fig. 3.47C). The question is what are the expected total deficit values for the Danube River annual flows? Eq. (3.144) is ready for the answer but first of all the standardized truncation level must be calculated according to Eq. (2.1), which yields

$$x_0 = \frac{8.70 - 8.61}{0.17} = 0.529$$

The substitution of this with other relevant numerical values into Eq. (3.144) leads to the total deficit amount in the logarithmic domain as

$$E(D) = \frac{\exp\left(-\frac{1}{2}0.529^2\right)}{0.7 \times 0.3 \times (2 \times 3.14)^2} + \frac{0.529}{0.7} = 0.861$$

The conversion of this value into the natural domain yields $\exp(0.861) = 2.36\,\mathrm{m^3/s}$

3.10.1.2 LOGARITHMIC NORMALLY DISTRIBUTED PROCESSES

This PDF has been applied in water resources system design for a long time to model daily streamflows, flood peak discharges, and annual, monthly, and daily rainfalls. It was adapted early in the statistical studies of hydrological data by Hazen (1914). The main reason for its adaptation is the skewness, which seems to represent the observed data more realistically than a symmetric PDF as in the previous example.

If any hydrologic variable, y, has log-normally PDF then $x = \ln(y)$ has normal (gaussian) PDF. The general form of a log-normal PDF is given by Benjamin and Cornell (1970) as

$$f(y) = \frac{1}{(2\pi)^{1/2}} \exp\left\{ -\frac{1}{2}\left[\frac{1}{\sigma_{Lny}}Ln\left(\frac{y}{\breve{m}_y}\right) \right]^2 \right\} \qquad 0 < y < \infty \qquad (3.145)$$

where $\breve{m}_y$ and σ_{Lny} are the two parameters of the distribution; namely, the median and the standard deviation, respectively. On the other hand, the relationship between parameters and the moments are given as

$$\breve{m}_y = m \, \exp\left(-\frac{1}{2}\sigma_{Lny}\right) \qquad (3.146)$$

$$\sigma_{Lny}^2 = Ln\left(V_y^2 + 1\right) \qquad (3.147)$$

and the coefficient of skewness, γ, is

$$\gamma = 3V_y + V_y^3 \qquad (3.148)$$

where m_y and V_y are the mean and coefficient of variation of the log-normally distributed random variable. From Eq. (3.138) one can obtain the truncated log-normal PDF as

$$f(y) = \frac{1}{p(2\pi)^{1/2}} \exp\left\{ -\frac{1}{2}\left[\frac{1}{\sigma_{Lny}} Ln\left(\frac{y}{\tilde{m}_y} \right) \right]^2 \right\} \quad y_0 < y < \infty \qquad (3.149)$$

where y_0 is the truncation level of log-normally distributed variate. The nth order moment can be easily obtained from Eq. (3.148) as

$$E(y^n) = \frac{\tilde{m}_y^n \exp\left(-\frac{1}{2}n^2\sigma_{Lny}^2 \right)}{p} \left\{ 1 - F\left[\frac{1}{\sigma_{Lny}} Ln\left(\frac{y_0}{\tilde{m}_y} \right) - n\sigma_{Lny} \right] \right\} \qquad (3.150)$$

where $F[.]$ is the area under the standardized normal distribution from $-\infty$ to the level of the brackets. Thus, the expected value about the origin becomes

$$E(y^n) = \tilde{m}_y \exp\left(-\frac{1}{2}n^2\sigma_{Lny}^2 \right) \frac{1 - F\left[\frac{1}{\sigma_{Lny}} Ln\left(\frac{y_0}{\tilde{m}_y} \right) - n\sigma_{Lny} \right]}{1 - F\left[\frac{1}{\sigma_{Lny}} Ln\left(\frac{y_0}{\tilde{m}_y} \right) \right]} \qquad (3.151)$$

It is possible to obtain the variance, $V(y)$, of deficits in the truncated log-normal distribution case by making use of the second-order moment for $n=2$ in Eq. (3.150). The calculations of $E(D)$, $V(D)$, and R in this case have been provided in Table 3.20. It is clear from this table that as the coefficient of skewness increases, $E(D)$ and $V(D)$ decrease.

Example 3.20 Truncated Normal PDF Drought Magnitude Calculations

It has been observed that an increase in the first-order autocorrelation coefficient, ρ (hence, autorun coefficient) causes the expected value of the drought duration to increase (see Eq. 3.34). Because the expected value of drought duration is explicitly related to expectation of the total deficit, as is obvious from Eq. (3.125), it is possible to conclude that ρ influences $E(D)$. The autocorrelation coefficient does not affect the PDF of observations. Thus, in the case of normal distribution, $E(D)$ can be obtained by substituting Eqs. (3.34) and (3.142) into Eq. (3.125), which leads to

$$E(D) = \frac{x_0}{(1-r)} + \frac{\exp\left(-\frac{1}{2}x_0^2 \right)}{p(1-r)(2\pi)^{1/2}} \qquad (3.152)$$

TABLE 3.20 The Various Run-Sum Properties of the Log-Normal Independent Process

q	p	$m_y = 1.00; \sigma_y = 0.597$ $V_y = 0.597; \gamma = 2.000$ $\sigma_{Lny} = 0.5521; \tilde{m}_y = 0.8586$				$m_y = 1.00; \sigma_y = 0.325$ $V_y = 0.325; \gamma = 1.000$ $\sigma_{Lny} = 0.3169; \tilde{m}_y = 0.951$				$m_y = 100.00; \sigma_y = 0.030$ $V_y = 0.01; \gamma = 0.030$ $\sigma_{Lny} = 0.1732; \tilde{m}_y = 0.9851$			
		y	E(S)	V(S)	R	y	E(S)	V(S)	R	y	E(S)	V(S)	R
0.10	0.90	0.42	14.97	205.28	0.99	0.63	16.84	100.14	0.995	0.79	18.19	95.81	0.998
0.20	0.80	0.53	8.43	28.14	0.97	0.73	9.12	24.42	0.991	0.85	9.54	22.41	0.997
0.40	0.60	0.75	5.16	7.31	0.94	0.88	5.17	5.46	0.987	0.94	5.12	4.64	0.995
0.50	0.50	0.86	4.55	4.69	0.92	0.95	4.40	3.26	0.979	0.98	4.24	2.62	0.992
0.60	0.40	0.98	4.23	3.20	0.91	1.03	3.91	2.02	0.963	1.03	3.67	1.16	0.990
0.80	0.20	1.36	4.12	1.61	0.85	1.24	3.43	0.77	0.959	1.14	3.00	0.51	0.990
0.90	0.10	1.74	4.08	1.11	0.78	1.43	1.77	0.41	0.926	1.23	2.56	0.25	0.872

TABLE 3.21 The Various Total Deficit Properties of the Normal Lag-One Markov Process

x_o	$\rho = 0.2$			$\rho = 0.4$			$\rho = 0.6$		
	r	E(S)	V(S)	r	E(S)	V(S)	r	E(S)	V(S)
−1.00	0.85	8.96	75.07	0.87	10.28	103.52	0.90	12.54	164.73
−0.50	0.73	3.72	12.63	0.76	4.36	19.07	0.81	5.42	34.21
−0.40	0.70	3.17	9.17	0.74	3.73	14.27	0.79	4.64	25.98
−0.30	0.67	2.74	6.84	0.72	3.23	10.47	0.77	4.03	20.25
−0.20	0.63	2.38	5.15	0.69	2.81	8.33	0.75	3.51	15.93
−0.10	0.60	2.08	3.95	0.66	2.48	6.35	0.74	3.07	12.78
0.00	0.56	1.83	3.07	0.63	2.16	5.15	0.70	2.70	10.37
0.10	0.53	1.62	2.41	0.60	1.91	4.14	0.68	2.39	8.52
0.20	0.49	1.44	1.93	0.57	1.70	3.37	0.63	1.98	6.44
0.30	0.46	1.29	1.57	0.50	1.40	2.40	0.61	1.78	5.49
0.40	0.52	1.16	1.29	0.49	1.37	2.32	0.58	1.60	4.72
0.50	0.39	1.06	1.07	0.48	1.23	1.96	0.57	1.45	4.11
1.00	0.24	0.51	0.34	0.33	0.79	1.14	0.46	0.97	2.41

If the generating mechanism of observations is assumed to be a standard form of the lag-one Markov process given in Eq. (3.11), then after the substitution of Eqs. (3.31), (3.35), and (3.41) into Eq. (3.129) and some tedious algebra, the variance, $V(D)$, can be derived as

$$V(D) = \frac{1}{(1-r)} \left[V(y) + \frac{rE^2(y)}{(1-r)} \right] + \frac{2V(y)\rho}{(1-\rho)} \left[\frac{r}{(1-r)} + \frac{(1-r)}{(1-\rho)(1-r\rho)} \right] \quad (3.153)$$

which in the case of independent process $(r = \rho)$ reduces to Eq. (3.131). The solutions to Eqs. (3.152) and (3.153) are presented in Table 3.21 for various ρ values. This table reveals that in the case of Markov processes, $E(D)$ and $V(D)$ increase but r decreases with an increase in ρ value.

APPENDIX 3.1 DROUGHT FEATURES SOFTWARE

```
function [C,S,SL,SM,SI,D,DL,DM,DI] = AnnualDroughtAnalysis(X,Xo,n)
% This program has been written by Zekâi Şen first in Fortran IV in
1977
% it has been converted to Matlab in 1999
% Drought analysis of a given sequence
```

```
% X is a given time series
% n is the number of observations
% Xo is the truncation level
% C is the crossing vector
% S (D)is the surplus (deficit) sum vector
% SL (DL) is the surplus (deficit) length vector
% SI (DI) is the surplus (deficit) intensity vector
% IMPORTANT NOTE: The negative (positive) values of D and S
distinguishes
% whether deficit or surplus according to the sign in the output
j=0;
for i=2:n
    sign =(X(i-1)-Xo)*(X(i)-Xo);
    if sign < 0
        j=j+1;
        C(j)=i-1;
    else
    end
end
j1=j-1;
    for i=1:C(1)
        surplus(i)=(X(i)-Xo);
    end
    S(1)=sum(surplus(1:C(1)));
    SM(1)=max(surplus(1:C(1)));
    SL(1)=C(1);
    SI(1)=S(1)/SL(1);
    m=1;
    for i=2:2:j1
        m=m+1;
        for k=C(i-1)+1:C(i)
            surplus(k)=(X(k)-Xo);
        end
        S(m)=sum(surplus(C(i-1)+1:C(i)));
        SM(m)=max(surplus(C(i-1)+1:C(i)));
        SL(m)=C(i+1)-C(i);
        SI(m)=S(m)/SL(m);
    end
        m=0;
    for i=1:2:j1
        m=m+1;
        for k=(C(i-1)+1):C(i)
            deficit(k)=(X(k)-Xo);
        end
        D(m)=sum(deficit((C(i-1)+1):C(i)));
```

```
         DM(m)=max(deficit((C(i-1)+1):C(i)));
         DL(m)=C(i+1)-C(i);
         DI(m)=D(m)/DL(m);
   end
          for i=(C(end)+1):n
              deficit(i)= (X(i)-Xo);
          end
%        D(m+1)=sum(deficit((C(end)+1):n));
         DD=sum(deficit(C(end):n));
         DL(m+1)=n-C(end);
         if DD < 0
             S(m+1)=DD;
             SM(m+1)=max(deficit((C(end)+1):n));
             SI(m+1)=S(m+1)/DL(m+1);
         else
%            DM(m+1)=max(deficit((C(end)+1):n));
%           . DI(m+1)=D(m)/DL(m);
         end
```

APPENDIX 3.2 INDEPENDENT BERNOULLI TRIALS SOFTWARE

```
function [Pr,Av,St] = BernoulliIndependentDrought(N,p)
% This program has been written by Zekâi Şen in 1979
% N = The number of observation
% Av= Average of the longest run
% St= Standard deviation of the longest run
% p = The probability of wet spell
%P(1:100,1:100)=0;
q = 1-p; % The probability of dry spell
    % For the first spell duration
P(1,1)=q;
P(1,2)=p;
    % For the subsequent spell durations
for i=2:N % Starts from the second run length
    P(i,1)=P(i-1,1)*q;
    for j=2:i % Ends at the previous spell than the last one
        k1=min(i-j+1,j-2);
        S1=0;
        if k1 >= 1
            for k=1:k1
                S1=S1+P(i-k-1,j-1)*q*p^(k-1);
            end
        else
        end
```

```
            S2=0;
            k2=min(i-j,j-1);
            if k2 >= 1
                for k=1:k2
                    S2=S2+P(i-k-1,j)*q*p^k;
                end
            else
            end
            P(i,j)=P(i-1,j)*q+(P(i-1,j-1)-S1)*p+S2;
        end
        P(i,i+1)=P(i-1,i)*p;
    end
    Pr=P(N,1:N+1);
    ProbabilitySum=sum(Pr)
    Sum1=0;
    Sum2=0;
    for i=1:N+1
        Sum1=Sum1+i*Pr(1,i);
        Sum2=Sum2+i^2*Pr(1,i);
    end
    Av=Sum1;
    St=sqrt(Sum2-Av^2);
end
```

APPENDIX 3.3 DEPENDENT BERNOULLI TRIALS SOFTWARE

```
function [Pr,Av,St] = BernoulliDependentDrought(N,p,ppp,pnn)
% This program has been written by Zekâi Şen in 1979
% N = The number of observation
% Av = Average of the longest run
% St = Standard deviation of the longest run
% p = The probability of wet spell
% ppp = wet spell followed by wet spell
% pnn = dry spell followed by dry spell
q = 1-p; % The probability of dry spell
    % For the first spell duration
Pn(1,1)=q;
Pp(1,2)=p;
    % Wet /dry( spell followed by dry (wet) spell
ppn=(1-ppp)*p/q;
pnp=(1-pnn)*q/p;
    % For the first two spell durations
Pn(2,1)=Pn(1,1)*pnn;
Pp(2,2)=Pn(1,1)*ppn;
```

```
Pn(2,2)=Pp(1,2)*pnp;
Pp(2,3)=Pp(1,2)*ppp;
    % For the subsequent spell durations
for i=3:N % Starts from the second run length
    Pn(i,1)=Pn(i-1,1)*pnn;
    for j=2:i % Ends at the previous spell than the last one
        k1=min(i-j,j-1);
        S1=0;
        if k1 >= 1
            for k=1:k1
                S1=S1+Pn(i-k-1,j-1)*ppn*ppp^(k-1);
            end
        else
        end
        S2=0;
        k2=min(i-j-1,j-2);
        if k2 >= 1
            for k=1:k2
                S2=S2+Pn(i-k-1,j)*ppn*ppp^(k-1);
            end
        else
        end
        Pp(i,j)=S1+S2;
        Pn(i,j)=Pp(i-1,j)*pnp+Pn(i-1,j)*pnn;
    end
    Pp(i,i+1)=Pp(i-1,i)*ppp;
end
Pr=P(N,1:N+1);
ProbabilitySum=sum(Pr)
Sum1=0;
Sum2=0;
for i=1:N+1
    Sum1=Sum1+i*Pr(1,i);
    Sum2=Sum2+i^2*Pr(1,i);
end
Av=Sum1;
St=sqrt(Sum2-Av^2);
end
```

APPENDIX 3.4 IDENTICALLY AND INDEPENDENTLY DISTRIBUTED VARIABLE SOFTWARE

```
function [P]=CrossingIdenticalIndependentProcess (p,n)
% This program calculates the number of crossings in independent
processes.
```

```
% Sen, Z. (1989. The theory of runs with application to drought
  prediction.
%                      Journal of Hydrology, Vol. 110, 383-391.
% p = proability of surpluss
% n = sample length (n > 4)
q=1-p;
P(1,1)=p;
P(1,2)=q;
P(2,1)=p*p+p*q+q*q;
P(2,2)=q*p;
P(3,1)=P(2,1)*q+p^3;
P(3,2)=(p*q+q*q)*p+q*p*q+q*p*p;
for i=4:n
    P(i,1)=P(i-1,1)*q+p^i;
    ju=round(i/2-0.001);
    for j=2:ju
        P(i,j)=P(i-2,j-1)*q*p+P(i-1,j)*q+(P(i-1,j)-P(i-2,j)*q)*p;
    end
in=i-2*ju;
    if in==0 P(i,ju+1)=(q*p)^ju;
    else
    P(i,ju+1)=P(i-2,ju)*q*p+P(i-1,ju+1)*q+P(i-1,ju+1)*p;
    end
end
end
```

APPENDIX 3.5 UN-IDENTICALLY AND INDEPENDENTLY DISTRIBUTED VARIABLE SOFTWARE

```
function [P]=CrossingUnidenticalProcess(p,n)
% This program calculates the number of crossings in independent
  processes.
% Sen, Z. (1989. The theory of runs with application to drought
  prediction.
%                      Journal of Hydrology, Vol. 110, 383-391.
% p = proability of surpluss
% n = sample length (n > 4)
q=1-p;
P(1,1)=p(1);
P(1,2)=q(1);
P(2,1)=p(1)*p(2)+p(1)*q(2)+q(1)*q(2);
P(2,2)=q(1)*p(2);
P(3,1)=P(2,1)*q(3)+p(1)*p(2)*p(3);
P(3,2)=(p(1)*q(2)+q(1)*q(2))*p(3)+q(1)*p(2)*q(3)+q(1)*p(2)*p(3);
for i=4:n
    P(i,1)=P(i-1,1)*q(i);
```

```
Mul1=1;
 for m=1:i
     Mul1=Mul1*p(i);
 end
     P(i,1)=P(i,1)+Mul1;
     ju=round(i/2-0.001);
     for j=2:ju
         P(i,j)=P(i-2,j-1)*q(i-1)*p(i)+P(i-1,j)*q(i)+(P(i-1,j)-
         P(i-2,j)*q(i-1))*p(i);
     end
     in=i-2*ju;
     if in==0
         Mul2=1;
         for m=1:2:ju
            Mul2=Mul2*q(m);
         end
         Mul3=1;
         for m=2:2:ju
             Mul3=Mul3*p(m);
         end
         P(i,ju+1)=Mul2*Mul3;
     else
         P(i,ju+1)=P(i-2,ju)*q(i-1)*p(i)+P(i-1,ju+1)*q(i)
         +P(i-1,ju+1)*p(i);
     end
end
```

REFERENCES

Aboammoh, A.M., 1991. The distribution of monthly rainfall intensity at some sites in Saudi Arabia. In: Environmental Monitoring and Assessment, vol. 17. Kluwer Academic Press, pp. 89–100.

Benjamin, J.R., Cornell, C.A., 1970. Probability Statistics, and Decision for Civil Engineers. McGraw-Hill, New York, 684 pp.

Beran, M.A., Rodier, J.A., 1985. Hydrological aspects of drought. UNBSCO-WMO, studies and reports in hydrology no. 39, UNESCO, Paris.

Chang, T.J., Kleopa, X.A., 1991. A proposed method for drought monitoring. Water Resour. Bull. 27 (2), 275–281.

Chow, V.T., 1964. The frequency formula for hydrologic frequency analysis. Trans. AGU 32, 231–237.

Cordery, I., McCall, M., 2000. A model for forecasting droughts from teleconnections. Water Resour. Res. 36 (3), 763–768.

Cramer, H., Leadbetter, M.R., 1967. Stationary and Related Stochastic Processes. John Wiley and Sons, New York, NY, 384 pp.

De Moivre, A., 1738. The Doctrine of Chances or a Method of Calculating Probabilities of Events in Play, second ed. (Reprinted by Frank Cass, London, 1967).

Downer, R., Siddiqui, M.M., Yevjevich, V., 1967. Applications of runs to hydrologic droughts. Hydrology paper 23, Colorado State University, Fort Collins, CO.

Dyer, T.G.J., Tyson, P.D., 1977. Estimating above and below normal rainfall periods over south Africa, 1972–2000. J. Appl. Meteorol. 16, 145–147.

Eljadid, A.G., 1997. Hydro-Meteorological Aspects and Water Resources Management in the Northern Part of Libya (Ph.D. thesis). Meteorology Department, Istanbul Technical University (Unpublished).

Eljadid, A.G., Şen, Z., 1997a. Regional wet and dry period statistics of Libyan rainfall. In: International Conference of Water Problems in the Mediterranean Countries, Cyprus. East Mediterranean University, Magosa.

Eljadid, A.G., Şen, Z., 1997b. Gamma distribution of the Libyan rainfall records. In: International Conference of Water Problems in the Mediterranean Countries, Cyprus. East Mediterranean University, Magosa.

Essenwanger, O., 1986. Elements of statistical analysis. In: World Survey of Climatology, vol. 1B. Elsevier, Amsterdam, 424 pp.

Feller, W., 1967. An Introduction to Probability Theory and Its Application. John Wiley and Sons, New York, 509 pp.

Feyerherm, A.M., Bark, L.D., 1967. Goodness of fit of a Markov chain model for sequences of wet and dry days. J. Appl. Meteorol. 6 (5), 770–773.

Feyerherm, A.M., Bark, L.D., 1973. Probability models for simulating temperatures and precipitations. In: Third Conference on Probability and Statistics in Atmospheric Science, Boulder, CO, pp. 248–249.

Gabriel, K.R., Neumann, J., 1957. On a distribution of weather cycles by length. Q. J. R. Meteorol. Soc. 83, 375–380.

Gabriel, K.R., Neumann, J., 1962. A Markov chain model for daily rainfall occurrence at Tel Aviv. Q. J. R. Meteorol. Soc. 88 (375), 90–95.

Gibbs, W.J., Maher, J.V., 1967. Rainfall deciles as drought indicators. Bulletin no. 45, Bureau of Meteorology, Melbourne.

Green, J.R., 1964. A model for rainfall occurrence. J. R. Stat. Soc. Ser. B Stat. Methodol. 26 (2), 345–353.

Guerrero-Salazar, P.L.A., Yevjevich, V., 1975. Analysis of drought characteristics by the theory of runs. Hydrology paper 80, Colorado State University, Fort Collins, CO.

Gumbel, E.J., 1963. Statistical forecast of droughts. Bull. Int. Assoc. Sci. Hydrol. 8, 5–23.

Hazen, A., 1914. Discussion on "flood flows" by W.E. Fuller. Trans. ASCE 77, 626–632.

Hisdal, H., Stahl, K., Tallaksen, L.M., Demuth, S., 2001. Have streamflow droughts in Europe become more severe or frequent? Int. J. Climatol. 21, 317–333.

Kendall, M.G., 1948. Rank Correlation Method, first ed. Griffin, London.

Kumar, V., Panu, U.S., 1997. Predictive assessment of severity of agricultural droughts based on agro-climatic factors. J. Am. Water Resour. Assoc. 33 (6), 1255–1264.

Linsley, R.K., 1982. Rainfall-runoff models—an overview. In: Singh, V.P. (Ed.), Proceedings of the International Symposium on Rainfall-Runoff Relationships. Water Resources Publications, Littleton, CO.

Llamas, J., Siddiqui, M.M., 1969. Runs of precipitation series. Hydrology paper 33, Colorado State University, Fort Collins, CO.

Lohani, V.K., Lognathan, G.V., 1997. An early warning system for drought management using the Palmer drought severity index. Nordic Hydrol. 29 (1), 21–40.

Millan, J., 1972. Statistical properties of runs as applied to hydrologic droughts. In: Floods and Droughts. Water Resources Publications, Colorado State University, Colorado, USA, pp. 627–636.

Millan, J., Yevjevich, V., 1971. Probabilities of observed droughts. Hydrology paper 50, Colorado State University, Fort Collins, CO.

Mood, A.M., 1940. The distribution theory of runs. Ann. Math. Stat. 11, 427–432.

Panu, U.S., Sharma, T.C., 2002. Challenges in drought research: some perspectives and future directions. Hydrol. Sci. J. 47, 19–29.

Parzen, E., 1962. Introduction to Stochastic Processes. Universal Book Stall, USA.

Piechota, T.C., Dracup, J.A., 1996. Drought and regional hydrologic variation in the United States: association with El Nino-Southern Oscillation. Water Resour. Res. 32, 1359–1373.

Rice, S.O., 1945. Mathematical analysis of random noise. Bell Syst. Tech. J. 24, 46–156.

Saldarriaga, J., Yevjevich, V., 1970. Application of run-lengths to hydrologic time series. Hydrology paper 40, Colorado State University, Fort Collins, CO.

Şen, Z., 1974. Small Sample Properties of Stationary Stochastic Processes and the Hurst Phenomenon in Hydrology. (Ph.D. thesis), University of London, Imperial College of Science and Technology. 256 pp. (Unpublished).

Şen, Z., 1976. Wet and dry periods of annual flow series. J. Hydraul. Div. 102 (HY10), 1503–1514 (ASCE, Proceeding Paper 12497).

Şen, Z., 1977. Run-sums of annual flow series. J. Hydrol. 35, 311–324.

Şen, Z., 1978. Autorun analysis of hydrological time series. J. Hydrol. 36, 75–85.

Şen, Z., 1980a. Statistical analysis of hydrological critical droughts. J. Hydraul. Div. 106, 99–115 (ASCE).

Şen, Z., 1980b. Regional drought and flood frequency analysis: theoretical considerations. J. Hydrol. 46, 265–279.

Şen, Z., 1980c. Critical drought analysis of periodic stochastic processes. J. Hydrol. 46, 25–263.

Şen, Z., 1980d. The numerical calculation of extreme wet and dry periods in hydrological time series. Hydrol. Sci. Bull. 25 (2), 135–142 (Bulletin des Sciences Hydrologiques).

Şen, Z., 1985. Autorun model for synthetic flow generation. J. Hydrol. 81, 157–170.

Şen, Z., 1989. The theory of runs with applications to drought prediction—comment. J. Hydrol. 110 (3–4), 383–390.

Şen, Z., 1990. Critical drought analysis by second order Markov chain. J. Hydrol. 129, 183–202.

Şen, Z., 2010. Fuzzy Logic and Hydrological Modeling. Taylor and Francis Group, CRC Press, New York, 340 pp.

Sharma, T.C., 2000. Drought parameters in relation to truncation levels. Hydrol. Process. 14 (7), 1279–1288.

Sheppard, W.F., 1918. On the calculation of the double integral expressing normal correlation. Proc. Camb. Philos. Soc. 19, 23–66.

Summer, G.N., 1988. Precipitation Process and Analysis. John Wiley and Sons, New York, 455 pp.

Tallaksen, L.M., 2000. Streamflow drought frequency analysis. In: Vogt, J.V., Somma, F. (Eds.), Drought and Drought Mitigation in Europe. Kluwer Academic Publishers, Netherlands, pp. 103–117.

Tallaksen, L.M., Hisdal, H., 1997. Regional analysis of extreme streamflow drought duration and deficit volume. In: Gustard, A., Blazkova, S., Brilly, M., Demunt, S., Dixon, J., van Lanen, H., Llasat, C., Mkhadni, S., Servat, E. (Eds.), FRIEND'97-Regional Hydrology: Concepts on model for Sustainable Water Resources Management. IAHS Publication No. 246.pp, 141–150.

Tase, N., 1976. Area-deficit-intensity characteristics of droughts. Hydrology paper 87, Colorado State University, Fort Collins, CO.

Theoharatos, G.A., Tselepidaki, I.G., 1990. The distribution of rainy days in the Aegean area. Theor. Appl. Climatol. 42, 111–116.

Todorovich, P., 1970. On some problems involving a random number of random variables. Ann. Math. Stat. 41 (3), 1059–1063.

Todorovich, P., Woolhiser, D.A., 1976. Stochastic structure of the local pattern of precipitation. In: Shen, H.W. (Ed.), Stochastic Approaches to Water Resources, vol. 2. Colorado State University, Fort Collins, CO.

Troccoli, A., Palmer, T.N., 2007. Ensemble decadal predictions from analyzed initial conditions. Philos. Trans. R. Soc. Lond. A 365, 2179–2191.

Wienser, E.M., 1965. Modified Markov probability models of sequences of precipitation events. Mon. Weather Rev. 93 (8), 511–516.

Yevjevich, V., 1963. Fluctuations of wet and dry years. Hydrology paper 1, Colorado State University, Fort Collins, CO.

Yevjevich, V., 1967. An objective approach to definition and investigation of continental hydrologic droughts. Hydrology paper 23, Colorado State University, Fort Collins, CO.

Zelenhasic, E., Salvai, A., 1987. A method of streamflow drought analysis. Water Resour. Res. 23 (1), 156–168.

Regional Drought Analysis and Modeling

Applied Drought Modeling, Prediction, and Mitigation

4.1 GENERAL

Droughts evolve over time, have a regional nature, may cover extensive areas, and continue for long time periods. Their assessment needs a regional context, more effective drought impact adaptation, mitigation, and preparedness studies ("see chapter: Climate Change, Droughts and Water Resources"). Regional drought properties can be identified by studying the spatial patterns at a set of sites (points). They are often based on single site event definitions where the areal aspect is included by studying the spatial pattern of point values and without introducing a separate regional drought event definition. Tallaksen and Hisdal (1997) and Hisdal et al. (2001) have advocated regional drought feature identifications. It is possible to study regional drought variables like the area coverage and total deficit over the drought area (Tase, 1976; Şen 1980a, 1998; Santos, 1983; Rossi et al., 1992; Vogt and Somma, 2000). The area affected by drought is rarely static during the course of the event. Different drought indices can be calculated for catchments or regions under "at-site" drought definitions, as they do not include the area covered by drought as a dynamic variable (varying in time) ("see chapter: Temporal Drought Analysis and Modeling"). As drought emerges and intensifies, its core area or epicenter shifts, expands, or contracts. A comprehensive drought early warning system is critical for tracking these changes in spatial coverage and severity (Panu and Sharma, 2002).

To assess the regional drought occurrence pattern, the study region must be divided into characteristically distinctive mutually exclusive subareas on the basis of prevailing meteorological, hydrological, atmospheric, geomorphologic, and social considerations. It is generally accepted that severe droughts arise as a result of apparently chance variations in the atmospheric circulation. Droughts are always initiated by a shortage of precipitation, and many of the traditional approaches to drought definition are restricted to rainfall analysis (Smith, 1965). Simple assessment of drought severity depends largely on the magnitude and regional extent of precipitation deficiencies from mean conditions. Because precipitation is the main source for water supply, it is one of the most dominant factors in the subdivision of the whole region into different drought potential subareas. Precipitation is the most easily available data compared to any other drought-effective variables. For the sake of simplicity, the precipitation is considered as one of the main variables, which gives rise to regional and temporal drought variations. Other major variables are deficits in water supply and soil moisture content, which vary regionally as well as temporally depending on human and agricultural activities and water availability.

Droughts have spatial extensiveness that may cover many hectares of areas, which are affected by severe drought; hence, they evolve gradually and regions of maximum intensity vary from season to season. In larger areas drought rarely, if ever, affects the entire region. For instance, during a severe drought only some percentage of the whole region or country, say 60% or 75%, may be covered by severe and extreme drought impacts. Spatial climatic diversity and the size of regions suggest that drought is likely to occur somewhere each year. On average, a certain percentage of a region is affected by severe to extreme drought annually. From a planning perspective, the spatial characteristics of drought have serious implications. Nations should determine the probability that drought may affect simultaneously all or a portion of a river basin and, accordingly, develop contingencies. It is important for central governments or local authorities to calculate the chances of a regional drought simultaneously affecting agricultural productivity, water resources and, consequently, food supplies.

Drought impacts are spread over a larger geographical area that may be under damages from other natural hazards such as floods, tropical storms, and earthquakes (see Fig. 1.1). The climate change impact in many areas is expected to increase the effect of drought areal damage more frequently than before. Concern exists that the threat of global warming may increase the frequency and severity of extreme climate events in the future (IPCC, 2001, 2007) ("see chapter: Drought Hazard Mitigation and Risk"). This will imply that the pressure on water resources for domestic, industrial, and especially agricultural activities may increase steadily, which may cause conflicts not only between the regions within a country but also between countries. Droughts are not absolutely existing natural and social features, but they have relativistic behaviors temporally and spatially depending on the local circumstances.

Water is a precious material not only in dry areas of the world but its importance also is increasing steadily by time and on an areal extent. Especially for agricultural planning, the temporal and more significantly areal pattern of rainy and nonrainy periods (days, months, years) occurrences and amounts become more important; therefore, their prediction models are in great demand. Information about rainy and nonrainy spell properties is very significant for planning and management purposes in hydrology, hydrogeology, meteorology, agricultural, and water engineering domains. Temporal drought behaviors are explained in the previous section and this chapter exposes areal drought patterns for theoretical and practical solutions. There are different and complementary solution techniques by various authors (Feyerherm and Bark, 1960; Gabriel and Neumann, 1962; Green, 1970; Hershfield, 1970; Şen, 1976, 1980b, 1998).

Due to different reasons, areal variation and variability of droughts affect socioeconomic consequences. In this chapter the probability and statistical methodologies are presented to depict the areal characteristics of droughts and, subsequently, simple but effective calculation methodologies are presented for rainy and nonrainy days.

4.2 REGIONAL NUMERICAL DEFINITION OF DROUGHTS

Instead of time intervals along the whole record length, in the study of areal and regional droughts basic units are subareas, which can be defined by overlaying a grid over the study area, and then each subarea can be regarded as a "wet" or "dry" spell unit as shown in Fig. 4.1.

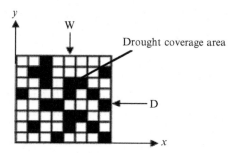

FIG. 4.1 Spatial dry units and drought coverage area.

Simultaneous occurrence of "dry" subareas over an area indicates the areal coverage of the drought phenomenon. In Fig. 4.1, there are $8 \times 8 = 64$ subareas of which 20 units are under dry condition. In case of a spatial drought, there are also "deficit" or "surplus" subareas and the projection of all the deficit subareas on a horizontal plane defines an irregular area that corresponds to drought coverage area as shown in Fig. 4.1.

Regional and spatial drought behavior features and prediction models are not well established in the literature as temporal drought descriptions and predictions. Areal drought coverage and its duration, in addition to magnitude and severity, are very important factors in drought planning and management work as well as for the assessment of drought mitigation impacts. Demuth and Stahl (2001) have provided some solutions on this aspect by studying the frequency, magnitude, and regional spread of droughts at different temporal and spatial scales. Theoretical development of areal drought coverage has been presented by Şen (1980b) for different drought descriptors. Many drought modeling techniques are available in different scientific journals but, unfortunately, their adaptation to practical applications has been lacking up to now.

4.3 TECHNIQUES TO PREDICT REGIONAL DROUGHTS

In general, among the techniques for predicting regional droughts are statistical regression, time series, stochastic (or probabilistic), and recently pattern recognition methods. These techniques require that a quantitative variable be identified to define drought with which to begin the process of prediction. In the case of agricultural drought, such a variable can be the yield (production per unit area) of the major crop in a region (Kumar and Panu, 1997; Boken and Şen, 2005). The crop yield in a year can be compared with its long-term average and drought intensity by considering the classifications as "nil," "mild," "moderate," "severe," or "disastrous," based on the difference between the current and the average yields.

4.3.1 Statistical Regression

Regression techniques estimate crop yields using yield-affecting variables. Usually, weather variables, which are routinely available for a historical period and significantly affect the yield, are included in a regression analysis. Regression techniques need to use weather data during a growing season for short-term estimates (Sakamoto, 1978; Idso et al., 1979; Slabbers and Dunin, 1981; Diaz, 1983; Cordery and Graham, 1989; Toure et al., 1995). Various researchers in different parts of the world have developed drought indices ("see chapter: Basic Drought Indicators") that can also be included along with the weather, such as the Western Canada Wheat Yield Model (Walker, 1989), a drought index using daily temperature and precipitation data, and advanced very high-resolution radiometer satellite data. The modified model improved the predictive power of the wheat yield significantly. The performance of the regression techniques has been found to have improved significantly by including satellite data-based variables.

The short-term estimates are available just prior to or around harvest time. Long-term predictions are required in predicting drought for next year so that long-term planning for tackling drought impacts can be initiated in time.

4.3.2 Time Series Analysis

Weather data during a growing season cannot be used for obtaining the long-term estimates simply because they are required prior to even sowing of the crop. As the yield is known to be influenced mostly by weather conditions during the growing season, it is a common practice to estimate yield using weather data. Attempts are limited to obtain long-term estimates that do not employ weather data.

As an alternative to weather data, annual time series of yield is utilized to obtain the long-term estimates by modeling the series. In a time series analysis, a variable, X_t, at time t, can be modeled and predicted as a function of time, $f(t)$, as

$$X_t = f(t) + \varepsilon_t \tag{4.1}$$

where ε_t refers to error, which is the difference between observations and the predicted values at the same time. Once a functional relationship is developed, then a forecast can be affected for any year ahead. Among the techniques for agricultural drought are linear and quadratic trends, simple exponential and double exponential smoothing, and simple and double moving averaging methodologies.

4.3.3 Pattern Recognition

Pattern recognition can be used to classify an object by analyzing the numerical data that characterize the object. Image processing, medical engineering, criminology, speech recognition, and signature identification all need pattern recognition techniques for classifying objects. Unfortunately, these techniques have not been used sufficiently for drought prediction tasks. In the case of agricultural drought, yield of a crop in the region plays the major role.

Although different pattern recognition techniques are available in the literature, only a few techniques are relevant in the case of agricultural drought (Boken et al., 2005). Potential variables that affect droughts can be derived from the combined use of weather and satellite data. Among such variables are average monthly temperature, total monthly precipitation, and soil moisture during the growing season.

4.3.4 Linear Discriminant and Nearest Neighbor Analysis

This can be applied on the subset of variables for their classification into a drought category. The whole set of data is divided into two parts as training and testing sets. The linear discriminant method can be employed for the training data set with the testing set for the classification performance of the same method. In the application of this method, it is assumed that in each category the values are distributed according to normal (Gaussian) PDF. Otherwise, instead a nonparametric technique of nearest neighbor approach may be employed. The use of this technique helps to classify a subset of variables into a drought or no drought category based on the category of the neighboring subsets (Şen, 2009).

4.3.5 Stochastic or Probabilistic Analysis

In the mathematical modeling of droughts, most often water records are taken as the basis where a time series of the precipitation or streamflow, $X_1, X_2, X_3, ..., X_n$ is truncated at a threshold value, X_0, as shown in Fig. 3.2. Accordingly, Bernoulli trials are attached to each time interval and the modeling of independent and dependent cases are explained already in Sections 3.7.1–3.7.2.

4.4 REGIONAL DROUGHT FEATURES

In chapter: Temporal Drought Analysis and Modeling, temporal characteristics of droughts are examined at a single site through various probabilistic and statistical models. It is not possible to generalize single point drought features to an area or region, but provided that a set of single drought characteristics locations are known, then a drought map can be drawn from information at these locations and areal drought feature can then be identified and interpreted. Such an approach is possible, but the main drawback is that at each location drought features are considered as independent from each other. At some locations there may be surpluses, whereas at others there may be deficits at simultaneous time intervals. This provides the ability of water transfer from surplus areas to deficit locations. Such water transfers are not related to drought occurrences only but also to human activities such as excess water exploitation due to domestic, agricultural, and industrial activities.

In areal drought assessments at a set of locations, joint drought occurrences are taken into consideration. This provides the opportunity to find the percentage of drought area (Tase, 1976; Şen, 1978). Areal drought features can be achieved either by probabilistic approaches or spatial data treatment processes such as Kriging (Matheron, 1963) or surface trend analysis. Although there are spatiotemporal drought applications, their practical uses are limited due to a set of restrictive assumptions (Şen, 2009).

The unprecedented increase in population, industry, and living standards in modern times gave rise to significant water shortages, which often creates drought periods when the shortages persist for lengthy periods of time. Although floods have been investigated empirically and theoretically by various researchers for a long time, the study of droughts is rather new. The pioneering work regarding the extreme values *s* is due to Dodd (1923), who studied the largest value other than normal PDF. However, Tippett (1925) succeeded in calculating the largest value probabilities observed in samples originating from normal PDF. Gumbel (1963) wrote a concise book, which was completely devoted to extreme value statistics. A different approach to flood frequency analysis was proposed by Todorovic (1970) on the basis of random number of random variables.

For the purpose of water resources system analysis, drought definition can be modified as (Palmer, 1965),

the interval of time generally of the order of several days, months or years in duration, during which the actual water supply at a given location rather consistently falls short of the demand

The classical approach to the drought problem began with the evaluation of the instantaneously smallest value by means of the theory of extremes (Gumbel, 1963). This approach does not tell us anything about drought duration. However, the first objective definition of drought was given by Yevjevich (1967) on the basis of run theory, the application of which has been performed by Downer et al. (1967), Saldarriaga and Yevjevich (1970), Millan and Yevjevich (1971), Guerrero-Salazar (1973), and Şen (1976, 1977), among other researchers.

All works mentioned heretofore are concerned with single-station investigations. However, it has been long accepted by various researchers that drought and flood analyses based on regional data are usually preferred to that developed for a single site, unless a very long record is available at this site. The main reason for this is the wide sampling variations at a single site. In the case of regional analysis such a variation is expected to diminish by exploiting all of the available data at multisites. The first study on the subject related to flood analysis was performed by Conover and Benson (1963) and further investigations were reported by Carrigan (1971). The basis of the method is the double ranking procedure in which individual sets of data from multistation areas are first ranked and then arranged in an order based on the ranking of the peak value of each set. This method, although it tells the greatest flood to be observed in a region, fails to specify the duration and areal coverage extent of the flood. Its failure can be circumvented by the theoretical methodology presented in this chapter. Similar studies can be performed for droughts, also.

The first study concerning regional droughts was performed by Tase (1976), who succeeded in determining experimentally (simulations) the area covered by a drought inside a fixed region, the total water deficit (areal magnitude of drought) below the demand level, and the maximum drought intensity. Due to the difficulties for the application of analytical methods in the investigation of area deficit-intensity characteristics of droughts, the experimental Monte Carlo or sample generation method has been preferred exclusively.

To assess a regional drought occurrence pattern, the study region must be divided into characteristically distinctive, but mutually exclusive, subareas on the basis of prevailing meteorological, atmospheric, geomorphologic, social, and other considerations (see Fig. 4.1). Drought is unique among atmospheric and other environmental hazards in that it creeps gradually on the afflicted area and, consequently, gives rise to social problems. It is generally accepted that severe droughts arise as a result of apparently chance variations of atmospheric circulation. Droughts are always started by a shortage of precipitation, and many traditional approaches to drought definition have been restricted to rainfall analysis (Smith, 1965). Simple assessment of meteorological drought severity depends largely on the magnitude and areal extent of precipitation deficiencies from the mean annual conditions. Because precipitation is the main source for water supply, it is one of the most dominant factors in subdivision of the whole region into different drought potentials. In addition, precipitation is the most easily available data compared to other drought-effective variables. For the sake of simplicity, precipitation is considered as one of the main variables that gives rise to regional and temporal variations of droughts. Other major variables are deficits in water supply and soil moisture content, which vary regionally as well as temporally depending on human and agricultural activities and water availability.

The relative significance of these major variables at various times and/or locations yield two hydrologically important events; namely, wet and dry spells. Dry spells correspond to water deficits, the persistence of which may lead to dangerous drought periods, as well as areal coverages; otherwise, water surpluses occur during wet periods.

Clustering of dry spells in a region is referred to as dry area; otherwise, wet area is valid. However, dry and wet areal spell configurations in a region give guidelines about regional water transportation by an interconnected network of channels, tunnels, or pipes.

Some versions of the aforementioned models are already applied in stochastic investigation of attachment and detachment of suspended solids to grains of deep-bed filter (Saatçi and Şen, 1982).

4.5 RANDOM DROUGHT COVERAGE AREAS

Precipitation, runoff, groundwater, evaporation, and soil moisture are hydrological quantities, which evolve continuously in space and time. Under given specific circumstances, each of these quantities has only a single value at each point in space and at each instant of time. Such a space–time distribution is referred to as a field of hydrometeorological quantity. As the quantity cannot be predicted with certainty, they are assumed to be random and it is necessary to study a new class of fields; namely, random fields (RFs). Quantitatively, a RF is characterized by the function, $\xi(x, y, z, t)$, called a random function at a point, with coordinates x, y, and z in addition to the time instant, t.

It is obvious that the concept of RF is a generalization of a stochastic process, for which time is the sole variable. The expression "the random function of the coordinate (x, y, z, t)" must be understood in the sense that at each point, (x, y, z, t) of four-dimensional space–time, the value $\xi(x, y, z, t)$ is a random variable and cannot be predicted exactly. In addition, the values $\xi(x, y, z, t)$ are subject to a certain law of probability. A complete description of a RF can be achieved by constructing all the finite dimensional PDFs of the field at different points in space. In practical applications, instead of these distributions the corresponding statistical moments might prove sufficient. In general, a RF has three types of moments.

1. Space moments, which are the time products of the values of the field at different points at a fixed time.
2. Time moments, which are the mean product of the values of the field at different times at a fixed point.
3. Space–time moments, which are the mean products from the values of the field at different points and times.

A RF at a fixed time with the given finite dimensional PDF is called homogeneous, if this density is invariant with respect to a shift in the system of points. The same RF is called statistically homogeneous and isotropic if the PDFs are invariant with respect to an arbitrary rotation of the system of points such as a solid body and to a mirror reflection of this system with respect to the arbitrary plane passing through the origin of the coordinate system. In other words, statistical moments depend upon the configuration of the system of points for which they are formed, but not upon the position of the system in space. Thus the correlation function depends only on the distance between two points but not on the orientation in space of the straight line that joints them (Yevjevich and Karplus, 1974; Şen, 1980b, 2009). A well-known time property of a RF is that it is stationary, and for the modeling of droughts and floods herein, the homogeneous-isotropic-stationary fields are considered.

For flood and drought analysis, the space components of the field are assumed to remain in the drainage plane. Furthermore, in practice, information about the RF can be sampled at a finite number of sites within the drainage basin. With fixed coordinates of sites relative to a reference system, the sampled hydrological phenomena constitute a multivariate stochastic process. However, the truncation of hydrologic variables at m different sites by a truncation level, ξ_0, yields two new variables at each site. These are the surplus and deficit as shown in Fig. 4.2. Surplus (deficit) occurs at ith site where $\xi(x, y, z, t) - \xi_0 > 0$ [$\xi(x, y, z, t) - \xi_0 \leq 0$].

It is obvious that as a result of truncation, in addition to the surplus/deficit magnitudes, their locations are over the truncation plane, and the numbers are also random. When the joint evolution of magnitudes and locations along the time axes are considered, they can be modeled by a RF.

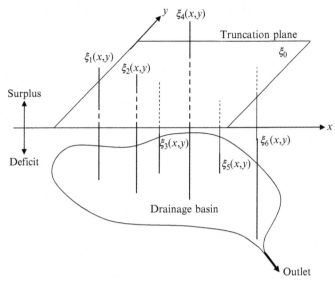

FIG. 4.2 Truncation of random field realizations over a drainage basin.

4.5.1 Regional Drought Descriptor Definitions

Drought and flood characteristics over an area can be determined objectively by several indices, which were proposed by Tase (1976) for droughts as the deficit (surplus) area, A_d (A_s), the total area deficit, (D), surplus, (S), and the maximum deficit (surplus) intensity, I_d (I_s). In the following theoretical treatments, the drought characteristics are examined only with deductions of the flood characteristics accordingly. For a given uniform truncation plane, ξ_0 (the same truncation level for all sites), these areal drought characteristics can be defined as follows:

$$A_d = \sum_{i=1}^{m} I_{(\xi_i \leq \xi_0)}(\xi_i) \tag{4.2}$$

$$D = \sum_{i=1}^{m} (\zeta_0 - \zeta_i) I_{(\zeta_i \leq \zeta_0)}(\zeta_i) \tag{4.3}$$

and

$$I_d = \xi_0 - \min(\xi_0, \xi_1, \xi_2, \ldots, \xi_m) \tag{4.4}$$

where $I_{(\xi_i \leq \xi_0)}(\xi_i)$ is an indicator function defined as

$$I_{(\xi_i \leq \xi_0)}(\xi_i) = \begin{cases} 1 & \text{if } \xi_i \leq \xi_0 \\ 0 & \text{if } \xi_i > \xi_0 \end{cases} \tag{4.5}$$

where m is the station number within the drainage basin. Similarly, for flood analysis,

$$A_s = \sum_{i=1}^{m} I_{(\xi_i > \xi_0)}(\xi_i) \tag{4.6}$$

$$S = \sum_{i=1}^{m} (\xi_i - \xi_0) I_{(\xi_i - \xi_0)}(\xi_i) \tag{4.7}$$

and

$$I_s = \max (\xi_0, \xi_1, \xi_2, ..., \xi_m) - \xi_0 \tag{4.8}$$

where $I_{(\xi_i > \xi_0)}(\xi_i)$ is an indicator function defined by

$$I_{(\xi_i > \xi_0)}(\xi_i) = \begin{cases} 0 & \text{if } \xi_i \leq \xi_0 \\ 1 & \text{if } \xi_i > \xi_0 \end{cases} \tag{4.9}$$

The deficit area does not yield any information about the distribution of deficits over the area considered, but it expresses the total number of deficits. It is an integer-valued random number with a minimum and maximum equal to 0 and m, respectively. On the other hand, it is straightforward to write

$$A_s + A_d = m \tag{4.10}$$

The total areal deficit does not represent the spatial distribution of deficits, but provides a means for measuring the volume of total water deficit in the area concerned. Its minimum value is equal to 0 and occurs when the whole area is under wet condition. It attains a maximum value when the whole area is simultaneously under a dry condition. It is intuitively evident from these two last statements that there is a strong correlation between A_d and D, as will be rigorously shown later in this chapter.

The maximum deficit intensity is the representative of the most severe local drought effect within the region, but it does not specify its whereabouts. An area with larger maximum deficit intensity is not necessarily more drought-stricken than other areas with relatively smaller maximum deficit intensity.

All the characteristics defined earlier are concerned with the state properties at a fixed time instant; therefore, they do not give any clue about the temporal drought or flood behaviors. However, at each time instant there exists one and only one value of these characteristics.

4.6 ANALYTICAL FORMULATION

Due to the difficulties for applying analytical models in a regional drought analysis, it is useful to consider first the simplified cases. The simple results may serve as a guide for investigation of more complex problems. The RF as defined previously is assumed to be constructed by spatially independent hydrologic variables at m sites. In addition, the PDF of the hydrologic variables are assumed to be independent. Hence, at a given uniform truncation plane level the probabilities of deficit and surplus at any site can be expressed as

$$q = F(\xi_0) = P(\xi \leq \xi_0) \tag{4.11}$$

and

$$p = 1.0 - F(\xi_0) = P(\xi > \xi_0) \tag{4.12}$$

Under the light of the aforementioned simplifications with Bernoulli trials at each site, the probability, $P(A_d = i)$, of i deficit subareas from the whole areal coverage of m subareas behaves as a binomial distribution.

$$P(A_d = i) = \binom{m}{i} q^i p^{m-i} \tag{4.13}$$

The mean and variance of this PDF are given by Benjamin and Cornell (1970) as follows:

$$\left. \begin{array}{l} E(A_d) = mq \\ \text{and} \\ V(A_d) = mpq \end{array} \right\} \tag{4.14}$$

In a similar way, the PDF and the various moments of the surplus area can be found. Eq. (4.14) shows that the mean deficit area increases linearly with the increase in the truncation plane level. This point was confirmed by Tase (1976) on the basis of Monte Carlo techniques. The PDF of the deficit area in Eq. (4.13) is in a sense independent from the underlying PDF of the hydrologic variable considered. In the case of sufficiently large m-values the binomial distribution in Eq. (4.13) converges to a normal PDF with mean mq and standard deviation $\sqrt{mpq}$. On the other hand, the number of sites can be regarded as a measure of the area size, because having practically independent events at distinct subareas they must be located at large distances from each other. The probability of drought covering the whole area is

$$P(A_d = m) = q^m \tag{4.15}$$

It is obvious from this expression that the smaller the area the greater the probability of the drought covering the whole region. Furthermore, this probability decreases rather rapidly with an increase in the areal size. Fig. 4.3 shows the areal drought change with the area size.

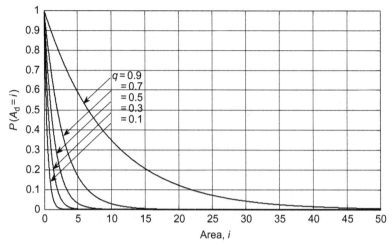

FIG. 4.3 Total area coverage probabilities versus areal size.

In some practical applications, the conditional PDF of A_d, given a group of randomly or symmetrically selected k deficit or surplus subareas, might have interest. For instance, if at one or more sites in a drainage basin there are various water resources alternatives such as reservoirs, groundwater storages, or possibilities of water transfer by interconnected network, then the probability of drought at these sites may become equal to zero. The probability that any k_1 of m sites have deficits at a truncation level ξ_0, corresponding to q, can be defined by virtue of independence principle as follows:

$$P\left(\zeta_1 \leq \zeta_0, \leq \zeta_2 \leq \zeta_0, \zeta_3 \leq \zeta_0, ..., \zeta_{k_1} \leq \zeta_0\right) = q^{k_1} \tag{4.16}$$

Assuming that any site is equally likely to experience a drought, the conditional probability that a certain group of sites has deficits (given that there exists actually a deficit area of size i in the whole area) can be written as

$$P\left(\xi_1 \leq \xi_0, \xi_2 \leq \xi_0, ..., \xi_{k_1} \leq \xi_0/A_d = i\right) = \left(\frac{i}{m}\right)^{k_1} \tag{4.17}$$

Hence, by consideration of the probability statement in Eq. (4.13), the joint probability can be written as

$$P\left(\xi_1 \leq \xi_0, \xi_2 \leq \xi_0, ..., \xi_{k_1} \leq \xi_0, A_d = i\right) = \left(\frac{i}{m}\right)^{k_1} \binom{m}{i} q^i p^{m-i} \tag{4.18}$$

The conditional probability of the deficit area, given that a certain group of sites has deficits, can be found by considering Eq. (4.16) as follows:

$$P\left(A_d = i/\xi_1 \leq \xi_0, \xi_2 \leq \xi_0, ..., \xi_{k_1} \leq \xi_0\right) = \left(\frac{i}{mq}\right)^{k_1} \binom{m}{i} q^i p^{m-i} \tag{4.19}$$

The conditional PDF in Eq. (4.19) has more information than the original distribution in Eq. (4.13). This fact can be shown objectively by considering the information content concept used by Matalas and Langbein (1962). The information content of a PDF is measured by the inverse of its variance. Therefore, the smaller the variance the greater the information content. To show that Eq. (4.19) contains more information than Eq. (4.13), let us simply assume that $k_1 = 1$. The mean and variance of PDF in Eq. (4.19) can then be obtained readily as follows:

$$E(A_d/\xi_1 \leq \xi_0) = mq + p \tag{4.20}$$

and

$$V(A_d/\xi_1 \leq \xi_0) = mpq - pq \tag{4.21}$$

respectively. Comparison of Eq. (4.21) with Eq. (4.14) shows that the conditional variance is always smaller than the original variance; hence, the aforementioned statement, concerning the information content, is valid. Similarly, the conditional probability of a deficit area, given that a group of k_2 has surpluses, can be calculated as

$$P\left(A_d = i/\xi_1 > \xi_0, \xi_2 > \xi_0, \dots, \xi_{k_2}^{2} > \xi_0\right) = \left(\frac{m-i}{mp}\right)^{k_2} \binom{m}{i} q^i p^{m-i} \tag{4.22}$$

This conditional probability statement has been solved for a given values of m, (10, 30, 50) at truncation level $q = 0.5$ and the resulting relationships are given in Fig. 4.4.

Furthermore, the conditional probability of a deficit area, given that a group of k_1 sites has deficits and another group of k_2 sites has surpluses, is

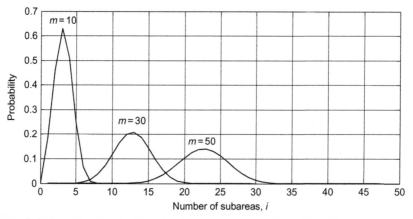

FIG. 4.4 Conditional probability change by subarea ($q = 0.5$ and $k_1 = 5$).

$$P\left(A_d = i/\xi_1 \leq \xi_0, \xi_2 \leq \xi_0, \dots, \xi_{k_1} \leq \xi_0, \xi_{k_1+1} > \xi_0, \xi_{k_1+2} > \xi_0, \dots, \xi_{k2} > \xi_0\right)$$
$$= \left(\frac{i}{mq}\right)^{k_1} \left(\frac{m-i}{mp}\right)^{k_2} \binom{m}{i} q^i p^{m-i} \tag{4.23}$$

The PDFs in Eqs. (4.19), (4.21), and (4.22) can be effectively employed to study regional drought configurations of a network design or some other hydrological problems.

On the other hand, Eq. (4.14) can be employed to evaluate the probability of more than a certain percentage of the whole area to be covered by a drought is

$$P(A_j \geq i) = 1 - \sum_{j=1}^{i} \binom{m}{j} q^j p^{m-j} \tag{4.24}$$

For large m values, the summation term on the right-hand side can be approximated by a normal PDF (see Fig. 4.5).

The PDF of the maximum deficit area, A_M, can be found provided that the deficit area evolutions along the time axis are assumed to be independent of each other. In general, for n time instances, the probability of the maximum deficit area to be less than or equal to an integer i-value can be written as

$$P(A_M \leq i) = P^n(A_d \leq i) \tag{4.25}$$

where from Eq. (4.13),

$$P(A_d \leq i) = \sum_{j=0}^{i} \binom{m}{j} q^j p^{m-j} \tag{4.26}$$

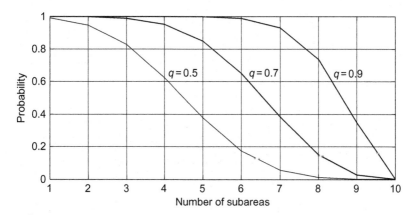

FIG. 4.5 Drought covered area versus probability for $m=10$.

of which the substitution into the previous expression leads to

$$P(A_M \leq i) = \left[\sum_{j=0}^{i} \binom{m}{j} q^j p^{m-j} \right]^n \tag{4.27}$$

However, it is well-known fact that

$$P(A_M = i) = P(A_M \leq i) - P(A_M \leq i - 1) \tag{4.28}$$

Substituting Eq. (4.27) into Eq. (4.28), one obtains

$$P(A_M = i) = \left[\sum_{j=0}^{i} \binom{m}{j} q^j p^{m-j} \right]^n - \left[\sum_{j=0}^{i-1} \binom{m}{j} q^j p^{m-j} \right]^n \tag{4.29}$$

Hence, the probability $P(A_M = m)$ of the whole area to be covered by a deficit can be obtained from Eq. (4.28) as

$$P(A_M = m) = 1 - (1 - q^m)^n \tag{4.30}$$

The numerical solutions of these two expressions are given in terms of the CDFs for a set of maximum areal coverages and time instances in Fig. 4.6 at a set of parameters ($q = 0.5$; $m = 100$; and $n = 2, 5, 10, 25, 50, 100$). Likewise, Fig. 4.7 presents maximum areal coverage CDFs for another set of parameters ($q = 0.1$, 0.3, 0.5, 0.7, 0.9; $m = 100$; and $n = 10$). The simple software for the numerical solution in MATLAB® is given in Appendix 4.1.

For a small region the number of stations is also small, in which case the probability in Eq. (4.30) is not zero, and there is a chance for the whole area to be covered by drought.

If desired, various statistical moments of the maximum deficit area can be found numerically. Similarly, the PDF of the minimum deficit, A_m, can be obtained after some algebra as

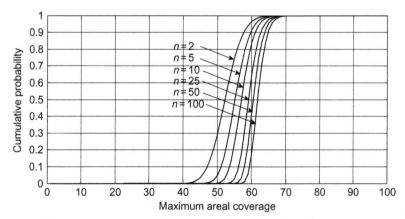

FIG. 4.6 Areal maximum probability coverages for a set of time instances.

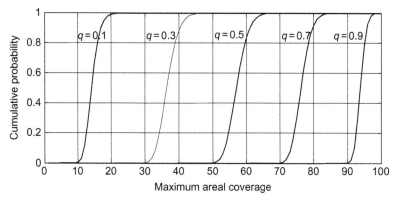

FIG. 4.7 Complete areal drought coverages for a set of truncation levels.

$$P(A_{\mathrm{m}} = i) = \left[\sum_{j=1}^{m} \binom{m}{j} q^j p^{m-j} \right]^n - \left[\sum_{j=i+1}^{m} \binom{m}{j} q^j p^{m-j} \right]^n \tag{4.31}$$

An interesting situation occurs when the probability of surpluses at different sites are different from each other. In such a situation, the RF is heterogeneous and anisotropic. Provided that the random events at distinct sites are mutually independent, then heterogeneous and anisotropic RFs are the result of the following three cases:

1. Identical PDFs at sites but nonuniform truncation level (ie, different truncations at different sites).
2. Nonidentical PDFs at sites and uniform truncation levels.
3. Nonidentical PDFs at sites and nonuniform truncation levels.

Let the surplus and deficit probabilities at m sites be given by p_i and q_i ($i = 1, 2, ..., m$), respectively. The probability of the deficit area being equal to i, $P(A_{\mathrm{d}} = i)$ can be evaluated by enumeration technique leading to

$$P(A_{\mathrm{d}} = i) = \frac{1}{i!} \sum_{\substack{j_1=1}}^{m} q_{j_1} \sum_{\substack{j_2=1 \\ i_2 \neq j_1}} q_{j_2} \cdots \sum_{\substack{j_1=1 \\ J_j \neq j_1 \\ l=1,2,...,i-1}}^{M} q_{j_i} \prod_{\substack{k=1 \\ k \neq j_1 \\ l=1,2,...,i}}^{m} p_k \tag{4.32}$$

or, succinctly,

$$P(A_{\mathrm{d}} = i) = \frac{1}{i!} \left(\prod_{k_1}^{i} q_{k_1} \sum_{\substack{k_{k_1}=1 \\ j_{k_1} \neq k_1 \\ l=1,2,...,j_{k_1}-1}}^{m} q_{k_1} \prod_{\substack{k_2=1 \\ k_1 \neq j_i \\ l=1,2,...,j_{k_1}}}^{m} p_{k_2} \right) \tag{4.33}$$

where the multiplication of i summations in the brackets includes all the possible combinations of i deficits at m sites, whereas the last multiplication term corresponds to possible surplus combinations. For identical PDFs, the terms in brackets simplify to $n(m-1) \ldots (m-i+1)q^i$ and the last multiplication to p^m, which yields Eq. (4.13). If required the maximum and minimum deficit areas of heterogeneous and anisotropic RFs can be found similar to Eqs. (4.29) and (4.31), but they have been avoided herein due to their large and complex expressions.

4.7 TOTAL AREAL DEFICIT (D)

The driving mechanism of the total areal deficit as defined in Eq. (4.3) involves two random phenomena: one is associated with the number of deficits (deficit area) and the other with their magnitudes (total deficit volume). In its probabilistic treatment the theory of sums of random number of random variables can be effectively used (Papoulis, 1965). The joint probability of the total areal deficit and deficit area can be written as follows (Şen, 1980b):

$$P(D \leq d, A_d = i) = P(D \leq d | A_d = i)P(A_d = i) \tag{4.34}$$

from which the marginal distribution of D follows.

$$P(D \leq d) = \sum_{i=0}^{m} P(D \leq d / A_d = i)P(A_d = i) \tag{4.35}$$

The substitution of Eq. (4.13) into this last expression leads to

$$P(D \leq d) = \sum_{i=0}^{m} \binom{m}{i} p^i p^{m-i} P(D \leq d / A_d = i) \tag{4.36}$$

The conditional probability of the total deficit area, given a specific deficit area, can be approximated by a normal PDF with mean and variance as follows:

$$\left. \begin{array}{c} \mu_D = i\mu_d \\ \text{and} \\ \sigma_D^2 = i\sigma_d^2 \end{array} \right\} \tag{4.37}$$

where μ_d and σ_d are the mean and variance of deficit at a single site. Downer et al. (1967) gave these parameters in terms of truncation level, ξ_0, by

$$\left. \begin{array}{c} \mu_d = \dfrac{\Phi(\xi_0)}{q} \\ \text{and} \\ \sigma_d = 1 + \dfrac{\xi_0}{q}\Phi(\xi_0) \end{array} \right\} \tag{4.38}$$

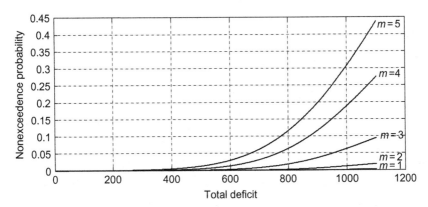

FIG. 4.8 Total areal deficit probability variations ($\xi_0 = 1$; $m = 5$).

where

$$\Phi(\xi_0) = \frac{1}{(2\pi)^{1/2}} \exp\left(-\frac{1}{2}\xi_0^2\right)$$

By virtue of the central limit theorem, one can write that

$$P(D \le d / A_j = i) = \frac{1}{\sqrt{2\pi}\sigma_D} \int_{-\infty}^{d} \exp\left[\frac{(D-\mu_D)^2}{2\sigma_D^2}\right] dD \quad (-\infty \le d \le \xi_0) \quad (4.39)$$

The substitution of Eq. (4.39) into Eq. (4.36) yields the complete form of the total areal deficit probability as

$$P(D \le d) = \frac{1}{\sqrt{2\pi\sigma_D^2}} \sum_{i=0}^{m} \binom{m}{i} q^i p^{m-i} \int_{-\infty}^{d} \exp\left[-\frac{1}{2}\left(\frac{D-\mu_D}{\sigma_D}\right)^2\right] dD \quad (-\infty \le d \le \xi_0)$$

$$(4.40)$$

This expression can be solved numerically on computers provided that the number of stations and truncation level are given. For this purpose a simple MATLAB program is written and given in Appendix 4.2. Fig. 4.8 indicates the total areal deficit variation probability.

4.8 MAXIMUM DEFICIT INTENSITY (I_d)

The RF of I_d has a similar mechanism to D as the maximum of the random number of random variables (Todorovic, 1970). Let us denote the probability of I_d to be less than or equal to s by $P(I_d \le s)$ and the sequence of events as $\{A_d = 0\}, \{A_d = 1\}, \ldots, \{A_d = m\}$, which represent a countable exhaustive set, Ω. Hence, it is possible to write for every $s > 0$,

$$P(I_d \leq s) = P(I_d \leq s, \Omega)$$

or by defining $\xi_i' = \xi_i - \xi_0$,

$$P(I_d \leq s) = P\left[\max \left(0, -\xi_1', -\xi_2', \ldots, -\xi_m' \right) \leq s, \bigcup_{i=0}^{m} (A_d = i) \right]$$

where $\cup$ denotes the union of events. Due to the mutual exclusiveness of events,

$$P(I_d \leq s) = \sum_{i=0}^{m} P\left[\max \left(0, -\xi_1', -\xi_2', \ldots, -\xi_m' \right), A_d = i \right] = P(A_d = 0)$$
$$+ \sum_{i=1}^{m} P\left[\max \left(0, -\xi_1', -\xi_2', \ldots, -\xi_m' \right), A_d = i \right]$$

If ξ_i', $(i = 1, 2, \ldots, m)$ represent a sequence of independent random variables with PDF $P\left(\xi_i' \leq s \right)$ then one can write that

$$P(I_d \leq s) = P(A_d = 0) + \sum_{i=1}^{m} \prod_{j=1}^{i} P\left(\xi_j' \leq s \right) P(A_d = j) \tag{4.41}$$

Furthermore, if ξ_i's are identically distributed, Eq. (4.41) becomes

$$P(I_d \leq s) = p^m + \sum_{i=1}^{m} P^i\left(\xi' \leq s \right) P(A_d = i) \tag{4.42}$$

The substitution of Eq. (4.12) into Eq. (4.42) leads to the simplest form of the maximum deficit-intensity PDF as

$$P(I_d \leq s) = p^m + \sum_{i=1}^{m} \binom{m}{i} q^i p^{m-i} P^i\left(\xi' \leq s \right) \tag{4.43}$$

The numerical solution of this last expression is given in Fig. 4.9 for $m = 5$ and the truncation level equal to 1.

4.9 AREAL JOINT DROUGHT PDF

In practical applications, the joint PDFs of drought and flood descriptors are more useful than the marginal distributions derived in the previous sections. The mutual relationships between the expectations of any two descriptors, especially, can be obtained easily from the relevant joint PDF and used in various engineering design projects. For instance, the relationship between the deficit area and the total areal deficit is very useful for predicting the total volume

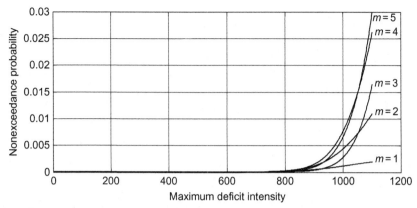

FIG. 4.9 Maximum deficit-intensity probability variations ($\xi_0=1$; $m=5$).

of water within a given drainage basin. Furthermore, these relationships effectively characterize the areal drought or flood structures as will be shown later in this chapter. The most important of these relationships is the one between the deficit area and total areal deficit. The joint PDF can be found by substituting Eqs. (4.13) and (4.37) into Eq. (4.34) (Şen, 1980b).

$$P(D \le d, A_d = i) = \frac{1}{\sqrt{2\pi\sigma_d^2}} \binom{m}{i} q^i p^{m-i} \int_{-\infty}^d \exp\left(-\frac{(D-i\mu_d)^2}{2i\sigma_d^2}\right) dD \quad (4.44)$$

The conditional PDF of D, given the deficit area has already been presented in Eq. (4.40) from which the conditional expectation of the total areal deficit can be found as

$$E(D/i) = i\mu_d$$

and unconditional expectation becomes

$$E(D) = E(A_d)\mu_d \quad (4.45)$$

This shows a perfect linear relationship between the two expectations. Consideration of Eq. (4.38) leads to

$$E(D) = E(A_d)\left[1 + \frac{\xi_0 \exp\left(-\frac{1}{2}\xi_0\right)}{q\sqrt{2\pi}}\right] \quad (4.46)$$

The slope of the straight line is dependent on the truncation level only. The variance of the total areal deficit is

$$V(D) = E(A_d) \left[1 + \frac{\xi_0 \exp\left(-\frac{1}{2}\xi_0\right)}{q\sqrt{2\pi}} \right]$$

$$- \left[V(A_d) - E(A_d^2) \right] \left[\xi_0 + \frac{\xi_0 \exp\left(-\frac{1}{2}\xi_0\right)}{q\sqrt{2\pi}} \right]^2 \qquad (4.47)$$

The maximum deficit intensity for a given deficit area can then be evaluated from the theory of classical extreme value statistics as

$$P(I_d \leq s, A_d = i) = P^i\left(\xi' \leq s\right) \qquad (4.48)$$

The joint PDF of the variables concerned is

$$P(I_d \leq s, A_d = i) = \binom{m}{i} q^i p^{m-i} P^i(\xi \leq s) \qquad (4.49)$$

For the regression function, the conditional expectation of I_d, given the deficit area, can be found from Eq. (4.49) as

$$E(I_d / A_d = i) = \int_{-\infty}^{\xi_0} P^i(\xi \leq s) ds$$

or unconditional expectation is

$$E(I_d) = \sum_{i=0}^{m} \binom{m}{i} q^i p^{m-1} \int_{-\infty}^{\xi_0} P^i(\xi \leq s) ds \qquad (4.50)$$

The ratio of Eq. (4.50) to Eq. (4.49) leads to

$$E(I_d) = \left[\frac{1}{mq} \sum_{i=0}^{m} \binom{m}{i} q^i p^{m-1} \int_{-\infty}^{\xi_0} P^i(\xi \leq s) ds \right] E(A_d) \qquad (4.51)$$

To find the expression relating to the expectation of the deficit area, the total areal deficit, and the maximum deficit intensity, the ratio of Eq. (4.46) to Eq. (4.51) gives

$$E(D) = \frac{E(A_d)E(I_d)\left[1 + \left((\xi_0 \exp\left(-(1/2)\xi_0^2\right))/(q\sqrt{2\pi})\right)\right]}{\sum_{i=0}^{m} \binom{m}{i} q^i p^{m-i} \int_{-\infty}^{\xi_0} P^i(\xi \leq s) ds} \qquad (4.52)$$

The substitution of Eq. (4.14) into this last expression gives the relationship between $E(D)$ and $E(I_d)$ as

$$E(D) = \frac{mq\left[1 + \left(\left(\xi_0 \exp\left(-(1/2)\xi_0^2\right)\right)/\left(q\sqrt{2\pi}\right)\right)\right]}{\sum_{i=0}^{m} \binom{m}{i} q^i p^{m-i} \int_{-\infty}^{\xi_0} P^i(\xi \leq s) ds} E(I_d) \qquad (4.53)$$

It is obvious from this expression that for a given truncation plane and area the total areal deficit increases proportionally with the maximum deficit intensity.

4.10 RAINY AND NONRAINY DAYS

In arid and semiarid regions of the world, water is an extremely scarce and valuable commodity. In these regions, existing resources are exploited mostly for domestic and agricultural activities. For effective planning of local agriculture, the analyses of temporal and regional wet- and dry-spell duration are necessary. The probability of wet- and dry-spell occurrences can be estimated usually with high reliability from available records at an individual site, but their regional distribution is rather difficult to assess. Wet and dry sequences are treated empirically or theoretically by different authors such as Feyerherm and Bark (1965), Gabriel and Neumann (1962), Green (1970), Şen (1976, 1980b, 1989), and Hershfield (1970). Large samples of temporal and spatial data are necessary to obtain estimates of regional wet and dry spells. Information about wet and dry spells provides insight into the various hydrologic, hydrometeorological, agricultural, and water engineering activities in an area.

The purpose of this section is to develop simple temporal and spatial models for estimating wet- and dry-spell duration distributions with applications in the southwestern part of Saudi Arabia based on daily data. Such a study serves as a guide for agricultural and groundwater recharge activities in an area where the rainfall amounts are rather insufficient for domestic and agricultural uses. Regional dry- and wet-spell durations within a 15-year period are mapped for the southwestern part of Saudi Arabia for summer and winter by Al-Yamani and Şen (1997).

Although Saudi Arabia is extremely dry, some parts of the country experience significant rainfall amounts (Şen, 1983; Nouh, 1987). From rainfall distribution analyses over the country, studied by Alqurashi (1981), it is noted that there are three rainy seasons; namely, winter, spring, and summer. Winter and spring rainfalls are caused by a combination of disturbances from the Mediterranean and Sudan troughs. They cover the northern half of the country. The summer rain is caused by the southwest monsoon air mass movement, which is more noticeable over the southern half of the country.

Southwestern Saudi Arabia is the selected area of study (Fig. 4.10). This area is mostly mountainous and receives the highest rainfall total in the country, with an average annual amount of about 450 mm.

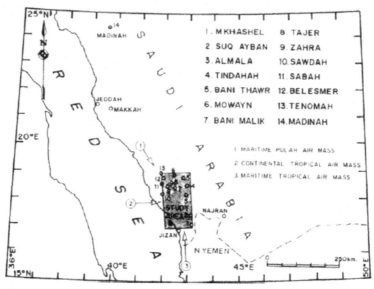

FIG. 4.10 Location map.

Any day with rainfall amount not equal to zero is considered as wet; otherwise, it is dry. An uninterrupted sequence of wet (dry) days preceded and succeeded by at least one dry (wet) day is referred to as a wet (dry) day spell. Previous studies have concentrated only on the rainfall amounts in the Kingdom of Saudi Arabia (Alqurashi, 1981; Şen, 1983; Nouh, 1987; Al-Yamani and Şen, 1992). In general, daily rainfall records from any rain gauge station in Saudi Arabia look like the one in Fig. 4.11.

Fig. 4.11 can be regarded as a representative sample for arid regions of the world with the following general features:

1. There are alternate occurrences of wet and dry spells along the whole record period.
2. Successive durations are randomly different from each other.
3. The total duration of wet spells is invariably smaller than total dry-spell duration as characteristics of arid regions.
4. By definition, the duration of any wet (dry) spell is equal to the number of uninterrupted sequence of wet (dry) days.
5. In the arid and semiarid regions, the smaller the wet-spell duration the more intensive the rainfall amounts. This is tantamount to saying that the overall maximum rainfall amount appears within comparatively short wet spells.
6. The number of wet spells is practically equal to the number of dry spells.

On the other hand, the following points are among some of the general concerns about the wet (dry) spells:

1. Wet day spells play an important role in agricultural planning and forest management.
2. It is possible to make quantitative description of drought and flood periods in addition to flash flood occurrence assessments and predictions.

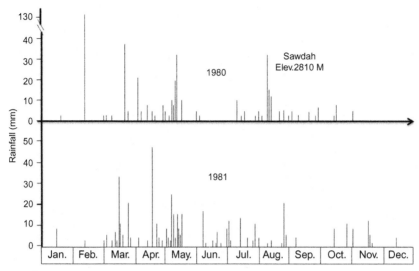

FIG. 4.11 Characteristic rainfall pattern for the study area.

3. Calculation of actual evaporation amounts on a daily basis. Such an approach is essential, especially in arid and semiarid regions.
4. Construction of rainfall and runoff relationships and their predictions to control surface flow discharges.
5. Operation of flood protection, water supply, and recreational dam storages.
6. Groundwater management studies need information on consecutive wet and dry day spells because groundwater abstraction is expected to increase during dry spells.

4.10.1 Daily Rainfall Data Statistics

The main purpose of this section is not a classical statistical analysis of actual rainfall amounts but rather the temporal as well as spatial behavior of dry- and wet-spell duration. Hence, each record is considered as being composed of binary events; that is, wet day (W) in the case of any rainfall occurrence and otherwise it is a dry day (D). Representation of a rainy day by +1 and a nonrainy day by −1 leads to a clear pattern of dry and wet spells as shown in Fig. 4.12.

The cluster of adjacent wet days (dry days) constitutes a wet spell (dry spell); notationally, WS (DS). Hence, the beginning and end states of any spell have transition states either in the form of WD or DW as implied in chapter: Temporal Drought Analysis and Modeling (Section 3.2). Prior to the extraction of dry-spell information, the available record is divided within each year into two equal 6-month periods, namely, the summer and winter seasons. The summer season starts from Apr. 1 and the winter season from Oct. 1. The purpose of making such a distinction is to investigate the wet- and dry-spell statistics in more detail. Furthermore, it has been shown by various authors (Şen, 1983; Al-Yamani and Şen, 1992) that the rainfall patterns during winter periods are significantly

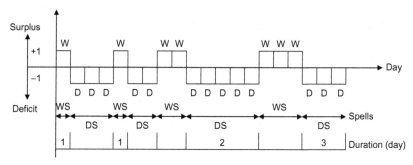

FIG. 4.12 Wet- and dry-spell representation.

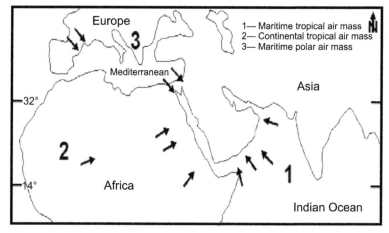

FIG. 4.13 Air masses movement over the Arabian Peninsula.

different from summer periods. Different air mass movements over the Arabian Peninsula are shown in Fig. 4.13.

During the summer season, for the whole region, the number of wet days is very small due to the prevailing advection of African continental warm air. This is also apparent from the observed, comparatively small amount of rainfall events during the summer season. On the other hand, the increase in the number of wet days in the area during the winter season is due to the Mediterranean-borne cold fronts, which invade the southern provinces of Saudi Arabia. Basic information concerning successive dry and wet spells of various durations is derived from a 15-year record of daily rainfall amounts.

Tables 4.1 and 4.2 include the number of successive wet and dry spells of given duration in days for winter and summer at a set of rainfall stations as in Fig. 4.10, respectively.

It should be born in mind that for each station the first line includes the number of wet spells, whereas the second line is for the number of dry spells. In order to further clarify these tables, let us give few practical examples. For instance, in Table 4.1 (winter) for a given 5-day duration, 2071 dry spells and only 2 wet

TABLE 4.1 Number of Successive Rainy and Nonrainy Days in Winter

Station Name	≥1	≥2	≥3	≥4	≥5	≥6	≥7	≥8	≥9	≥10	≥11	≥12	≥13	≥14	≥15
															Duration (Day)
M'khashel	170	50	18	6	2										
	2571	2416	2289	2174	2071	1974	1885	1800	1726	1655	1567	1523	1462	1404	1350
Suq Ayhan	198	61	22	9	6	4	3	2	1						
	2537	2360	2228	2124	2008	1896	1803	1715	1636	1559	1428	1419	1355	1294	1239
Almala	439	189	89	40	23	11	6								
	2296	2028	1854	1651	1511	1394	1287	1197	1112	1031	960	894	834	780	737
Tindahah	173	63	22	9	3	2	1								
	2561	2424	2300	2187	2086	1995	1910	1823	1749	1673	1598	1530	1463	1397	1333
Bani Thawr	161	34	12	6	3	1									
	2577	2417	2291	2178	2072	1973	1884	1795	1714	1635	1561	1501	1434	1370	1307
Mowayn	211	83	39	19	8	7	6	5	4	3	2	1			
	1526	2370	2227	2102	1990	1883	1795	1708	1626	1546	1469	1395	1325	1259	1196
Bani Malik	131	41	12	7	4	2	1								
	2603	2485	2377	2279	2168	2106	2028	1951	1830	1813	1747	1685	1624	1566	1513

Continued

TABLE 4.1 Number of Successive Rainy and Nonrainy Days in Winter—cont'd

Station Name		Duration (Day)													
	≥1	≥2	≥3	≥4	≥5	≥6	≥7	≥8	≥9	≥10	≥11	≥12	≥13	≥14	≥15
Tajer	216	56	22	9	4	1									
	2519	2337	2180	2045	1919	1805	1702	1707	1516	1433	1354	1287	1204	1135	1069
Zahra	244	67	23	4	1										
	2490	2286	2118	1980	1854	1738	1631	1527	1429	1336	1255	1176	1104	1034	967
Sawdah	496	232	125	70	43	24	13	7	3	1					
	2234	1951	1718	1535	1378	1240	1119	973	909	872	742	671	606	543	486
Sabah	254	78	25	7	1										
	2481	2284	2122	1980	1851	1737	1601	1529	1432	1343	1259	1181	1110	1045	981
Belesmer	423	186	85	41	27	21	14	11	9	7	5	3	2	1	
	2311	2045	1868	1682	1541	1421	1316	1118	1052	948	868	794	732	673	
Tonomah	570	281	150	90	54	33	21	13	7	3	1				
	2163	1858	1625	1442	1293	1168	1064	976	897	827	762	701	645	591	545
Madinah	85	19	6	1											
	2650	2561	2492	2391	2313	2234	2158	2087	2016	1947	1850	1815	1751	1691	1631

TABLE 4.2 Number of Successive Rainy and Nonrainy Days in Summer

Station Name		≥1	≥2	≥3	≥4	≥5	≥6	≥7	≥8	≥9	≥10	≥11	≥12	≥13	≥14	≥15
									Duration (Day)							
M'khashel		392	137	51	21	8	4	1								
		2346	2062	1859	1687	1558	1452	1358	1274	1203	1138	1080	1031	984	942	900
Suq Ayhan		252	63	15	3											
		2492	2297	2131	1991	1867	1758	1664	1578	1505	1463	1407	1356	1308	1261	1214
Almala		323	105	32	16	8	3									
		2422	2188	2002	1851	1724	1619	1530	1447	1372	1302	1240	1183	1130	1080	1039
Tindahah		96	17	6	4	2	1									
		2649	2555	2475	2402	2331	2264	2201	2139	2081	2025	1968	1920	1870	1822	1735
Bani Thawr		81	25	11	6	5	4	3	2	1						
		2660	2591	2525	2464	2407	2351	2298	2247	2200	2154	2103	2062	2017	1973	1929
Mowayn		104	23	4												
		2638	2547	2464	2391	2323	2259	2196	2138	2087	2038	1992	1947	1904	1861	1818
Bani Malik		87	23	4												
		2658	2576	2907	2441	2378	2320	2267	2218	2170	2125	2081	2038	1996	1956	1916

Continued

TABLE 4.2 Number of Successive Rainy and Nonrainy Days in Summer—cont'd

Station Name	Duration (Day)														
	≥1	≥2	≥3	≥4	≥5	≥6	≥7	≥8	≥9	≥10	≥11	≥12	≥13	≥14	≥15
Tajer	108	19	4	1											
	2636	2520	2428	2342	2266	2194	2127	2064	2004	1945	1886	1829	1774	1720	1668
Zahra	100	22	5												
	2645	2552	2466	2388	2315	2247	2185	2128	2069	2016	1965	1915	1866	1818	1720
Sawdah	497	187	77	36	14	7	4	3	2	1					
	2248	1921	1668	1469	1307	1168	1053	953	868	797	734	679	629	584	547
Sabah	102	20	9	2											
	2642	2544	2461	2372	2286	2203	2125	2054	1983	1918	1857	1796	1738	1681	1628
Belesmer	131	42	18	12	9	7	5	4	3	2	1				
	2614	2511	2422	2341	2271	2204	2138	2075	2017	1962	1909	1859	1810	1762	1714
Tonomah	229	75	32	20	15	12	9	7	5	3	1				
	1517	2351	2218	2093	1980	1882	1792	1760	1631	1556	1456	1417	1353	1292	1233
Madinah	34	12	4	1											
	2710	2672	2635	2600	2568	2537	2507	2478	2449	2420	2392	2364	2336	2308	2280

spells occurred at M'khashel. However, for the same duration during summer, Table 4.2 shows at the same station 8 wet and 1558 dry spells. Station to station comparisons from these tables led to the following important conclusions:

1. The longest wet spell is observed in Belesmer, which lies in the Scarp-Hijaz Mountains. Table 4.1 shows that in the winter season only this station had one 14-day wet spell, which is rather an extreme event for Saudi Arabia. However, during the summer period, the same station has also one of the longest wet spells, but for 11 successive days only (Table 4.2).

2. The next most persistent wet spell in the winter season appears at Mowayn with 12-days. However, this station has the lowest maximum wet-spell duration in summer. Hence, the difference between summer and winter spells is very significant in depicting different air masses in winter and summer. The climatological investigation in the area shows that in summer, the monsoon flow affects Mowayn, but in winter relatively moist Mediterranean air masses enter the area.

3. Contrary to the aforementioned two stations, M'khashel, Bani Thawr, and Tenomah experience longer wet spells in summer than in winter. Tenomah and Bani Thawr stations are in the north of the study area where Mediterranean air (maritime polar) has an influence, whereas M'khashel, in the very south, gets monsoonal (maritime tropical) air during the summer monsoon. Orographic conditions are also important for the observed spatial distribution of precipitation in the study area.

4. Column-wise comparisons of wet and dry spells between the two tables for fixed duration reveal significant changes in the regional rainfall pattern. For instance, Sawdah has a maximum number of wet spells of only 1 day during summer and winter, while Madinah has the lowest number of wet spells of only 1-day duration.

5. The arid nature of the study region is apparent from the comparison of wet and dry spells, as the number of dry spells for a fixed duration is invariably greater than the number of wet spells of the same duration. Further insight into the precipitation regime can be understood from the conventional summary statistics of wet and dry daily rainfall sequences as presented in Tables 4.3 and 4.4.

On the other hand, a comparison of mean and mode values at each station leads to the following classification:

1. The daily rainfall distributions comply with a negatively skewed distribution in stations where the mode value is smaller than the mean. All stations in this study have pronounced negatively skewed distributions (negative exponential or log-normal distributions) (Şen, 1983) during winter and summer. The skewness of the distribution function is evidenced by the use of relative error (α) based on conventional summary statistics; namely, the mean ($\bar{x}$) and mode (m) values as

$$\alpha = 100 \frac{\bar{x} - m}{\bar{x}} \tag{4.54}$$

TABLE 4.3 Summary Statistics of Successive Rainy Days

Station Name	Winter			Summer			Annual		
	Mean	Mode	St. Dev.	Mean	Mode	St. Dev.	Mean	Mode	St. Dev.
M'Khashel	49.2	6.0	70.1	87.7	8.0	142.4	39.5	50.0	18.1
Suq Ayban	34.0	4.0	64.4	83.2	3.0	115.4	30.0	18.0	16.4
Almala	113.7	23.0	156.7	81.3	8.0	124.2	50.7	45.0	12.8
Tindahah	39.0	3.0	63.0	21.0	2.0	37.2	17.9	26.0	8.5
Bani Thawr	36.2	3.0	62.3	15.3	4.0	25.7	16.1	17.0	8.7
Mowayn	32.3	5.0	61.0	43.7	4.0	53.1	21.0	17.0	9.8
Bani Malik	28.3	4.0	47.4	37.0	3.0	44.2	14.5	10.0	10.5
Tajer	50.5	4.0	81.2	33.0	1.0	50.6	21.5	22.0	7.8
Zahra	67.8	4.0	101.9	42.3	5.0	50.7	23.0	20.0	7.7
Sawdah	101.4	13.0	156.4	82.8	4.0	156.6	66.1	63.0	19.0
Sabah	75.0	7.0	104.6	33.3	2.0	46.4	23.6	20.0	9.8
Belesmer	59.6	9.0	115.8	21.3	5.0	38.2	36.9	45.0	12.6
Tenomah	111.2	21.0	174.1	37.1	9.0	67.1	53.3	55.0	24.2
Madinah	27.8	1.0	38.9	12.5	1.0	14.9	7.9	8.0	5.4

TABLE 4.4 Summary Statistics of Successive Dry Days

Station Name	Winter			Summer			Annual		
	Mean	Mode	St. Dev.	Mean	Mode	St. Dev.	Mean	Mode	St. Dev.
M'Khashel	1859.0	1726	381.1	1391.6	1274	435.1	325.9	316	18.2
Suq Ayban	1778.1	1636	405.8	1686.1	1578	394.6	335.4	348	16.4
Almala	1304.2	1112	478.3	1541.9	1447	421.7	314.4	302	12.9
Tindahah	1868.7	1749	383.3	2165.1	2139	275.2	347.4	353	8.5
Bani Thawr	1847.1	1714	395.6	1165.9	1147	229.9	349.1	363	8.9
Mowayn	1761.2	1626	414.2	2173.5	2138	257.8	343.6	339	10.4
Bani Malik	1898.7	1880	342.9	2243.3	2218	232.5	351.0	355	10.3
Tajer	1673.9	1516	450.5	2093.5	2064	299.7	343.0	344	7.9
Zahra	1595.0	1429	472.8	2156.3	2128	273.9	341.8	345	8.1
Sawdah	1128.9	909	537.9	1108.9	953	519.1	298.9	296	19.4
Sabah	1596.6	1432	467.8	2419.2	2125	323.7	341.3	346	10.1
Belesmer	1302.8	1118	499.4	2107.3	2757	280.8	328.4	330	12.6
Tenomah	1103.8	897	486.9	1767.3	1710	398.4	311.8	310	24.3
Madinah	2107.8	2016	326.1	2483.7	2428	135.9	357.5	363	5.6

Substitution of relevant parameters into this equation indicates that α is more than 5% for daily rainfall amounts. However, when the records are considered on an annual basis such a difference reduces to less than 5%, as observed at several stations (eg, Bani Thawr, Tajer, Tenomah, and Madinah).

2. Overall comparisons between dry- and wet-spell statistics indicate that invariably dry-spell statistics are larger than wet spells. This implies, physically, the general arid region character of the study area. In addition, dry-spell duration standard deviations are comparatively very large, which indicates significant random variations. Consequently, reliable calculation of dry-spell duration predictions presents difficulties using classical statistical techniques. It is, therefore, advisable to investigate the spatiotemporal variations of dry and wet spells, which are discussed in the following sections.

4.10.2 Temporal Variation

To identify the temporal variation of dry and wet spells, their durations are plotted against numbers of occurrences on the semilogarithmic scale for each station during summer and winter seasons separately. Examples of the results are presented in graphical form on semilogarithmic paper as in Fig. 4.14.

It is possible to deduce the following significant conclusions from the visual inspections of these graphs.

1. There is an inverse relationship between the spell numbers and durations. This suggests that short spells occur more frequently than long spells.
2. In all the graphs, on semilogarithmic paper, there is no curvature and, in fact, the majority has a single straight-line relationship. This means that the rainfall generation mechanism (cyclonic, convection, or orographic) is rather uniform, at least during the available record length over the study area.

    Generally, the straight-line equation on semilogarithmic paper can be written as

$$D = a + b\log N \qquad (4.55)$$

where D is a specified duration and N is the number of spells. The constants "a" and "b" must be determined from the simple regression line fit to scatter diagrams in Fig. 4.14.

On the semilogarithmic paper, "a" and "b" represent the intercept and the slope of the straight line. However, physically, "a" corresponds to the average maximum possible spell duration to occur at the station, whereas "b" indicates the severity of the spell. The steeper the line (ie, the greater the value of "b") the more persistent the impact of such a spell. In other words, in such a situation wet and dry-spell durations do not occur randomly. The values of "a" and "b" for single straight-line cases are shown in Tables 4.5 and 4.6.

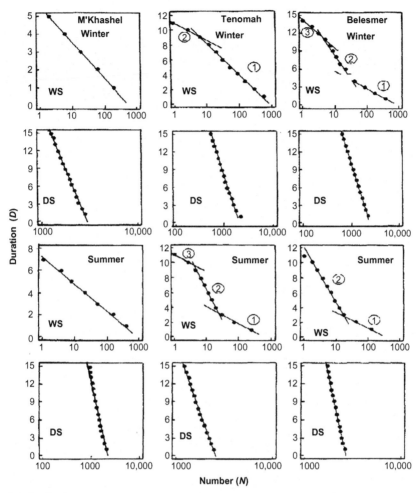

FIG. 4.14 Duration and number of dry and wet spells in summer and winter seasons. *WS*, Wet period; *DS*, dry period.

3. Although all the summer or winter dry-spell duration–number relationships appear as straight lines, some of the wet spells show broken lines. For instance, winter season wet periods at stations Suq Ayban, Tindahah, Bani Thawr, Mowayn, Bani Malik, Tajer, and Tenomah, as well as summer seasons at Tindahah and Sawdah stations, have break points between two straight-line portions. Such a change in the slope implies local climate variation in time due to monsoon effects. In their quantitative representations two consecutive straight lines can be written as

$$D = a_1 + b_1 \ \log N \quad \text{for} \ N < N_1^* \tag{4.56}$$

$$D = a_2 + b_2 \ \log N \quad \text{for} \ N > N_1^* \tag{4.57}$$

TABLE 4.5 Model Parameters for Window Season

| | | Winter | | | |
| | | WP | | DP | |
Station Name	Line No.	a	b	a	b
M'Khashel	1	5.62	−2.09	173.44	−50.70
Suq Ayban	1	5.78	−2.09	156.20	−45.60
	2	9.32	−5.43		
Almala	1	9.37	−3.23	96.90	−28.80
Tindahah	1	6.17	−2.32	172.10	−50.30
	2	7.06	−4.10		
Bani Thawr	1	4.27	−1.48	177.20	−48.40
	2	6.13	−2.78		
Mowayn	1	7.59	−2.87	150.10	−43.90
	2	12.71	−6.99		
Bani Malik	1	5.02	−1.87	206.60	−60.03
	2	7.08	−3.69		
Tajer	1	4.99	−1.71	131.10	−38.40
	2	6.13	−2.25		
Zahra	1	5.05	−1.66	119.00	−34.70
Sawdah	1	10.60	−3.45	73.10	−21.62
Sabah	1	5.20	−1.68	120.13	−25.20
Belesmer	1	8.74	−2.96	91.10	−26.90
	2	16.90	−8.36		
	3	14.71	−4.35		
Tenomah	1	12.40	−4.21	80.50	−24.10
	2	11.10	−2.65		
Madinah	1	4.07	−1.58	228.03	−66.30

where $a_1, b_1, a_2,$ and b_2 are parameters that can be determined from the available data through the regression technique. N_1^* is the threshold number for the break points.

4. Belesmer in winter and summer as well as Tenomah and Bani Thawr in summer have three straight lines for wet spells. They are located close to one another (Fig. 4.10) and to the north. According to Şen (1983), this region

TABLE 4.6 Model Parameters for Summer Season

| Station Name | Line No. | Winter | | | |
| | | WP | | DP | |
		a	b	a	b
M'Khashel	1	7.17	−2.39	37.20	−9.15
Suq Ayban	1	4.79	−1.56	154.00	−45.20
Almala	1	7.10	−2.51	131.13	−38.80
Tindahah	1	5.99	−3.32	279.10	−81.33
	2	6.07	−3.74		
Bani Thawr	1	5.31	−2.65	349.12	−102.08
	2	11.80	−5.72		
	3	8.99	−3.33		
Mowayn	1	3.87	−1.41	300.20	−87.70
Bani Malik	1	3.69	−1.26	324.20	−94.33
Tajer	1	3.94	−1.47	248.00	−72.30
Zahra	1	4.07	−1.54	280.10	−81.50
Sawdah	1	8.00	−2.62	76.90	−23.10
	2	10.20	−4.92		
Sabah	1	4.55	−1.80	228.41	−66.52
Belesmer	1	5.84	−2.30	266.06	−77.70
	2	12.90	−8.29		
	3	11.00	−3.33		
Tenomah	1	6.45	−2.33	157.10	−46.10
	2	15.00	−8.46		
	3	10.99	−2.09		
Madinah	1	4.07	−1.98	651.10	−189.00

is influenced by Mediterranean and monsoonal air movements during winter. Their reliable estimations will require three subsequent straight-line equations,

$$D = a_1 + b_1 \ \log N \ \text{ for } \ N < N_2^* \tag{4.58}$$

$$D = a_2 + b_2 \ \log N \ \text{ for } \ N_2^* < N < N_3^* \tag{4.59}$$

$$D = a_3 + b_3 \ \log N \ \text{ for } \ N \geq N_2^* \qquad (4.60)$$

where N_2^* and N_3^* are the lower and upper threshold (break point) values that can be read directly from the scatter diagrams. The numerical values of parameters as well as threshold values are presented in Tables 4.5 and 4.6.

4.10.3 Spatial Variation

It is possible to prepare a set of maps showing the spatial variability of wet- or dry-spell durations individually. This is not a practical solution because there will be many maps and, more importantly, their use will depend on prefixing the numbers for wet- or dry-spell durations. In addition, it is not possible to switch from number to duration or vice versa.

The technique proposed in this section is rather straightforward and does not require mapping each time. Based on the information available in Tables 4.5 and 4.6, it is possible to construct maps for the regional variations of regression parameters.

The major problem is the incorporation of single, double, or triple straight-line cases into a unique map. In the case of dry spells, where all stations have only one intercept, the regional map of intercept variation can be drawn without any ambiguity. Figs. 4.15 and 4.16 show regional patterns for summer and winter seasons, respectively.

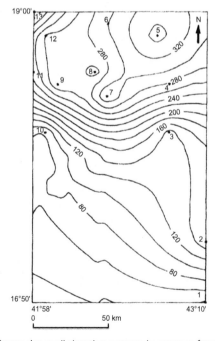

FIG. 4.15 The maximum dry-spell duration pattern in summer (number of days).

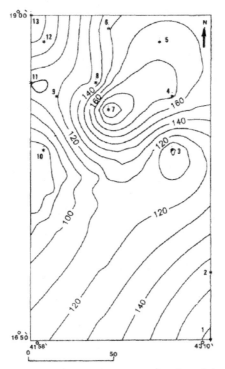

FIG. 4.16 The maximum dry-spell pattern in winter (number of days).

In fact, these maps represent physically the maximum expected dry-spell duration within the study area at any desired location. It is obvious from Fig. 4.16 that the longest dry-spell duration appears in the northern parts of the study area, in general, and in the northeast in particular.

It should be remembered that the northeastern locations are close to the Rub-Al-Khali desert, which has the severest dry conditions within the Kingdom of Saudi Arabia the whole year round. However, a secondary maximum of dry-spell duration is observed in the southeast (Fig. 4.15). The maximum durations in winter are shorter than in summer. The climatological reason for such a shift is the Mediterranean cold air mass reaching the area, especially during Nov. and Dec. (see Fig. 4.16).

Finally, regional patterns of drought severity during summer and winter seasons are shown in Figs. 4.17 and 4.18, respectively. These maps show similar patterns to their respective counterparts in Figs. 4.15 and 4.16. It is also possible to prepare similar maps for the wet spells, which are most useful in agricultural, irrigational, and groundwater recharge activities.

In order to assess the reliability of proposed interpolation technique, a 2-day wet-spell duration frequency map is presented in Figs. 4.19 and 4.20.

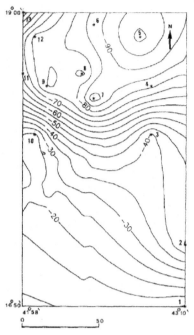

FIG. 4.17 Dry-spell severity pattern in summer factor to Eq. (4.55).

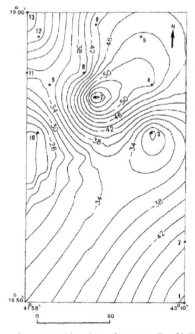

FIG. 4.18 Dry-spell severity pattern in winter factor to Eq. (4.55).

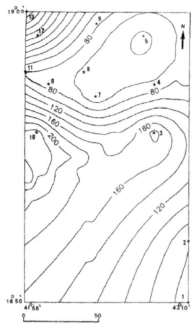

FIG. 4.19 Two-day wet-spell duration historic pattern.

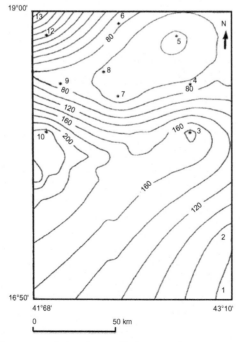

FIG. 4.20 Two-day wet-spell duration prediction pattern.

For the construction of these figures, it is necessary to execute the following steps in sequence:

1. Decide on the duration in days; here, a 2-day wet-spell duration has been adapted.
2. Decide on the locations in the study area.
3. Read off "*a*" and "*b*" values from maps in Figs. 4.15–4.18 for these locations.
4. Calculate the duration of wet (dry) spell using the appropriate equation of Eqs. (4.55)–(4.60). For instance, if the station has only one straight-line relationship between N and D, then Eq. (4.55) leads to

$$N = \exp\left(\frac{D-a}{b}\right) \tag{4.61}$$

As a and b are available for any point within the study area from Figs. 4.15 and 4.16, their substitutions into Eq. (4.61) together with any desired value of D gives the desired value of the number of occurrences. The calculated wet-spell frequency map for $D=2$ is given in Fig. 4.20. This figure has almost the same pattern as Fig. 4.19. This point shows the efficiency of the spatiotemporal calculation method presented in this section.

After all that has been explained earlier, it is obvious that dry- and wet-spell durations, especially on the basis of daily rainfall records, provide useful knowledge for agriculture, irrigation, groundwater recharge, and different hydrological activities.

4.11 DOUBLE-LOGARITHMIC METHOD FOR DETERMINATION OF MONTHLY WET AND DRY PERIODS

In arid regions of the world, although rainfall occurrences and amounts are random and sporadic, there is some regularity in the way rainfall systems evolve. In such regions, dry spells occur rather persistently, leading to extreme drought periods that may cause damage to water resources if necessary measures are not taken based on the reliable prediction of these periods. On the other hand, wet spells are also significant in arid regions because they are related directly to groundwater recharge and to flash floods, which are the major natural water supply for the whole region. Herein, a graphical procedure is presented for assessment of wet- and dry-spell duration based on the monthly rainfall amounts, in order to depict the change of wet and dry period durations with their number within a given record length. A linear regression formulation is described on a log–log plot between spell durations and numbers (Bazuhair et al., 1997). Because the longest durations of wet and dry periods occur once in the whole record length, their predictions can be found from the double-logarithmic graphs corresponding to occurrence number one.

The alternating nature of wet and dry periods of any rainfall record, especially as they occur in arid regions of the world, is presented in Fig. 4.21. In a hydrometeorological context, wet and dry periods are most often referred to interchangeably as rainy and nonrainy spells, respectively.

The methodology developed is applied to monthly rainfall sequences at 16 observation stations representatively distributed within different climatological regions of the Kingdom of Saudi Arabia, as shown in Fig. 4.22 (see also Table 4.7).

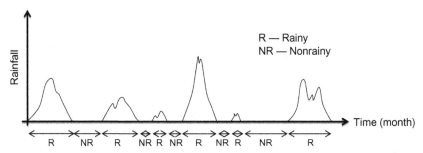

FIG. 4.21 Schematic wet and dry periods of a rainfall record in arid regions.

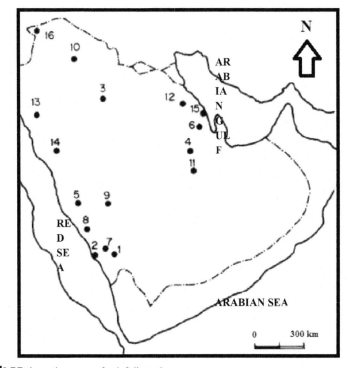

FIG. 4.22 Location map of rainfall stations.

TABLE 4.7 Rainfall Record Characteristics in Saudi Arabia

Number	Station Name	Period	Yearly Average (mm)	Wet Period (%)	Dry Period (%)	Elevation (m)
1	Al-Amir	1966–86	33	79	21	2100
2	Qamah	1965–86	24	31	69	20
3	Ba'Qaa	1967–86	9	47	53	755
4	Harad	1966–86	18	49	51	300
5	Turabah	1965–86	17	43	57	1126
6	Hofuf	1967–86	15	45	55	160
7	Abha	1965–86	30	87	13	2200
S	Beljurshi	1965–86	38	86	14	2400
9	Ranyah	1965–86	21	54	46	810
10	Sakakah	1965–86	11	45	55	574
11	Yatrib	1968–86	11	36	64	119
12	Sarar	1968–86	16	49	51	75
13	Al-Ula	1966–86	17	36	64	650
14	Mosajid	1966–86	18	39	61	471
15	Qatif	1967–86	18	40	60	5
16	Qurrayat	1967–86	9	46	55	549

The western part is along the Red Sea, which is affected in the south by the monsoonal circulation during the spring and summer seasons (Fig. 4.13). To the north, this region is under the influence of the Mediterranean climate in the winter season. The stations are mostly located along the Red Sea escarpment, which runs roughly parallel to the Red Sea coast. Some of the stations are located on flat coastal areas, which show extreme dry periods to the east. However, the stations on the escarpment receive comparatively higher rainfall amounts more frequently; therefore, their dry durations are shorter. Some relevant and important quantitative characteristics of these stations such as the serial number, name, record duration, yearly average, total wet and dry duration percentages and elevations are shown in Table 4.7. It should be noticed that there are a few stations at over 2000 m above mean sea level; therefore, their wet durations are expected to be significantly long.

Information about the statistical behavior of each station is given in Table 4.8 terms of monthly means (first line) and standard deviations (second line). It is obvious that, on average, the dry spell prevails from Jun. to Oct. inclusive, except at high-elevation stations. It is equally obvious that high averages

TABLE 4.8 Monthly Means and Standard Deviations

Station Number	Month											
	J	F	M	A	M	J	J	A	S	O	N	D
1	20.0	24.5	47.0	56.0	55.0	17.5	38.0	05.6	07.4	03.9	12.0	09.0
	23.5	35.0	61.0	52.5	43.0	14.7	21.0	04.5	08.0	06.5	21.0	14.5
2	23.0	07.0	02.8	05.8	02.8	00.2	04.5	01.0	01.3	02.5	08.2	05.0
	31.5	09.0	09.0	12.5	06.5	01.0	29.5	02.5	04.5	06.0	22.0	21.5
3	40.0	06.7	22.0	14.0	07.4	00.0	00.0	00.0	00.0	02.7	20.3	04.0
	55.5	18.0	30.5	19.5	09.0	01.0	00.5	00.0	00.0	06.5	22.0	14.5
4	13.0	10.0	08.3	10.0	02.2	00.0	00.0	00.0	00.0	00.0	00.0	01.8
	23.5	16.0	12.0	12.5	03.5	00.0	01.5	00.0	00.5	00.5	00.5	02.5
5	12.0	08.0	22.6	33.0	13.0	00.0	01.9	04.9	01.3	00.0	08.1	00.7
	17.5	17.0	25.0	51.0	16.0	00.0	05.0	09.0	03.5	02.5	15.4	17.5
6	16.2	11.3	15.0	15.0	01.2	00.0	00.0	00.0	00.0	00.0	02.5	04.6
	26.0	15.0	19.0	19.5	02.0	00.0	00.0	00.0	00.0	01.0	05.5	05.5
7	33.4	27.0	49.0	56.0	44.0	15.0	44.0	40.0	07.3	10.6	16.5	08.0
	26.3	57.8	67.0	43.9	35.5	12.0	19.4	16.0	08.0	18.0	38.0	18.0
8	71.6	29.1	31.0	56.0	25.0	12.0	22.0	28.0	08.7	22.0	50.0	44.0
	77.4	41.5	37.0	89.5	25.0	12.0	27.5	32.0	12.0	20.0	50.0	61.0

Continued

TABLE 4.8 Monthly Means and Standard Deviations—cont'd

Station Number						Month						
	J	F	M	A	M	J	J	A	S	O	N	D
9	06.1	04.5	39.0	46.0	10.9	00.0	01.5	00.0	00.0	01.6	04.6	02.4
	08.5	11.5	27.0	49.0	22.5	00.0	05.5	00.0	00.0	06.5	13.0	05.5
10	11.0	05.7	07.9	08.0	05.2	00.0	00.0	00.0	00.0	03.6	19.0	03.9
	16.5	08.0	17.5	18.0	09.0	00.0	00.0	00.0	00.0	09.0	28.5	05.5
11	10.0	05.5	03.5	12.3	01.2	00.0	00.0	00.0	00.0	00.5	06.1	09.0
	10.5	14.0	06.5	26.0	02.5	00.0	00.0	00.0	00.0	02.0	07.5	12.5
12	13.0	12.0	18.0	18.0	04.0	00.0	00.0	00.0	00.0	04.4	12.8	15.0
	21.5	17.0	24.0	27.5	05.5	00.5	01.5	00.0	00.0	14.0	18.0	07.0
13	06.4	04.9	08.8	16.5	02.8	00.0	00.0	01.3	00.0	01.5	08.9	05.4
	09.0	05.5	17.5	20.0	05.5	00.5	00.0	04.0	00.0	05.0	10.0	09.0
14	20.0	09.3	12.0	15.2	03.3	00.0	00.0	02.1	03.2	00.0	16.0	06.0
	22.5	12.0	22.0	27.5	09.0	01.5	01.5	05.5	04.5	01.0	17.5	39.0
15	17.5	14.0	13.6	13.4	00.0	00.0	00.0	00.0	00.0	00.0	10.0	11.0
	17.5	29.0	20.5	18.5	01.5	00.0	00.0	00.0	00.0	00.5	11.5	23.5
16	08.7	05.6	10.0	03.5	02.9	00.0	00.0	00.0	00.0	03.3	07.5	04.5
	07.5	08.0	15.0	06.0	05.5	00.0	00.0	00.0	01.0	05.0	08.3	05.5

are associated with high standard deviations and vice versa. The maximum averages and standard deviations occur mainly during Apr., when there is an unstable transition period from the Mediterranean to the monsoonal effect. The second month with the same pattern is Jan., corresponding to the start of the Mediterranean climate effect.

For the Kingdom of Saudi Arabia, whenever there is appreciable rainfall in a month, it is considered a wet (rainy) month, otherwise zero monthly rainfall amounts indicate a dry (nonrainy) month. A succession of uninterrupted wet (dry) months preceded and succeeded by at least one dry (wet) month is referred to as a wet (dry) spell and the number of months in such a spell is its period (ie, duration or length).

In order to present the application of graphical methodology in finding the durations of wet and dry spells as well as their numbers, monthly rainfalls at Al-Amir station from 1966 to 1979 are given in Table 4.9. Given the aforementioned definition it is obvious that there is a comparatively much smaller number of dry than wet spells. Consequently, longer wet duration spells occur more frequently.

The sequence of monthly rainfall amounts are converted into wet- and dry-spell numbers of fixed durations along the available record. If the interest lies in the number of wet and dry periods of, say, a 2-month (ie, two successive months) duration only, then calculations from Table 4.9 show that there are 49 and 12 cases, respectively, for these durations within the whole record. Similarly, for wet or dry periods of 6 uninterrupted consecutive months, 10 and 1 occurrences are found within the Al-Amir monthly record. Hence, for all the possible monthly durations the whole results are presented in Table 4.10.

As expected, in general, a first glance at Table 4.10 shows that there is an inverse relationship between the duration (period) and number of wet and dry months. Specifically, Fig. 4.23 shows such a relationship on a log–log graph for the Al-Amir station.

The slope of the straight line is in reverse relationship with the duration of wet or dry periods. The steeper the slope, the shorter the duration change. It is possible to obtain a preliminary prediction of the possible longest wet- or dry-spell duration, on the average, from Fig. 4.23 at Al-Amir station, corresponding to occurrence number 1. Hence, at Al-Amir the expected longest wet- and dry-spell durations appear as intercepts of the straight lines on the horizontal axis in Fig. 4.23 as 33 and 7 months, respectively. Similar calculations are carried out for other stations considered in this section, and Table 4.11 summarizes the straight-line intercept and slope values for monthly wet- and dry-spell durations.

Fig. 4.24 represents the wet and dry spell period–number of occurrences charts for all the stations.

The following significant interpretations can be drawn from Table 4.11 and the corresponding graphs in Fig. 4.24:

1. The station location is referred to as humid if the wet period straight line is significantly above the dry line (Al-Amir in Fig. 4.23; Abha and Beljurshi in

TABLE 4.9 Al-Amir Station Monthly Rainfall Amounts

Year	Month											
	J	F	M	A	M	J	J	A	S	O	N	D
1979	49.0	00.0	25.5	02.5	65.5	25.5	12.0	138.0	24.3	20.4	00.0	18.0
1978	29.2	30.4	07.0	27.2	61.0	04.0	72.4	59.0	10.3	00.0	00.0	01.5
1977	29.8	02.5	20.0	19.5	66.9	07.3	33.9	108.0	00.0	16.2	00.0	00.0
1976	06.4	00.0	102.3	37.8	45.5	11.8	20.7	49.9	00.0	07.0	35.0	00.0
1975	13.0	11.5	09.5	157.0	08.0	21.2	29.1	90.9	07.9	00.0	03.5	00.0
1974	00.0	00.0	212.0	18.5	56.2	38.0	17.8	116.0	16.5	00.0	00.0	54.8
1973	26.3	00.0	00.0	08.2	35.7	10.6	18.7	34.6	01.0	02.5	07.0	09.5
1972	08.2	05.2	103.0	141.0	50.3	14.5	69.9	38.8	00.0	06.5	03.0	12.2
1971	00.0	00.0	00.0	129.0	180.0	08.7	00.0	48.6	00.0	00.0	00.0	00.0
1970	40.3	00.0	91.1	00.0	00.0	16.5	30.2	21.8	16.8	00.0	00.0	00.0
1969	77.5	98.1	63.5	34.1	74.1	10.1	30.2	18.1	10.1	00.0	00.0	00.0
1968	00.0	76.4	00.0	66.6	58.6	55.7	09.7	096.3	00.0	02.1	22.4	00.0
1967	00.0	00.0	30.7	75.9	70.3	11.5	35.3	29.4	19.7	00.0	70.5	00.0
1966	00.0	119.0	00.0	65.6	14.5	16.3	11.6	27.9	00.0	00.0	29.3	00.0

TABLE 4.10 Wet and Dry Periods and Their Duration

| | Number | |
Periods (Months)	Wet	Dry
1	120	49
2	49	12
3	30	5
4	21	1
5	12	1
6	10	1
7	9	0
8	7	0
9	4	0

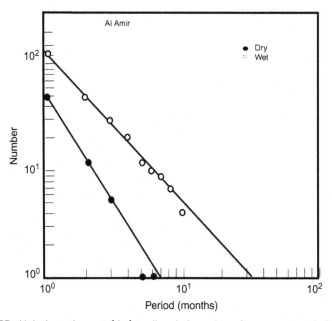

FIG. 4.23 Al-Amir station wet (dry) spell period–number of occurrences relationship.

Fig. 4.24). Otherwise, the location is under an arid condition (Harad, Yatrib, Al-Ula, Mosajid, and Qamah in Fig. 4.24). Semiaridity prevails if these two lines are close to each other (Baqa'a, Turabah, Hofuf, Ranyah, Sakakah, Sarar, Qatif, and Qarrayat in Fig. 4.24). Hence, out of the 16 rainfall stations

TABLE 4.11 Intercept and Slope Values for Various Stations

Number	Station Name	Intercept (Month)		Slope (1/Month)	
		Wet	Dry	Wet	Dry
1	Al-Amir	65	7	1.25	2.00
2	Qamah	9	38	2.20	1.53
3	Ba'Qaa	24	35	1.55	1.39
4	Harad	16	52	1.44	1.26
5	Turabah	17	27	1.65	1.53
6	Hofuf	30	40	1.39	1.32
7	Abha	90	5	1.19	2.29
8	Beljurshi	100	6	1.20	2.12
9	Ranyah	21	17	1.60	1.70
10	Sakaka	32	55	1.40	1.25
11	Yatrib	18	55	1.58	1.25
12	Sarar	32	42	1.39	1.26
13	Al-Ula	6	35	2.38	1.49
14	Mosajid	12	26	1.80	1.50
15	Qatif	27	46	1.32	1.40
16	Qurrayat	26	34	1.44	1.39

considered in this paper, 3, 8, and 5 are humid, semihumid, and arid locations, respectively. Furthermore, it is important to notice that humid locations are situated in the southwestern part of the Kingdom of Saudi Arabia, where the Asir Mountains reach elevations up to 3000 m above mean sea level. This area experiences orographic rainfall due to the topographic and monsoon effects.

2. There are no crossings of wet and dry straight lines. This is tantamount to saying that wet and dry periods are always shorter or longer than each other.

3. The slope of straight lines indicates the intensity of wet and dry periods. The smaller the slope, the shorter and more persistent the wet or dry periods. For instances, among the plots in Fig. 4.24, Beljurshi station has the least wet-period slope, which implies that this station has the longest and most persistent durations of wet spells. Similarly, the shortest dry period occurs at the Abha location because it has the minimum slope (1.19 per month) value among the other stations (see Table 4.11). However, this slope is not practically different from the one for Beljurshi station.

4.12 POWER LAW IN DESCRIBING TEMPORAL AND SPATIAL PRECIPITATION PATTERN

Wet- and dry-spell properties of monthly rainfall series at a set of meteorology stations can be examined by plotting successive wet and dry month durations versus their number of occurrences on the double-logarithmic paper. Straight-line relationships on such graphs show that power laws govern the pattern of successive persistent wet and dry monthly spells. Functional power law relationships between the number of dry and wet spells for a given monthly period can be derived from the available monthly precipitation data. The probability statements for wet and dry period spells are obtained from the power law expression, the comparison of which at distinct sites provides useful interpretation about the temporal and spatial rainfall pattern.

Concern about the variability of weather and climate has grown in recent years. The severest adverse effect of climatic change, however, could be felt in rainfall amounts. Any substantial rainfall change either in amount or timing or both in the rainfall regime for a region could be catastrophic (Tegart et al.,

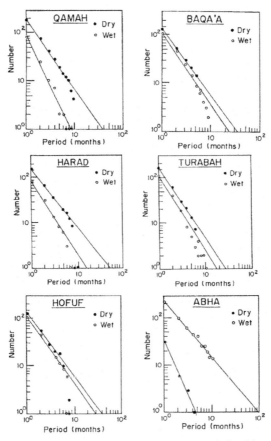

FIG. 4.24 Wet (dry) spell period–number of occurrences relationships.

(Continued)

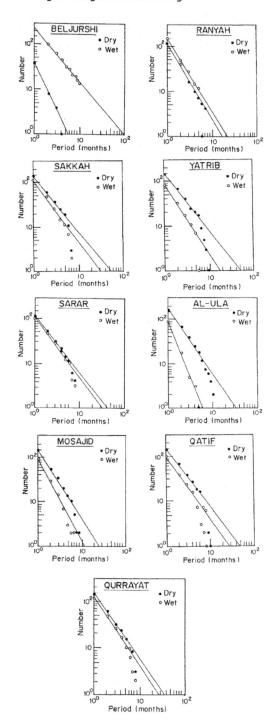

FIG. 4.24—CONT'D

1990). For example, inadequate rainfall in a growing season might cause a crop failure, which in turn can lead to famine for many regions (Todorov, 1985).

Apart from studying the amount of precipitation, it is also of hydrological and agricultural importance to consider the length of rainy periods. The duration of these periods is measured by the number of successive rainy months in a given record. Geographical distribution of the number of rainy days and the frequency of consecutive rainy days appearance in the area of the Aegean Sea are examined by Theoharatos and Tselepidaki (1990).

A change in the distribution of precipitation toward a decrease in the number of rainy days or months, as well as a significant shortening in the length of the rainy season, have been mentioned by many authors who have analyzed Sudano-Sahelian precipitation data (Olaniran, 1991). Any decrease in rainfall or its concentration into a shorter rainy period could reduce crop yields. It is also noted that the area of precipitation over the globe decreases even though global mean precipitation increases (Noda and Tokioka, 1989). Convective changes imply an increase in more intense local rain storms, with less persistent rainfall events associated with local convective storms (Mitchell et al., 1990).

Many studies have been performed to examine the trends and variability of the monthly or annual precipitation totals in various regions of the world (Todorov, 1985). In general, the history of each station shows that there are neither location changes nor environmental effects that might have caused heterogeneity involvements in the rainfall series. This allows one to interpret any regional change in the climatic regions as well as to compare them for any temporal differences. In addition, the same methodology is applied to monthly wet and dry periods within two nonoverlapping consecutive half (NCH) periods each as 1931–60 and 1961–90 for detecting any possible temporal change. The main reason for the comparison, simply between two 30-year periods, is due to extensive industrial and urban developments as well as the increase in environmental problems in the last 30 years. However, it is also possible to examine the changes that take place between overlapping periods of varying lengths of various running means. In the following discussions the first NCH period will be referred to as "first half (FH)" and the subsequent period as "second half (SH)." This is a similar procedure to the innovative trend analysis already mentioned in chapter: Temporal Drought Analysis and Modeling and by Şen (2012, 2014). The change of monthly rainy and nonrainy periods at the stations within two NCH periods are, therefore, determined and compared to search for any possible characteristic systematic variations such as trends.

4.12.1 Description of Data Preparation

To effectively explain the power law procedure, a database is adapted from Turkish monthly total precipitation records for identification of possible climatic or artificially induced trends. The calculations and interpretations are performed for Adana, Ankara, and İzmir cities as already mentioned in chapter: Basic Drought Indicators, Section 2.17. In the following sequence, two more cities are added; namely, Kars city in the eastern part of Turkey and Samsun city in the northern part of Turkey along the Black Sea coastal area (see Fig. 2.20).

In this data set, the stations have various periods of records. For conclusive results in a climatological work, it is necessary to have long enough time series. Additionally, for exploring the changes in wet and dry spells, the data series should have stationary characteristics. In the power law technique application to monthly total precipitation data, the following steps should be followed:

1. At each station, the median value of monthly precipitation totals is calculated for the NCH periods (see Table 4.12).

2. Median values are subtracted from the corresponding monthly precipitation totals. These differences can be thought of as a first-order stationary climatological time series because periodicity in the average fluctuations has been removed. In the first-order stationary time series, negative and positive values form a continuous subseries of alternate random dry and wet periods. Randomness herein implies that occurrences of wet- and dry-spell duration do not affect each other (Şen, 1980b).

3. Numbers of various length dry and wet periods are determined for FH and SH periods. Numerical values of wet- and dry-spell numbers of a predetermined time scale in months are presented in Tables 4.13 and 4.14, respectively.

It is worth noting that although the time scale is a selected deterministic variable, the corresponding number of spells of this duration is a random variable. The probabilistic behavior of wet and dry spells has already been treated analytically by Şen (1976, 1980a, 1989). However, herein the randomness will be investigated on double-logarithmic paper, which is the simplest tool for the probabilistic behavior of the phenomenon concerned.

According to calculated values in Table 4.13, for instance, if 3-month length is considered, then there are 36 wet-run cases within Adana station data during the FH period. Likewise, the number of dry-run periods of the same length is given for the Adana FH period as 21 in Table 4.14. The simple comparison of 3-month dry and wet spells indicates that during the FH period in Adana station, the occurrence of wet periods are more persistent than dry periods.

Furthermore, it is obvious from Table 4.13 that longer wet spells appear in the SH period in Adana and Ankara stations as 13 months and in Kars station again in the SH period but for 11 months. It is well known in Turkey that the first two cities experienced unprecedented industrial development over the last 30 years. On the contrary, in Samsun station the longest wet spell appears during the FH period for 11 months. Temporally, this implies that although Adana, Ankara, and Kars stations become wetter in the SH period, Samsun station has just the opposite occurrence.

Again in Table 4.14, Samsun station reflects its distinctive characteristic for dry spells within the FH period with a very long dry period of 14 months, while others remain at 7 and 8 months. So as far as dry spells are concerned, all the stations exhibit more or less the same dry-spell duration lengths during FH and SH periods, but at Samsun the dry-spell length during the FH period is almost twice that of the SH period. This may be a clear indication of a dry-spell trend at this station.

TABLE 4.12 Medians of Each Month of Different Time Intervals

Station	Period	Months											
		1	2	3	4	5	6	7	8	9	10	11	12
Adana	1931–60	104.3	67.3	54.3	36.1	38.6	13.9	0.1	0.0	7.9	29.3	56.9	83.0
	1961–90	93.4	89.4	57.7	48.0	35.6	18.0	0.0	0.0	9.6	35.7	68.1	110.1
	1931–90	97.4	84.3	57.6	40.1	35.6	16.7	0.0	0.0	8.6	33.8	60.3	92.3
Ankara	1931–60	32.5	29.9	28.4	36.3	46.5	23.2	5.4	3.4	7.8	19.8	34.2	38.7
	1961–90	42.0	29.3	33.5	48.0	51.8	26.7	9.1	6.0	13.2	20.3	25.8	44.0
	1931–90	36.3	29.5	30.2	39.8	48.9	23.3	7.3	4.0	9.2	19.8	27.5	41.4
Izmir	1931–60	134.9	91.3	49.0	33.5	32.2	4.3	0.2	0.0	4.4	38.0	75.6	123.8
	1961–90	131.3	86.6	70.8	39.8	19.6	3.7	0.0	0.0	2.7	26.3	82.4	133.0
	1931–90	134.8	89.1	64.5	38.1	27.3	3.7	0.0	0.0	-3.5	30.0	78.6	130.2
Kars	1931–60	27.0	25.3	30.0	42.0	88.5	71.8	52.1	37.3	26.0	30.0	28.0	1.40
	1961–90	17.2	22.6	27.7	40.6	75.8	73.6	47.2	34.2	18.5	34.8	19.6	19.9
	1931–90	20.9	23.6	26.3	41.4	81.7	71.8	48.5	36.3	23.6	34.0	22.5	20.5
Samsun	1931–60	76.2	64.8	66.6	55.2	39.8	32.5	24.0	20.1	46.6	69.9	88.5	70.8
	1961–90	52.8	43.1	54.8	52.0	42.7	40.5	23.1	27.4	51.3	72.3	75.9	68.1
	1931–90	66.4	53.9	62.3	53.5	41.2	37.2	23.8	22.4	48.1	70.6	84.8	69.2

TABLE 4.13 Numbers of Wet Runs During FH and SH Periods

Month	Adana FH	Adana SH	Ankara FH	Ankara SH	İzmir FH	İzmir SH	Kars FH	Kars SH	Samsun FH	Samsun SH
1	198	210	189	181	196	210	180	180	180	180
2	71	78	67	65	69	82	65	64	60	60
3	36	39	26	22	28	39	28	28	28	28
4	17	25	12	9	15	23	15	17	13	10
5	8	13	5	5	9	11	5	8	6	6
6	5	9	2	2	2	7	2	4	4	2
7	2	7	1	1	1			2	3	1
8	1	4	1	1		2		2	2	2
9	1	3	1	1	1			2	2	
10	1	1	1					2	1	
11		1	1					2	1	
12		1	1							
13		1	1							

TABLE 4.14 Numbers of Dry Runs During FH and SH Periods

Month	Adana FH	Adana SH	Ankara FH	Ankara SH	İzmir FH	İzmir SH	Kars FH	Kars SH	Samsun FH	Samsun SH
1	162	150	183	179	164	150	180	180	180	180
2	52	46	64	58	51	49	62	60	62	68
3	21	18	21	28	21	17	29	31	22	23
4	8	6	13	15	6	7	16	12	13	10
5	3	5	6	5	3	4	6	10	8	2
6	3	2	4	1	2	2	3	5	3	1
7	2	1	3		1	1	1	3	3	1
8			2					1	1	
9									1	
10									1	
11									1	
12									1	
13									1	

4.12.2 Methodology

The essence of the power law technique is based on the examination of some small-scale property resemblance to the moderate and large-scale behaviors within the whole record period. In order to search for any regional climatic change over Turkey, dry and wet periods of two NCH periods at different stations are compared with each other. In this kind of study, starting from the smallest time scale, say a month, one can investigate the number of the nonoverlapping wet (dry) periods throughout the whole record length. Likewise, the counting process is repeated for other time scales, as already shown in Tables 4.13 and 4.14. If the time scale and corresponding number of dry (wet) periods are shown by $S_d(S_w)$ and $n_d(n_w)$, respectively, it is logical to deduce that with increase in $S_d(S_w)$, $n_d(n_w)$ will decrease. The decrease would not be a linear one, but expected to appear as a power function expressed mathematically, in general, as

$$S_d = \frac{a_d}{n_d^{-b_d}} \quad \text{or} \quad S_w = \frac{a_w}{n_w^{-b_w}} \tag{4.62}$$

where a_d, a_w, b_d, and b_w are parameters depending on the characteristics of dry- or wet-spell phenomena concerned. These parameters can be estimated from a double-logarithmic plot of selected time scale versus a corresponding number of dry (wet) periods. Taking logarithms of both sides leads to

$$\log S_d = \log a_d - b_d \log n_d \quad \text{or} \quad \log S_w = \log a_w - b_w \log n_w \tag{4.63}$$

On double-log graph paper, these relationships appear as straight lines with slopes giving estimations of b_d and b_w. The intercept values of S_d and S_w for $n_d = n_w = 1$ provide the estimates of a_d and a_w (see Fig. 4.25).

Power law expressions can be converted to the probability statements for the occurrences of wet- and dry-spell durations greater than or equal to a given duration. For this purpose, it is sufficient to divide both sides of the power law by the number of months within the whole record length, which is equal to $12 \times 60 = 720$ for the five stations considered in this study. It is, therefore, possible to write the probability of wet-, W, or dry-, D, spell duration as $P(W \geq d)$ or $P(D \geq d)$ to be more than a prespecified monthly duration, d, as

$$P(W \geq d) = \frac{a_w}{N} d^{-b_w} \tag{4.64}$$

and

$$P(D \geq d) = \frac{a_d}{N} d^{-b_d} \tag{4.65}$$

where N is the number of total months, that is, $N = 720$.

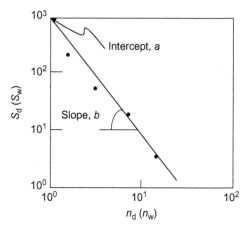

FIG. 4.25 Power law graphs on double-logarithmic scale.

The following points should be kept in mind while interpreting the figures on double-log paper:
1. If the scatters of points are close to a single line within ±10% significance level, then the wet (dry) spell has homogeneous regime. In this case, the phenomenon concerned occurs under uniform environmental conditions. There is no significant climatic change during the period concerned.
2. Any systematic deviation from the straight line implies the evolution of phenomena under nonuniform conditions. Such deviations could be as a result of any short-duration climatic change or anthropogenic processes due to industrial and urban development. Nonuniformity should be interpreted here as the involvement of more than one external effect in the evaluation of the rainfall event.
3. If the double-logarithmic characteristic of a climatic variable at a location is shown on the same figure for two NCHs, it is then possible to infer whether there are temporal climatic changes. For example, if the fitted lines of these two normal periods are close or parallel to each other without any crossing, then there is no evidence for climatic change. Crossing of the lines for NCH periods raises the question of a climatic change. Of course, herein, crossing implies that the FH and SH periods have different slopes.

4.12.3 Graphical Analysis of Numbers of Wet and Dry Months

Double-logarithmic paper plots of numerical values given in Tables 4.13 and 4.14 are represented in Figs. 4.26–4.30 for five stations. Figs. 4.26–4.30A and B exhibit wet (dry) spell number variation with periods of the spell. The best regression linear lines are fitted by least squares technique to the scatter of points separately for wet and dry spells. The slope and intercept values for each NCH periods' wet and dry spells are calculated from these figures and presented in Table 4.15.

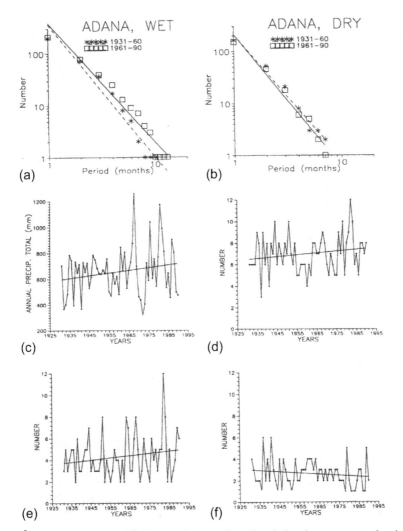

FIG. 4.26 Adana station: (A) Change of wet-spell number during the two nonoverlapping consecutive half (NCH) periods with period length; (B) change of dry-spell number during NCH periods with period length; (C) annual total precipitation time series; (D) yearly 1-month wet-spell numbers as a function of time; (E) time series of the longest successive wet month duration; and (F) time series of the longest successive dry month duration.

These parameters help to make future predictions of wet- or dry-spell lengths from Eq. (4.64) or (4.65). For instance, answers to the questions of wet-spell period occurrence of at least 12-month length after the SH period at Adana station can be calculated by using Eq. (4.64) with necessary parameters from Table 4.15 as

$$P(W \geq 12) = \frac{386.68}{720} \times 12^{-2.28} = 1.77 \times 10^{-3}$$

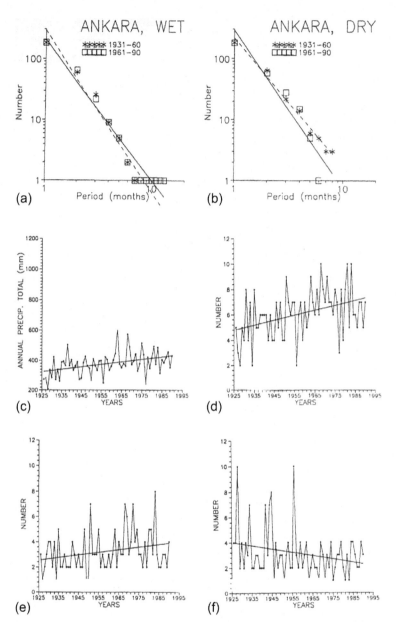

FIG. 4.27 Ankara station: (A) Change of wet-spell number during the two NCH periods with period length; (B) change of dry-spell number during NCH periods with period length; (C) annual total precipitation time series; (D) yearly 1-month wet-spell numbers as a function of time; (E) time series of the longest successive wet month duration; and (F) time series of the longest successive dry month duration.

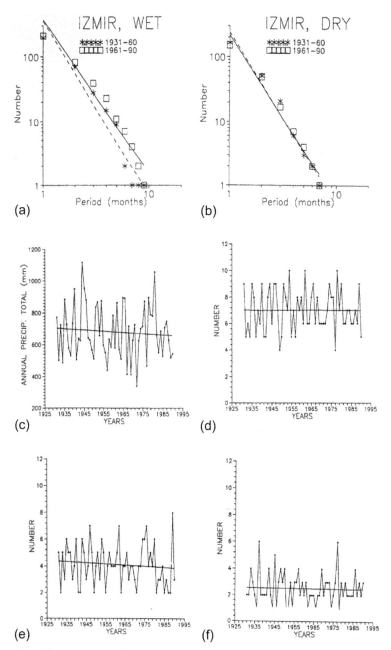

FIG. 4.28 İzmir station: (A) Change of wet-spell number during the two NCH periods with period length; (B) change of dry-spell number during NCH periods with period length; (C) annual total precipitation time series; (D) yearly 1-month wet-spell numbers as a function of time; (E) time series of the longest successive wet month duration; and (F) time series of the longest successive dry month duration.

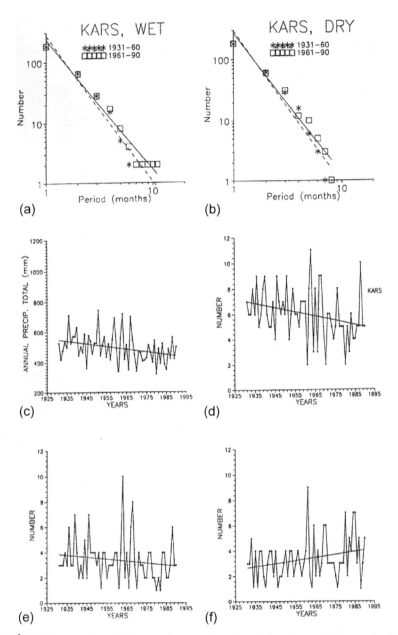

FIG. 4.29 Kars station: (A) Change of wet-spell number during the two NCH periods with period length; (B) change of dry-spell number during NCH periods with period length; (C) annual total precipitation time series; (D) yearly 1-month wet-spell numbers as a function of time; (E) time series of the longest successive wet month duration; and (F) time series of the longest successive dry month duration.

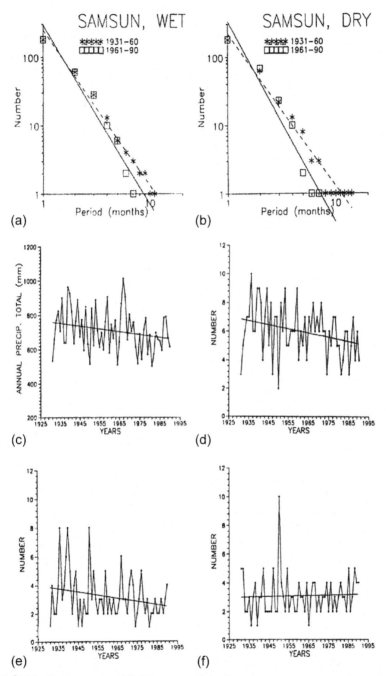

FIG. 4.30 Samsun station: (A) Change of wet-spell number during the two NCH periods with period length; (B) change of dry-spell number during NCH periods with period length; (C) annual total precipitation time series; (D) yearly 1-month wet-spell numbers as a function of time; (E) time series of the longest successive wet month duration; and (F) time series of the longest successive dry month duration.

TABLE 4.15 Slopes (*b*) and Intercepts (*a*)

	Wet Spell				Dry Spell			
	a		*b*		*a*		*b*	
Stations	FH	SH	FH	SH	FH	SH	FH	SH
Adana	360.96	382.68	2.55	2.28	209.23	209.55	2.38	2.53
Ankara	288.68	203.79	2.65	2.31	215.71	289.36	2.10	2.60
Izmir	353.85	369.10	2.68	2.36	241.41	215.71	2.67	2.56
Kars	267.32	229.43	2.40	2.13	296.28	269.14	2.52	2.31
Samsun	249.52	249.52	2.28	2.28	291.01	327.89	2.23	2.29

The same question for 6-month length leads to $P(W \geq 6) = 6.60 \times 10^{-3}$. Likewise, for 1-month duration $P(W \geq 1) = 0.511$, which is equal to the actually observed value.

The following interpretations and conclusions result from the visual inspection and interpretation of these double-logarithmic plots:

1. In general, there are no significant differences between wet- and dry-spell behaviors on double-logarithmic paper except in Samsun, where rather considerable derivations are observed for dry spells (Fig. 4.30B) between the NCH periods. In fact, the crossing of two straight lines is an indication of heterogeneity, most likely in the form of a trend as in Fig. 4.30D. This heterogeneity is not due to station location change but might be a result of local climatic change in the vicinity of this region. Furthermore, at other stations the numbers of successive wet month periods is longer during the FH in comparison with the SH (Figs. 4.26–4.30A),

2. The changes in the series of consecutive dry months are not so distinct for Adana (Fig. 4.26B). There is no difference in dry months between two NCH periods at Izmir (Fig. 4.27B). This is a clear indication that on the basis of 30-year data, dry spells at these stations did not change temporally.

3. It is clear from Figs. 4.26–4.30A and B that the differences between the numbers of wet and dry runs of two NCH periods appear, especially for rather long wet and dry periods. In the SH period, the length of periods longer than 3 months have distinctly changed.

4. Comparisons of the annual precipitation totals with those of the wet and dry double-logarithmic plots (Figs. 4.26A–C, respectively) lead to quite uncommon expectations. Generally, an increasing (decreasing) trend in the series of the annual precipitation totals (Figs. 4.26–4.30C) does not usually coincide with an increase (decrease) in the recent wet (dry) periods (Figs. 4.26–4.30A and B). For example, let us consider Izmir; while Fig. 4.28A shows an increase in the numbers of the wet periods and Fig. 4.28B shows no change in the numbers of dry periods, the annual precipitation totals steadily decrease with time (Fig. 4.28C).

TABLE 4.16 Slopes of Linear Trend

	Slopes			
Station	One-Month Wet	Max-Wet	Max-Dry	Annual Precipitation Total
Adana	0.017	0.021	−0.011	2.080
Ankara	0.039	0.021	−0.027	1.639
İzmir	−0.001	−0.008	−0.002	−0.789
Kars	−0.034	−0.016	0.024	−1.832
Samsun	−0.029	−0.023	0.003	−1.665

5. Table 4.16 indicates that Adana and Ankara have positive trends in their annual precipitation totals (Figs. 4.26 and 4.27A), the numbers of 1-month wet (Figs. 4.26 and 4.27D) and the longest period of successive wet months (Figs. 4.26 and 4.27E).

At these two rainfall stations, the longest periods of successive dry months are, therefore, decreasing. In other words, it can be said that there has been an increasing trend in the wet periods and precipitation amounts. Especially at Ankara station, the number of the 1-month-duration wet spells is increasing rapidly. It indicates that the persistence of monthly precipitation amounts below median (ie, short droughts) decreases substantially. However, the rates of increase in the largest number of the successive wet months in a year are not parallel to the rates of decrease in the largest number of the successive dry months (Figs. 4.26–4.30 also compare the estimated slopes of linear changes in max-wet and max-dry periods from Table 4.15

6. Table 4.16 and Figs. 4.28–4.29C show that the annual precipitation totals for the whole period have decreasing trends at Izmir, Kars, and Samsun stations. The largest portion of this decrease occurs in the numbers of 1-month-duration wet spells at Kars and Samsun (Figs. 4.28–4.30D), and of the longest period of successive wet months at İzmir (Fig. 4.28E).

Throughout the study, the significance of a trend is tested by considering their slope values at the 5% significance level. Significant trends are found in Figs. 4.27C–F, 4.29C, D, and 4.30C, D.

Finally, Fig. 4.31 shows the part of a year that these changes took place. The monthly medians at Adana and İzmir increase in wet part of the year (winter +spring) and they do not change significantly in the rest of the year.

Accordingly, this may explain why the numbers of dry periods do not change significantly in Adana and İzmir (Figs. 4.26 and 4.27B). In contrast to these sites at Samsun, the monthly medians decrease mostly in winter and spring months, but increase slightly in the rest of the year. These are shown in Fig. 4.30A and B, as a decrease both in the numbers of wet and dry periods.

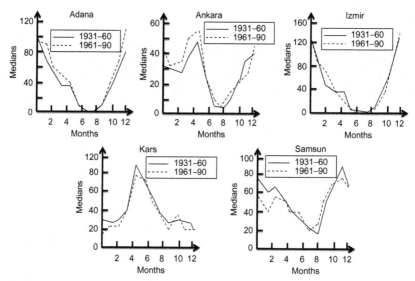

FIG. 4.31 Monthly medians for NCH period of five stations.

A simple power law relationship is proposed and implemented for describing the temporal and spatial precipitation pattern in Turkey. For this purpose, analyses of 60-year original raw precipitation records at five Turkish sites are carried out by considering simple double-logarithmic graph paper representation. Most of the results obtained on this paper could not be extracted solely by classical trend analysis of precipitation time series.

As is expected for temperate areas, evidence from five stations suggests that the rainfall regimes of these sites are mostly dominated by the numbers of wet and dry months whose period lengths are shorter than a meteorological season.

With the exception of Samsun, all stations show on double-logarithmic plots that the number of wet periods that are longer than 3 months have increased over the 1961–90 period. However, the increase in the number of extreme wet or dry periods is much greater. Although the classical trend analysis studies of precipitation series have not shown any distinctive spatial trend patterns, the double-logarithmic graphs indicate rather uniform spatial pattern with no significant regional changes.

While there is a decrease in the number of dry periods at Kars and no change occurs at Izmir, the other three sites selected show a decrease with increasing wet periods. The rates of these increases in the largest number of the successive wet months in a year, are not usually comparable to decreases in the largest number of the successive dry months in a year. As in the example of Turkey, this may indicate that in temperate areas, the precipitation changes are governed mostly by a 1-month-long dry or wet spells. This may indicate that monthly data are insufficient for the resolution of wet (dry) period effects; therefore, daily data must be used in future studies in order to identify this point. The methodology developed in the form of power law is applicable to data anywhere in the world.

APPENDIX 4.1 AREAL MAXIMUM PROBABILITY COVERAGES

```
function [sum,CPD] = MaximumDeficitArea(q,m,t)
% This program is written by Zekâi Şen on 31 December 2014
% q : Truncation level
% t : Time sequence
% m : Total areal coverage
p=1-q;
f=factorial(m);
sum(1)=0;
for i=2:m
    sum(i)=sum(i-1)+((f/(factorial(i)*factorial(m-i))))*q^i*p^(m-i);
end
for i=2:m
    P(i)=sum(i).^t-sum(i-1).^t;
end
CPD=cumsum(P);
end
```

APPENDIX 4.2 TOTAL AREAL DEFICIT

```
function [P]= TotalArealDeficit(t,n)
% This program is written by Zekâi Şen on 27 December 2014
% t     : Truncation level
% n     : Number of deficit locations
% m     :
q=normcdf(t,0,1);
p=1-q;
dm=normpdf(t,0,1)/q; % Deficit mean value
ds=1+t*dm;
f=factorial(n);
x=-10:0.01:t;
sum=0;
for i=1:n
    Dm=i*dm; % Deficit area mean value
    Ds=sqrt(i)*ds; % Deficit area standard deviation
    y=normcdf(x,Dm,Ds);
    sum=sum+((f/(factorial(i)*factorial(n-i))))*q^i*p^(n-i);
    P=sum*y;
    plot(P,'k')
    hold on
end
end
```

REFERENCES

Alqurashi, M.A., 1981. Synoptic climatology of the rainfall in the southwest region of Saudi Arabia. M.A. Thesis, Western Michigan Univ, USA, 97 pp.

Al-Yamani, M., Şen, Z., 1992. Regional variation of monthly rainfall amounts in the Kingdom of Saudi Arabia. J. King Abdulaziz Univ. Fac. Earth Sci. 6, 113–133.

Al-Yamani, M., Şen, Z., 1997. Spatiotemporal dry and wet spell duration distributions in Southwestern Saudi Arabia. Theor. Appl. Climatol. 57, 165–179.

Bazuhair, A.S., Al-Gohani, A., Şen, Z., 1997. Determination of monthly wet and dry periods in Saudi Arabia. Int. J. Climatol. 17, 303–311.

Benjamin, J.R., Cornell, C.A., 1970. Probability, Statistics and Decision for Civil Engineers. McGraw-Hill, New York, 684 pp.

Boken, V., Şen, Z., 2005. Techniques to predict agricultural droughts. Monitoring and Predicting Agricultural Droughts. Oxford University Press, UK, pp. 40–65 (Chapter 4).

Boken, V.K., Craqcknell, A.P., Heathcore, R.H., 2005. Monitoring and Predicting Agricultural Drought. A Global Study. Oxford Press, USA, 472 pp.

Carrigan Jr., P.H., 1971. A flood-frequency relation based on regional record maxima. Professional paper 434-F, U.S. Geological Survey, Reston, VA.

Conover, W.J., Benson, M.A., 1963. Long-term flood frequencies based on extremes of short-term records. Short papers in geology, hydrology and topography: professional paper 450-E, U.S. Geological Survey, Reston, VA, pp. E159–E160.

Cordery, I., Graham, A.G., 1989. Forecasting wheat yields using water budgeting model. Aust. J. Agric. Res. 40, 715–728.

Demuth, S., Stahl, K. (Eds.), 2001. Assessment of the regional impact of droughts in Europe. Final report to the European Union, ENV-CT97-0553. Institute of Hydrology, University of Freiburg, Germany.

Diaz, H.F., 1983. Some aspects of major dry and wet periods in the contiguous United States, 1895–1981. J. Clim. Appl. Meteorol. 22, 3–16.

Dodd, E.L., 1923. The greatest and the least variate under general laws of error. Trans. Am. Math. Soc. 25, 525–539.

Downer, R., Siddiqui, M.M., Yevjevich, V., 1967. Applications of runs to hydrologic droughts. Hydrology paper no. 23, Colorado State University, Fort Collins, CO.

Feyerherm, A.M., Bark, L.D., 1960. Goodness of fit of a Markov chain model for sequences of wet and dry days. J. Appl. Meteorol. 6, 770–773.

Feyerherm, A.M., Bark, D., 1965. Statistical methods for precipitation pattern. J. Appl. Meteor. 4, 320–328.

Gabriel, K.R., Neumann, J., 1962. A Markov chain model for study for daily rainfall occurrence at Tel-Aviv. Q. J. R. Meteorol. Soc. 88, 90–95.

Green, J.R., 1970. A generalized probability model for sequences of wet and dry days. Mon. Weather Rev. 98 (3), 238–241.

Guerrero-Salazar, P.L.A., 1973. Statistical Modeling of Droughts. Ph. D. Dissertation, Colo. State Univ., Fort Collins, Colo., Unpublished Ph. D. Thesis.

Gumbel, E.J., 1963. Statistical forecast of droughts. Bull. Int. Assoc. Sci. Hydrol. 8, 5–23.

Hershfield, D.M., 1970. Parameter estimation for wet-day sequences. J. Am. Water Resour. Assoc. 7 (3), 441–446.

Hisdal, H., Stahl, K., Tallaksen, L.M., Demuth, S., 2001. Have streamflow droughts in Europe become more severe or frequent? Int. J. Climatol. 21, 317–333.

Idso, S.B., Hatfield, J.L., Jackson, R.D., Reginato, R.J., 1979. Grain yield prediction: extending the stress-degree-day approach to accommodate climatic variability. Remote Sens. Environ. 8, 267–272.

IPCC, 2001. Climate change 2001: impacts, adaptation and vulnerability. Report of working group II for the third assessment report of the IPCC, summary for policymakers, http://www.ipcc.ch/pub/wg2SPMfinal.pdf.

IPCC, 2007. The physical science basis. IPCC fourth assessment report working group I report, Cambridge University Press, New York.

Kumar, V., Panu, U.S., 1997. Predictive assessment of severity of agricultural droughts based on agro-climatic factors. J. Am. Water Resour. Assoc. 33 (6), 1255–1264.

Matalas, N.C., Langbein, W.B., 1962. Information content of the mean. J. Geophys. Res. 67 (9), 3441–3448.

Matheron, G., 1963. Principles of geostatistics. Econ. Geol. 58, 1246–1266.

Millan, J., Yevjevich, V., 1971. Probabilities of observed droughts. Hydrology paper no. 50, Colorado State University, Fort Collins, CO.

Mitchell, J.F.B., Manabe, S., Meleshko, V., Tokioka, T., 1990. Equilibrium climate change—and its implications for the future. In: Houghton, J.T., Jenkins, G.J., Ephraums, J.J. (Eds.), Climate Change: The IPCC Scientific Assessment. Cambridge University Press, Cambridge, pp. 137–164.

Noda, A., Tokioka, T., 1989. The effect of doubling the CO_2 concentration on convective and no convective precipitation in a general circulation model coupled with a simple mixed layer ocean model. J. Meteorol. Soc. Jpn. 67, 1057–1067.

Nouh, M.A., 1987. Analysis of rainfall in the south-west region of Saudi Arabia. Proc. Inst. Civ. Eng., Part 2 83, 339–349.

Olaniran, O.J., 1991. Rainfall anomaly patterns in dry and wet years over Nigeria. Int. J. Climatol. 11, 177–204.

Palmer, W.C., 1965. Meteorological drought. US Weather Bureau research paper no. 45, 58 pp.

Panu, U.S., Sharma, T.C., 2002. Challenges in drought research: some perspectives and future directions. Special Issue: Towards integrated water resources management for sustainable development, Hydrol. Sci. J. Sci. Hydrol. 47, S19–S30.

Papoulis, A., 1965. Probability, Random Variables, and Stochastic Processes. McGraw Hill, New York, 583 p.

Rossi, G., Benedini, M., Tsakiris, G., Giakoumakis, S., 1992. On regional drought estimation and analysis. Water Resour. Manage. 6, 249–277.

Saatci, A.M., Şen, Z., 1982. A stochastic model of attachment and detachment in fixed beds. In: Proceedings of the Third World Filtration Congress, Downingtown, Pennsylvania, USA.

Sakamoto, C.M., 1978. The Z-index as a variable for crop yield estimation. Agric. Meteorol. 19, 305–313.

Saldarriaga, J., Yevjevich, V., 1970. Application of run lengths to hydrologic series. Hydrology paper no. 40, Colorado State University, Fort Collins, CO.

Santos, M.A., 1983. Regional droughts: a stochastic characterization. J. Hydrol. 66, 183–211.

Şen, Z., 1976. Wet and dry periods of annual flow series. J. Hydraul. Div. ASCE 102 (HY10), 1503–1514, Proc. Paper 12457.

Şen, Z., 1977. Run-sums of annual flow series. J. Hydrol. 35, 311–324.

Şen, Z., 1978. Autorun analysis of hydrologic time series. J. Hydrol. 36, 75–85.

Şen, Z., 1980a. Statistical analysis of hydrologic critical droughts. J. Hydraul. Div. ASCE 106, 99–115, Proc. Pap. 14134.

Şen, Z., 1980b. Regional drought and flood frequency analysis. J. Hydrol. 46, 258–263.

Şen, Z., 1983. Hydrology of Saudi Arabia, Water Research in the Kingdom of Saudi Arabia, Management, Treatment and Utilization, vol. 1. College of Eng., King Saud Univ., Riyadh.

Şen, Z., 1989. The theory of runs with applications to drought prediction—comment. J. Hydrol. 110 (3–4), 383–390.

Şen, Z., 1998. Average areal precipitation by percentage weighted polygon method. J. Hydrol. Eng. 3 (1), 69–72.

Şen, Z., 2009. Spatial Modeling Principles in Earth Sciences. Springer, New York, 351 pp.

Şen, Z., 2012. Innovative trend analysis methodology. J. Hydrol. Eng. 17 (9), 1042–1046.

Şen, Z., 2014. Trend identification simulation and application. J. Hydrol. Eng. 19 (3), 635–642.

Slabbers, P.J., Dunin, F.X., 1981. Wheat yield estimation in northwest Iran. Agric. Water Manage. 3, 291–304.

Smith, K., 1965. Principles of Applied Climatology. McGraw-Hill Book Co., London, 233 pp.

Tallaksen, L.M., Hisdal, H., 1997. Regional analysis of extreme streamflow drought duration and deficit volume. In: Gustard, A., Blazkova, S., Brilly, M., Demuth, S., Dixon, J., van

Lanen, H., Llasat, C., Mkhandi, S., Servat, E. (Eds.), FRIEND'97-Regional Hydrology: Concepts and Models for Sustainable Water Resource Management, pp. 141–150, IAHS Publ. No. 246.

Tase, N., 1976. Area-deficit-intensity characteristics of droughts. Hydrology paper no. 87, Colorado State University, Fort Collins, CO.

Tegart, W.J., Sheldon, G.W., Griffiths, D.C. (Eds.), 1990. Climate Change: The IPCC Impacts Assessment. Australian Gov. Pub. Services, Canberra.

Theoharatos, G.A., Tselepidaki, I.G., 1990. The distribution of rainy days in the Aegean area. Theor. Appl. Climatol. 42, 111–116.

Tippett, L.H.C., 1925. On the extreme individuals and the range of samples from a normal population. Biometrika 17, 364–387.

Todorov, A.V., 1985. Sahel: the changing rainfall regime and the "Normals" used for its assessment. J. Clim. Appl. Meteorol. 24, 97–107.

Todorovic, P., 1970. On some problems involving random number of random variables. Ann. Math. Stat. 41 (3), 1059–1063.

Toure, A., Major, D.J., Lindwall, C.W., 1995. Comparison of five wheat simulation models in Southern Alberta. Can. J. Plant Sci. 75, 61–68.

Vogt, J.V., Somma, F. (Eds.), 2000. Drought and Drought Mitigation in Europe. Kluwer Academic Publishers, The Netherlands.

Walker, G.K., 1989. Model for operational forecasting of western Canada wheat yield. Agric. For. Meteorol. 44, 339–351.

Yevjevich, V., 1967. An objective approach to definition and investigation of continental hydrologic droughts. Hydrology paper no. 23, Colorado State University, Fort Collins, CO.

Yevjevich, V., Karplus, A.K., 1974. Area-time structure of the monthly precipitation process. Hydrology paper no. 64, Colorado State University, Fort Collins, CO.

Spatiotemporal Drought Analysis and Modeling

5.1 GENERAL

The questions "How extensive is the wet or dry spell areal coverage?" and "What are their effects?" are very common in governmental and farmer circles. "What portion of the industrial, agricultural, and cultural activities will be

Applied Drought Modeling, Prediction, and Mitigation

affected by possible future dry or wet spells?" is another question that must be answered based on various information sources concerning regional weather, climatology, and water resources conditions. For instance, the meteorological information may provide a useful source of data only for the assessment of questions such as "What is the crop moisture probability in the soybean producing regions?"

A dry (wet) spell may be regarded as a normal, recurrent feature of climate for virtually all climatic regions. It is not restricted only to low-precipitation regions of the world, but occasionally high precipitation areas may also experience wet and dry spell areal and temporal occurrences. Therefore, it is necessary to distinguish a dry (wet) spell from the aridity (humidity) point of view, which is restricted to low (high) rainfall regions and it is a permanent climatic feature like it is in the arid (humid) regions of the world. The character of a dry (wet) spell is not only temporal as most people understand, but more severely spatially relating unique meteorological, hydrological, agricultural, and socioeconomic characteristics of the region concerned. Dry (wet) spells are related to relatively long-term average conditions of balance between precipitation and evapotranspiration in a particular area. Droughts triggered by dry spell persistence, especially, differ from other natural hazards in several ways. It is a creeping phenomenon that makes its onset and end difficult even to feel. The effects of drought accumulate slowly over a considerable period of time, and may continue for many years after the termination of the event. On the other hand, drought impacts are less obvious and spread over a larger area than are damages that result from other natural hazards such as floods, earthquakes, volcanoes, and so on. Consequently, dry (wet) spell impact quantification and provision are far more difficult tasks than other natural hazards. Therefore, it is necessary to seek help from probabilistic and statistical approaches and their overall combinations from the stochastic evaluation methodologies for quantification purposes of the spatiotemporal predictions (Şen, 1978). These techniques are helpful in cases where the lack of a precise and objective definition occurs in specific situations. The lack of a precise and objective definition as well as incomplete data, especially, are the main obstacles to proper understanding of dry and wet spells and their modeling and, subsequently, this has led to indecision and incapability in actions against the droughts or floods on the part of managers, policy makers, and others. The main dry (wet) spell causes are concerned more with the effects on the surface or subsurface water supply in addition to streamflow, reservoir, lake, and groundwater levels, rather than with precipitation shortfalls. During agricultural drought periods, competition for water escalates and significant conflicts can arise between water users. In general, upstream changes in land use such as deforestation and changes in cropping patterns may alter runoff and, especially, infiltration rates, which may lead to frequent and severe agricultural drought or flood occurrences. On the other hand, the socioeconomic dimension of wet and dry spells associates the supply and demand of some economic good with elements of meteorological, hydrological, and agricultural phenomena.

So far, there is no procedure for accurately predicting the time occurrence of wet and dry spell (drought) durations with their regional extensions. Although in the past various empirical approaches have been employed for drought estimations, they all ended in surprising failures. For instance, multiple regression analysis or Monte Carlo simulation techniques are used to answer questions concerning regional and temporal wet and dry spell frequencies. In modern times, drought estimations are sought on the basis of objective and systematic scientific procedures and here probability theory and statistics provide convenient procedures for wet and dry spell predictions. These techniques, in general, are used for depicting the quantitative relationships between weather variables and wet and dry duration characteristics.

The main purpose of this chapter is to present a rigorous analytical methodology to model theoretically the spatiotemporal heterogeneous probabilities in a given region during a certain time duration. This empowers meteorologists and hydrologists alike to model areal and temporal wet or dry spell occurrences even for the case of the most complicated set of heterogeneous probabilities.

5.2 SPATIOTEMPORAL DROUGHT MODELS

For regional drought analysis as shown in Fig. 4.1, the whole area can be divided into a set of mutually exclusive subareas. In general, any drought analysis starts with precipitation data; therefore, for regional meteorological drought analysis, spatial precipitation distribution provides information by convenient methodology. There are different sets of areal average precipitation calculation methodologies such as arithmetic average (Thiessen, 1911), polygon, isohyetal mapping, weighted averaged polygon method (Şen, 1998), and others. In general, regional drought analysis of precipitation starts with a consideration of deviations from the regional overall climatic behaviors (ie, averages) (see Fig. 4.2).

In an approach to investigate spatiotemporal drought analyses, two distinctive models can be considered. The first model relies on the regional dry and wet spell occurrences only. The regional wet and dry spell probabilities are denoted by p_r and q_r, respectively, and they are mutually exclusive,

$$p_r + q_r = 1 \tag{5.1}$$

This model assumes that once a subarea is hit by a dry spell, it remains under this state in subsequent time instances. Therefore, as time passes, the number of dry spell hit subareas steadily increases until the whole region is covered by drought. This model is referred to as the regional persistence model (Şen, 1980a). The application of such a model can be valid in a region for drought periods only. When the whole region is under drought effect this model ceases to function. However, the same model can be applied during wet periods in order to assess the time persistence of wet area over a region. Application of this model is convenient for arid regions where long dry periods exist.

The second model takes into account the regional as well as temporal occurrence probabilities of wet and dry spells. The temporal probabilities of wet p_t and dry q_t spell occurrences are mutually exclusive; therefore,

$$p_t + q_t = 1.0 \tag{5.2}$$

Initially, the region is thought to be subjected to regional drought and, subsequently, dry spells in subareas are subjected to temporal drought effects. The model is referred to as a multiseasonal model because it can be applied for a duration, which may include several dry and wet periods (Şen, 1980b). Although Lee et al. (1986) suggested multiyear drought durations analysis, their arguments were based on several hazard function models, which were examined with regard to their ability to represent the duration-dependent termination rate of drought data set. The view taken in this chapter is entirely different and based on the objective probability models as spatiotemporal concepts.

5.2.1 Persistent Regional Drought Model

Let us assume that the region under consideration is subdivided into a set of n subareas as in Fig. 4.1. Each one of the subareas is already subject to dry or wet spell effect randomly by time. An independent Bernoulli process is thought to prevail during Δt time interval and the total areal drought coverage is denoted by A_d. In such a case, the probability of n_1 subareas, $P_{\Delta t}(A_d = n_1)$, to have a dry spell during the first time interval, Δt, can be calculated similar to the Bernoulli events as has already been stated in Eq. (4.12) (Feller, 1967),

$$P_{\Delta t}(A_d = n_1) = \binom{m}{n_1} q_r^{n_1} p_r^{m-n_1} \tag{5.3}$$

Hence, the percentage of drought-stricken area is $100 n_1/m$, where m is the number of available subareas within the study region. In the next time interval, there remains $(m - n_1)$ subareas that are subject to dry spell hit. Because each dry spell along the time axis is independent even from Eq. (5.3), the number of dry spell hit areas during the second time interval, $n_2 > n_1$ and its probability is,

$$P_{2\Delta t}(A_d = n_2) = \sum_{n_1=0}^{n_2} \binom{m}{n_1} \binom{m - n_1}{n_2 - n_1} q_r^{n_1} q_t^{n_2 - n_1} p_r^{m - n_1} p_t^{m - n_2} \tag{5.4}$$

Herein, n_2 indicates the total number of dry spell hit areas at the end of the next time interval. This last expression can be written in the form of a recurrence formulation by consideration of Eq. (5.3) as

$$P_{2\Delta t}(A_d = n_2) = \sum_{n_1=0}^{n_2} P_{\Delta t}(A_d = n_1) P(A_d = n_2 - n_1) \tag{5.5}$$

Here, $P(A_d = n_2 - n_1)$ is the probability that $(n_2 - n_1)$ basic subareas are hit by a dry spell among the available $(m - n_1)$ potential subareas during the second time interval. Similar to Eq. (5.5), the drought probability for other time intervals, say ith time interval, can be written for $n_i > n_{i-1}$ as

$$P_{i\Delta t}(A_d = n_i) = \sum_{n_{i-1}=0}^{n_i} P_{(i-1)\Delta t}(A_d = n_{i-1})P(A_d = n_i - n_{i-1})$$ (5.6)

This expression can be reduced to simple drought analysis for $i=1$ (Şen, 1980b). The solution of this simple case was given numerically after extensive Monte Carlo simulations by Tase (1976). It is possible to obtain from Eq. (5.6) that probability after a certain time, $i\Delta t$, is equal to j less than this subarea,

$$P_{i\Delta t}(A_d \leq j) = \sum_{k=0}^{j} P_{i\Delta t}(A_d = k)$$ (5.7)

5.2.2 Multiseasonal Regional Drought Model

If n_1' dry spell subareas are available at the begining of any time instant, then there are n_1 subareas for possible dry spell hit; therefore, the conditional probability can be written as follows:

$$P_{\Delta t}\left(A_d = n_1' | A_d = n_1\right) = \binom{m}{n_1} \binom{n_1}{n_1 - n_1'} q_r^{n_1} p_r^{m-n_1} q_t^{n_1'} p_t^{n_1 - n_1'}$$

or succinctly,

$$P_{\Delta t}\left(A_d = n_1' | A_d = n_1\right) = P_{\Delta t}(A_d = n_1) \binom{n_1}{n_1 - n_1'} q_t^{n_1'} p_t^{n_1 - n_1'}$$ (5.8)

Here, always $n_1 \geq n_1'$ and the difference $j = n_1 - n_1'$ gives the possible distinctive transition number. By taking Eq. (5.8) as the basis after the end of the same time interval, the appearence of n_1' dry spell subareas hit probability can be expressed after some algebraic manipularion as

$$P_{\Delta t}\left(A_d = n_1'\right) = \sum_{k=0}^{m=n_i'} P_{i\Delta t}\left(A_d = k + n_i'\right) \binom{k - n_i'}{k} q_t^k p_t^{n_i'}$$ (5.9)

Similarly, the probability after $i\Delta t$ time interval becomes

$$P_{i\Delta t}\left(A_d = n_i'\right) = \sum_{k=0}^{m=n_i'} P_{i\Delta t}\left(A_d = k + n_i'\right) \binom{k + n_i'}{k} q_t^k p_t^{n_i'}$$ (5.10)

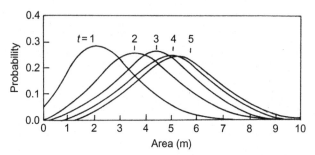

FIG. 5.1 Multiseasonal drought model.

The validity of this expression has been verified numerically by Şen (1980b) and the solutions are presented in Fig. 5.1 for $m = 10$, $p_r = 0.7$, $p_t = 0.8$ and for $i = 1, 2, 3, 4$, and 5 time intervals. Although there is some skewness in each PDF, but they can be considered as approximately symmetrical.

Another example for a multiseasonal drought model is to trace the whole drought area along the time axis. In this case, from Eq. (5.3) one can calculate the drought probability for the first time interval. At the end of the second time interval, on the condition that at the previous time interval there were n_1 subareas with dry spell, the probability of j subarea to be dry can be given as

$$P_{2\Delta t}(A_d = j | A_d = n_1) = P\Delta t(A_d = n_1)\binom{n_1}{j}q_t^j p_t^{n_1 - j}\qquad(5.11)$$

The probability out of n_1 subareas, j subareas to be dry, that is $(n_1 - j)$ subareas to be dry, can be written as

$$P_{2\Delta t}(A_d = j) = \sum_{k=0}^{m-j} P_{\Delta t}(A_d = k + j)\binom{k+j}{j}q_t^j p_t^k$$

In general, for the ith time interval one can write

$$P_{i\Delta t}(A_d = j) = \sum_{k=0}^{m-j} P_{(i-1)\Delta t}(A_d = k + j)\binom{k+j}{j}q_t^j p_t^k\qquad(5.12)$$

Fig. 5.2 gives the numerical solution of this expression for a set of parameters as $m = 10$, $p_r = 0.7$ and $p_t = 0.5$. In this figure, the PDFs are also slightly skewed.

5.2.3 Drought Parameters

Even though the PDFs as derived in the previous sections are important, in practical work, their first- (arithmetic average, expectation) and second-order moment (variance) play significant roles in the description of regional droughts.

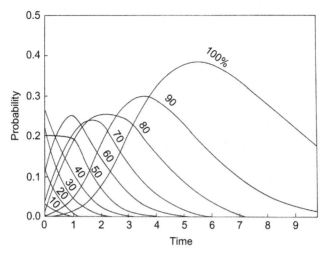

FIG. 5.2 Probability distribution functions for a multiseasonal drought model.

Given m subareas in a region, the expected value of a drought after $i\Delta t$ time interval can be calculated as

$$E_i(A_d) = \sum_{k=0}^{m} kP_{i\Delta t}(A_d = k) \tag{5.13}$$

Similarly, the variance of drought-stricken area can be calculated from the following expression:

$$V_i(A_d) = \sum_{k=0}^{m} k^2 P_{i\Delta t}(A_d = k) - E_i^2(A_d) \tag{5.14}$$

After the substitution of Eq. (5.6) into Eq. (5.12) one can obtain the expected value of the whole drought-stricken area as follows:

$$E_i(A_d) = mp_r \sum_{k=0}^{i-1} q_r^k \tag{5.15}$$

or succinctly as

$$E_i(A_d) = m\left(1 - q_r^i\right) \tag{5.16}$$

The percentage (probability) of the area subject to drought, P_A^i, can be obtained by dividing both sides of this last expression by m.

$$P_A^i = \left(1 - q_r^i\right) \tag{5.17}$$

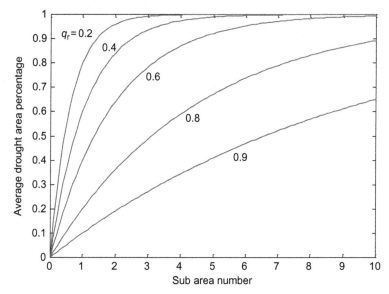

FIG. 5.3 Drought area coverage percentage.

Fig. 5.3 provides the change of average drought area percentage with the number of subareas, given the probability of dry spell probability, q_r.

Eq. (5.17) yields for $i=1$ the regional drought variability $P_A^1 = p_r$. On the other hand, theoretically as $i \to \infty$, which means that the whole region is covered by the drought, then $P_A^\infty = 1$. In the case of i subareas simultaneously under the drought conditions out of m subareas, the probability falls between these two extremes, $p_r \leq P_A^i \leq 1$. In a region as the drought coverage area increases, the average drought area reaches its maximum value in a relatively shorter time duration. One can obtain from Eq. (5.17) that

$$i = \ln\left(1 - P_A^i\right) / \ln q_r \tag{5.18}$$

This expression helps to predict the average time duration for a given drought coverage percentage of the region. Fig. 5.4 presents the variation of i with p_r.

In addition to all that has been stated earlier, in practical studies the maximum drought probability can be calculated as $1/m$ or, preferably, as $1/(m+1)$ (see Eq. 2.24). The substitution of this condition into Eq. (5.18) leads to the following expression:

$$i = \ln\left(1 - P_A^i\right) / \left[\ln\left(1/(m+1)\right)\right] \tag{5.19}$$

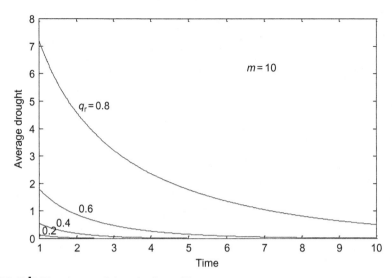

FIG. 5.4 The change of drought time with p_r.

This expression does not provide information about the shape of the drought hit area but about its extent. This was already stated earlier by Tase and Yevjevich (1978).

Another important measure for the areal drought assessment is the variance of drought hit area. Generally, the smaller the variance the smaller the drought hit area. The variance of persistent regional drought model can be calculated from Eqs. (5.17) and (5.14), after some simple algebra, as,

$$V_i(A_d) = m\left(1 - q_r^t\right)q_r^t \tag{5.20}$$

One can find the percentage variance of drought-stricken area by dividing both sides of this expression by m. The probability of i subareas to cover the drought area after T time duration can be obtained as

$$P_V^{i,T} = \left(1 - q_r^i\right)q_r^T \tag{5.21}$$

Combining this last expression with Eq. (5.17) provides the relationship between the regional percentage drought area average and percentage variance as

$$P_V^{i,T} = P_A^i q_r^T \tag{5.22}$$

Fig. 5.5 indicates the change of regional drought area percentage by i number of subarea given the dry spell probability as $q_r = 0.7$.

Another application of the area–time model can be achieved by considering subarea dry spell probability, q_r, and time transition probability, p_t. It is rather

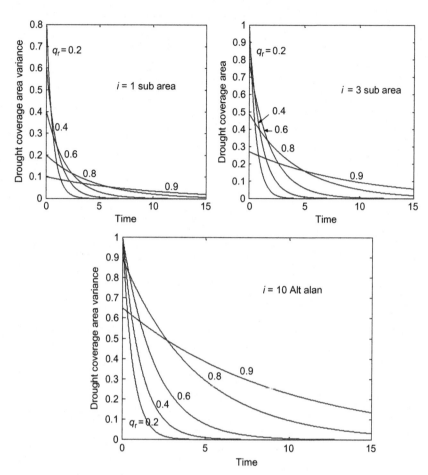

FIG. 5.5 Percentage variance variation of areal drought.

complicated to derive a formulation for any time interval, but one can do it for the first time interval as

$$E_1(A_d) = m(1 - q_r)(1 - p_t) \quad (5.23)$$

This expression shows the areal and temporal drought effects on the regional drought. Fig. 5.6 indicates the change of dry spell time probability with the average areal drought coverage for a set of given q_r values.

Finally, according to a simple area–time model, the variance of areal drought again for the first time instant becomes

$$V_1(A_d) = E_1(A_d)(q_r + p_t p_r) \quad (5.24)$$

The numerical solution of this expression is given in Fig. 5.7 for different q_r values.

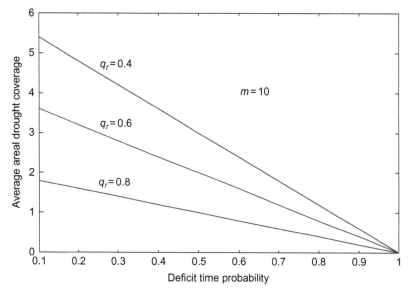

FIG. 5.6 Average areal drought coverage variation for the first time interval.

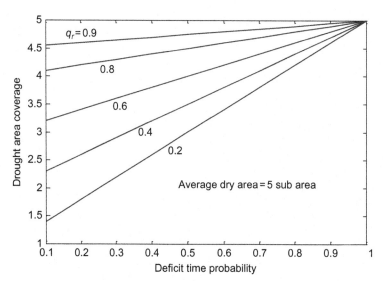

FIG. 5.7 Drought area average variance variation with the dry spell probability.

5.3 DROUGHT SPATIOTEMPORAL MODELING

Regional water exchange is relatively new in the hydrologic literature and it can be defined in general terms as water exchange between drought-stricken and water surplus subject areas. The water transfer between different regions can be achieved by means of bilateral interregional water transmission networks

(Takeuchi, 1974). By simultaneous consideration of all the water resources and consumption in a region, the planner would like to find answers to various questions such as

1. Which areas of the region are subject to droughts?
2. What is the extent of areal drought?
3. What is the drought duration (dry period) or wet period duration at any site?
4. What are the possibilities of alleviating drought by transmitting water from other alternatives?
5. To what extent it is possible to alleviate drought both in duration and intensity?
6. What is the percentage of failure of all the water resources considered together to meet areal demand?

Some of these questions can be answered objectively by the methodology presented in this section, provided that the necessary basic information such as spatial and temporal data on hydrologic variables is available.

In order to measure the effectiveness of the regional water exchange, Takeuchi (1974) used the sum of ranges of multiple streamflows, which is an extension of a simple range for a single reservoir investigated by various researchers (Hurst, 1951; Feller, 1951; Anis and Lloyd, 1953; Yevjevich, 1967; Şen, 1977). This sum of ranges of multiple streamflows is not capable to account for drought durations but for intensities only. On the other hand, the theory of runs was employed to investigate various problems in water resources systems design by Yevjevich (1967), Saldarriaga and Yevjevich (1970), Downer et al. (1967), and Şen (1976, 1977). The regional water exchange can be controlled effectively by joint run properties of hydrologic variables at a number of sites within the region (Şen, 1978).

5.3.1 Multivariate Runs

In a regional drought analysis, run properties of an isolated site is not informative; therefore, joint properties of more than one site must be established. The runs of bivariate processes were described by Yevjevich (1967) without any analytical treatment, but later Salazar and Yevjevich (1975) presented methodologies useful for the analysis of joint run properties of stationary processes at two sites only.

Let the concurrent observations at two sites of a catchment area be given as two time series, $X_{11}, X_{12}, ..., X_{1n}$ and $X_{21}, X_{22}, ..., X_{2n}$ together with their respective truncation levels X_{10} and X_{20}, which are not necessarily equal. A truncation of these observations at given levels generates four mutually exclusive but collectively exhaustive events at a given time instant, i, as

1. Joint deficits at two sites ($X_{1i} \leq X_{10}$ and $X_{2i} \leq X_{20}$).
2. Deficit $X_{1i} \leq X_{10}$ and surplus $X_{2i} > X_{20}$.
3. Surplus $X_{1i} > X_{10}$ and deficit $X_{2i} \leq X_{20}$.
4. Joint surpluses at two sites ($X_{1i} > X_{10}$ and $X_{2i} > X_{20}$).

Successive occurrences of joint deficits will be referred to as a twofold negative run, and the length of such a run is equal to the concurrent negative runs at two

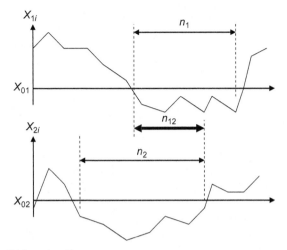

FIG. 5.8 Twofold run-lengths.

sites. For instance, in Fig. 5.8 n_1 and n_2 are the concurrent individual negative runs at two distinct sites, but n_{12} is a twofold negative run-length whose length is equal to the common part of n_1 and n_2. Similarly, as for the bivariate case, n-fold negative runs may be defined for multivariate hydrologic variables at n sites. As will be shown later, n-fold run statistics (especially expectation) provide a very useful tool in measuring the average spatiotemporal drought coverage.

5.3.1.1 DISTRIBUTION OF N-FOLD RUN-LENGTH

There are two different approaches in the application of the theory of runs to hydrology problems: the integration approach and the combinatorial approach (Salazar and Yevjevich, 1975).

The first approach refers to runs of infinite populations and in the case of stationary-ergodic processes any run is synonymous with the first run. Thus the PDF of the first n-fold run-length is sufficient to describe the joint run-length properties of series at n distinct sites.

Given n stationary and ergodic series $X_{1i}, X_{2i}, \ldots, X_{ni}$ with corresponding n different truncation levels X_{0i} ($i = 1, 2, \ldots, n$), sequences of simple Bernoulli trials at the ith site have simple probabilities, which can be written similar to Eqs. (3.12) and (3.13) as

$$q_i = P(X_i \le X_{0i}) = \int_{-\infty}^{X_{0i}} f(X_i) \mathrm{d}X_i \tag{5.25}$$

and complementarily,

$$p_i = P(X_i > X_{0i}) = 1 - q_i = \int_{X_{0i}}^{+\infty} f(X_i) \mathrm{d}X_i$$

Hereafter, q_is will be referred to as truncation levels in probability terms. When n simultaneous series are considered then the probability of n-fold negative run-length, n_n, and occurrence of length k can be defined generally as

$$P(n_n = k) = P \begin{pmatrix} X_{11} \leq X_{01}, X_{12} \leq X_{01}, \ldots, X_{1k-1} \leq X_{01}, \ B_k^{(1)}, \\ X_{21} \leq X_{02}, X_{22} \leq X_{02}, \ldots, X_{21k-1} \leq X_{02}, \ B_k^{(2)}, \\ \cdots \\ X_{n1} \leq X_{0n}, X_{n2} \leq X_{0n}, \ldots, X_{n1k-1} \leq X_{0n}, \ B_k^{(n)}), \end{pmatrix} \quad (5.26)$$

where $B_k^{(i)}$ shows a Bernoulli event at ith site and kth time instant. By definition, in order to obtain the event ($n_n = k$), the necessary and sufficient condition is to have at least one of the Bernoulli events as a surplus. Each line in Eq. (5.26) gives an event, which results in a negative run-length of length k at an individual site. In general, the solution of Eq. (5.26) can be achieved when multivariate PDF of nk observations is known. Depending on serial and mutual structures of n series, Eq. (5.26) can be simplified to exactly or approximately manageable forms, whichever is possible.

5.3.1.1.1 n Series Serially and Mutually Independent

This is the simplest case where the hydrologic variables in a region show neither significant temporal nor spatial correlations. Llamas and Siddiqui (1969) and Salazar and Yevjevich (1975) obtained simple expressions for the bivariate case ($n = 2$) only. If n series are spatially and temporally independent, then Eq. (5.26) becomes

$$P(n_n = k) = \prod_{i=1}^{n} \prod_{j=1}^{k-1} P(X_{ij} \leq X_{0i}) \prod_{m=1}^{n} P[B_k^m] \quad (5.27)$$

where the right most multiplication term is the probability of occurrence with at least one surplus in n trials, and it can be written as

$$\prod_{m=1}^{n} P[B_k^{(m)}] = 1 - P(B_k^0) \quad (5.28)$$

where $P(B_k^0)$ is the probability of no surplus in n series at the kth time instant. By considering Eqs. (5.25) and (5.26), Eq. (5.27) can be rewritten after some algebra as

$$P(n_n = k) = \left(\prod_{i=1}^{n} q_i \right)^{k-1} \left(1 - \prod_{i=1}^{n} q_i \right) \quad (5.29)$$

In general, this last expression is a geometric distribution with parameter

$$Q = \prod_{i=1}^{n} q_i \tag{5.30}$$

For $n=1$ and $n=2$, Eq. (5.29) reduces to the probability of negative run-lengths in univaiate and bivariate cases previously given by Feller (1951).

It follows from Eq. (5.29) that the occurrence of a dry period at any site has greater probability than all sites considered together. Thus, the joint operation of water resources systems at distinct sites reduces the risk of a dry period at single sites. Such a joint operation necessitates regional water exchange between available water resources alternatives. In relation to risk analysis, the designer is interested in knowing the exceedance probability of a particular dry period in a region, which turns out to be

$$P(n_n > k) = \left(\prod_{i=1}^{n} q_i \right)^{k-1} \tag{5.31}$$

Table 5.1 gives run-lengths for a given risk and truncation levels at three sites considered individually and jointly.

This table reveals the fact that the larger the number of water resource units, the smaller the duration of joint wet period. A basic assumption implicit in this last statement is that if there exists surpluses at some sites at a particular time instant, then the sum of these surpluses is enough to offset deficits at other sites at the same time instant. However, such an ideal situation may not happen in practice; even so, the statement is always valid.

The probability of an n-fold positive run-length, n_p, with length k can be obtained similar to Eq. (5.29) as

$$P(n_p = k) = \left(\prod_{i=1}^{n} p_i \right)^{k-1} \left(1 - \prod_{i=1}^{n} p_i \right) \tag{5.32}$$

This is useful for assessing the joint reliability of n water structures, if each were subjected to an independent sequence of hydrologic variables such as floods, rainfall, and so forth. Furthermore, Eq. (5.32) can be interpreted as the nonfailure probability of n interconnected water structures over k time periods. Thus, the probability of failure, p_f, becomes

$$p_f = 1 - \left(\prod_{i=1}^{n} p_i \right)^{k-1} \left(1 - \prod_{i=1}^{n} p_i \right) \tag{5.33}$$

This last expression verifies quantitatively two important conclusions by Yevjevich (1967).

TABLE 5.1 Threefold Negative Run-Length for Given Risk and Truncation Level

Truncation Level			Risk, $P(k > j)$								
			0.01				0.50				
1	2	3	1	2	3	Joint	1	2	3	Joint
0.700	0.800	0.900	13.91	21.63	44.71	7.72	2.94	4.10	7.58	2.01
0.800	0.900	0.990	21.63	44.71	459.21	14.00	4.10	7.58	69.96	3.05
0.900	0.990	0.999	44.71	459.21	4603.9	40.56	40.56	69.96	693.80	6.95

1. With the passage of time, the probability of system failure increases because for a fix n, p_f, increases with k.
2. The larger the number of constructed structures, the greater becomes the probability for one of them to fail, because for fixed k, p_f increases with n.

Thus Eq. (5.32) can be used effectively as a measure of regional reliability (safety) of a system with n components to function for at least k years without failure. After simple algebric treatment the expectation and variance of n-fold negative run can be found from Eq. (5.29) as

$$E(n_n) = \frac{1}{\left(1 - \prod_{i=1}^{n} q_i\right)} \qquad (5.34)$$

and

$$V(n_n) = \frac{\prod_{i=1}^{n} q_i}{\left(1 - \prod_{i=1}^{n} q_i\right)^2} \qquad (5.35)$$

respectively. The statistics, especially $E(n_n)$, may have several applications in water engineering, one of which is the measure of serial and mutual independencies of n sequences. Table 5.2 gives $E(n_n)$ values for various numbers of sites at a truncation level equal to respective medians.

If there are $n = 10$ sites and the average of n-fold negative run-lengths is close to 1, then the sequences might be expected to be temporally and spatially independent. On the other hand, if $E(n_n)$ is the sum of univariate dry periods of streamflows without a regional exchange scheme and $E^0(n_n)$ is the joint dry period of all streamflows within a regional exchange system, then the dry period reduction can be defined as

$$\alpha_n = 1 - \frac{E(n_n)}{E^0(n_n)} \qquad (5.36)$$

or more specifically,

$$\alpha_n = 1 - \frac{\left(1 - \prod_{i=1}^{n} q_i\right)^{-1}}{\sum_{i=1}^{n}(1 - q_i)^{-1}} \qquad (5.37)$$

TABLE 5.2 Expected Values of n-Fold Negative Run-Lengths

	Number of Sites, n					
	1	2	3	4	5	10
$E(n_n)$	2.00	1.33	1.14	1.06	1.03	1.00

In Eq. (5.36) $E(n_n) < E^0(n_n)$ and for small $E(n_n)$, α_n approaches unity, indicating significant reduction in dry periods. Obviously, the degree of reduction in dry periods depends on the operation of regional water exchange. In the aforementioned approach, economic considerations are ignored completely and thus the maximum reduction in dry periods may not conform with an economical solution.

Example 5.1

In an area there are two sites where the observations are serially and mutually independent from each other, and they are truncated at different levels. Hence, calculate drought durations at 0.01, 0.05, 0.25, 0.50, 075, 0.95, and 0.99 risk levels.

As serial and mutal independences are valid, then according to Eq. (5.31) the risks can be calculated. If the two truncation levels at these two sites are equal to each other as $q_1 = q_2$, then the calculations are shown in Table 5.3.

5.3.1.1.1.1 Run-Sums. In general, the sum of deficits (surpluses) over a dry (wet) period is referred to as negative (positive) run-sum. In a similar manner, n-fold run-sums may be defined as run-lengths. For the univariate case, a general expression of expected positive run-sum was given by Şen (1977) as

TABLE 5.3 Dry Period Lengths for Given Truncation and Risk Levels

Truncation Level, $q_1 = q_2$	Simple Risk, $P(k > j)$						
	0.01	**0.05**	**0.25**	**0.50**	**0.75**	**0.95**	**0.99**
0.1	2.00	1.65	1.30	1.15	1.06	1.01	1.00
0.2	2.43	1.93	1.43	1.21	1.09	1.01	1.00
0.3	2.91	1.74	1.57	1.29	1.12	1.02	1.00
0.4	3.51	2.63	1.75	1.38	1.15	1.03	1.00
0.5	4.32	3.16	2.00	1.50	1.20	1.04	1.00
0.6	5.50	3.93	2.35	1.68	1.28	1.05	1.00
0.7	.45	5.19	2.94	1.97	1.40	1.07	1.01
0.8	11.31	7.71	4.10	2.55	1.64	1.11	1.02
0.9	22.85	15.21	7.58	4.29	2.36	1.2	1.04
0.99	230.10	150.03	69.97	35.48	15.31	3.55	1.50
0.999	2302.43	1498.11	69.80	346.40	144.77	26.63	6.02

$$E(S) = \sum_{i=1}^{\infty} E(S|i)P(n_n = i) \tag{5.38}$$

This expression can be extended to covar n series as

$$E(S) = \sum_{j=1}^{n} \sum_{i=1}^{\infty} E(S|i,j)P(n_n = i|j) \tag{5.39}$$

where j and i are counters referring to site number and length of a temporal negative run-length, respectively. $P(n_n = i\,|\,j)$ is the conditional probability of a negative run-length of length i given jth site. After some tedious algebra, the expected value of n-fold negative run-sums turns out to be

$$E(S) = E(n_n) \sum_{j=1}^{m} E(d_j) \tag{5.40}$$

where d_j shows deficit at the jth site, and $E(d_j)$ can be evaluated from the appropriate truncated distribution of original PDF at a truncation level, x_{0i} (see Section 3). The expectation in Eq. (5.40) can be used as a descriptor of regional drought intensity. The larger the expectation, the greater the regional drought intensity.

5.3.1.1.2 n Series Serially İndependent but Mutually Dependent

A similar procedure can be employed as for the serially and mutually independent series. Single events, $(X_{ij} \leq X_{0i})$ in the lines of Eq. (5.26), are independent, whereas events in the same column are dependent. Eq. (5.26), can be rewritten in its transposed form as

$$P(n_n = k) = P \begin{pmatrix} X_{11} \leq X_{01}, X_{21} \leq X_{02}, \ldots, X_{n1} \leq X_{0n}, & B_k^1 \\ X_{12} \leq X_{01}, X_{22} \leq X_{02}, \ldots, X_{n2} \leq X_{0n}, & B_k^2 \\ \ldots \\ X_{1k-1} \leq X_{01}, X_{2K-1} \leq X_{02}, \ldots, X_{nk-1} \leq X_{0n}, & B_k^n \end{pmatrix} \tag{5.41}$$

where now lines include dependent events and each line is independent from the others; therefore, Eq. (5.41) factorizes to

$$P(n_n = k) = \prod_{i=1}^{k-1} P(X_{1i} \leq X_{01}, X_{2i} \leq X_{02}, \ldots, X_{ni} \leq X_{0n})P(B_k^1, B_k^2, \ldots, B_k^n) \tag{5.42}$$

To further simplify this expression, the dependence between two successive events is assumed to be Markovian; thus, using the results from Section 3.6.1, it can be rewritten shortly as

$$P(n_n = k) = \left(\prod_{i=1}^{n} r_{i-1,i} \right)^{k-1} \left(1 - \prod_{i=1}^{n} r_{i-1,i} \right) \tag{5.43}$$

where $r_{0,1} = q$ and $r_{i-1,i}$ are the autorun coefficients defined as autorun coefficient ("see chapter: Temporal Drought Analysis and Modeling")

$$r_{i=1,i} = \int_{-\infty}^{X_{0i-a}} \int_{-\infty}^{X_{oi}} f(X_{i-1}, X_i) dX_{i-1} dX_i \tag{5.44}$$

This can be numerically calculated either on a computer or from standard bivariate normal tables. If all the series had the same truncation levels and first-order serial correlation coefficients, then $r_{i-1,i} = r$ are the same and Eq. (5.43) could be written in its simplest form as

$$P(n_n = k) = r^{n(k-1)}(1 - r^n) \tag{5.45}$$

where r is a function of both correlation coeffcient, ρ, and truncation level, q ("see chapter: Temporal Drought Analysis and Modeling, Eq. (3.30)"). In this particular case, the risk of a regional drought exceeding a preselected duration can be found from Eq. (5.43) as

$$P(n_n > k) = \left(\prod_{i=1}^{n} r_{i-1,i} \right)^{k-i} \tag{5.46}$$

5.3.1.1.3 n Series Serially Dependent but Mutually Independent
For this case the analysis is similar to the case of serially independent but mutually dependent series. The only difference is that the joint probabilities must take into account the serial dependencies in n series. Hence, Eq. (5.26) can be simplified to

$$P(n_n = k) = \prod_{i=1}^{k-1} P\left(X_{i1} \leq X_{0i}, X_{i2} \leq X_{0i}, \ldots, X_{ik-1} \leq X_{0i}, B_k^i \right) \tag{5.47}$$

A further simpification of the multivariate probability in this expression is possible when the successive observations in a series are correlated according to the lag-one Markov process. On the basis of wet and dry period analysis of stationary processes (Şen, 1976), Eq. (5.47) can be written succinctly as

$$P(n_n = k) = \prod_{i=1}^{n} q_i r_i^{k-2} \left(1 - \prod_{i=1}^{n} r_i \right) \tag{5.48}$$

Here, $r_j = P(X_{ij} \leq X_{0i} \mid X_{ij-1} \leq X_{0j})$ and can be calculated from standard bivariate normal tables. From this last expression the risk of a regional drought period to exceed a given duration k becomes

$$P(n_n > k) \prod_{i=1}^{n} q_i r_i^{k-2} \qquad (5.49)$$

The greater the serial correlations the larger the r_is and, accordingly, the greater the risk of a regional drought.

5.3.1.1.4 *n* Series Serially and Mutually Dependent

This is the most complicated of all the cases and requires nk-dimensional joint PDF. Because this case is also frequently encounted in practice, approximate analytical solutions are presented herein. It can be considered as a combination of the two previous cases. Thus, to simplify the general expression in Eq. (5.26), again the Markovian property is assumed not only for the serial correlations but also for mutual correlations, as a result of which Eq. (5.26) yields

$$P(n_n = k) = \prod_{i=1}^{n} r_i^{k-1} r_{i-1,i} \left(1 - \prod_{i=1}^{n} r_i\right) \qquad (5.50)$$

On the other hand, the risk of a regional drought to be greater than a preselected k period is

$$P(n_n > k) = \prod_{i=1}^{n} r_i^{k-1} r_{i-1,i} \qquad (5.51)$$

which is a function of temporal as well as spatial correlations. A comparison of Eqs. (5.31), (5.46), (5.49), and (5.51) for a set of the same parameters shows the expected fact that n serially and mutually dependent series cause droughts of k-period duration greater than the other three cases. In this case the regional droughts are longer than they are in other situations.

5.4 REGIONAL WET AND DRY SPELL ANALYSIS WITH HETEROGENEOUS PROBABILITY OCCURRENCES

This section provides regional probabilistic wet and dry period areal coverage modeling, which is useful for temporal and spatial wet or dry risk predictions and parameter assessments. The basis of the methodology is mutually exclusive and independent subareal (site) dry and wet spell heterogeneous probabilities. Derived expressions are general and reduce to a simple Bernoulli (homogeneous) trial case as available in chapter: Temporal Drought Analysis and Modeling.

Panu and Sharma (2002) gave perspectives and challenges for future drought studies and they also undertook a critical review of the existing literature. Mishra and Singh (2010) provided an extensive review of drought phenomenon by considering many journal papers. Furthermore, Mishra et al. (2009) investigated alternative renewable process and run theory for drought interval time distribution, mean drought interarrival time, and joint PDF. Chung and Salas (1999) proposed low-order discrete autoregressive moving average and transition probabilities of drought events. Specifically, researchers center their attention on the occurrence of drought events, particularly their duration, by using the concept of runs (Mood, 1949; Feller, 1967). The probability distribution of drought occurrence, expected values and variances of first arrival and interarrival times of drought events, and the associated risks are derived in this section. Recently, another approach appears as the copula functions, which are applied in bivariate drought case drought duration and severity frequency analyses (Lee et al., 2012).

In all the previous studies the spatial subarea (site) spells are assumed as homogeneous (equal probability of occurrences). The main purpose of this section is to provide dry and wet spell duration assessments based on temporal regional probabilistic modeling, where each subarea is assumed to have different wet and dry spell occurrence probabilities (ie, heterogeneous probabilities). The general spatiotemporal drought model is developed for this purpose, based on the assumptions that regional dry and wet spell occurrences are mutually exclusive, independent, but heterogeneous. In the literature the simplest model of homogeneity exists with all other assumptions remaining the same.

5.4.1 Dry and Wet Spell Features

Droughts are extensive regional covers in continental scales. They strike not only one country, but several countries or subareas (sites) within the same country in different proportions. In the literature, almost all the theoretical studies are confined to temporal assessment ("see chapter: Temporal Drought Analysis and Modeling") with few studies concerning areal coverage ("see chapter: Temporal Drought Analysis and Modeling").

To model the coverage of occurrence patterns the study area can be divided into characteristically distinctive and mutually exclusive subareas on the basis of prevailing rainfall considerations. It is not necessary that each subarea be equal in size. It is generally accepted that severe dry and wet spells arise as a result of apparent chance variations in atmospheric circulation. For instance, dry spells are always initiated by a shortage of precipitation, and many of the traditional approaches to drought definition concentrate on rainfall analysis only (Smith, 2001).

Simple assessment of drought severity depends largely on the magnitude and regional extent of precipitation deficiencies from mean climatic conditions. Precipitation is the most easily available data compared to any other effective variables. For the sake of simplicity, precipitation is considered as the main

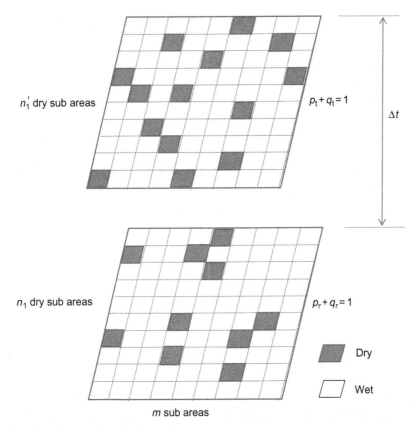

FIG. 5.9 Spatiotemporal wet and dry pattern grids.

variable, which gives rise to regional and temporal variations of wet and dry spells. Fig. 5.9 indicates subareas along the time axis. At initial time instant, there are n_1 dry spells, but after one time step, Δt, there are $n_{1'}$ dry spells.

5.4.2 Regional Models

By considering explanations given for Fig. 5.9, in case of heterogeneous subareal probabilities the general form of Eq. (5.3) was derived by Şen (2014) as

$$P_{\Delta t}(A_d = n_1 \mid m) = \frac{1}{n_1!} \underbrace{\sum_{\substack{j_1=1}}^{m} \sum_{\substack{j_2=1 \\ j_2 \neq j_1}}^{m} \cdots \sum_{\substack{j_{n_1}=1 \\ j_{n_1} \neq j_k \\ k=1,2,\ldots,n_1-1}}^{m}}_{n_1-\text{fold}} p_{r_{j_1}} p_{r_{j_2}} \ldots p_{r_{j_k}} \underbrace{\prod_{\substack{j_k=1 \\ j_k \neq j_l \\ l=1,2,\ldots,n_1}}^{m} q_{r_{j_k}1}}_{(m-n_1)-\text{fold}}$$

$$(5.52)$$

where $p_{r_{j_i}}$ is the regional probability of dry spell hit of subarea, j_i. Herein, indice i takes its value according to arrangements beneath the summation and multiplication signs.

The n_1-fold summation terms imply the exhaustive alternatives of dry periods and the $(m-n)$-fold multiplication term is for collective wet spells at the remaining subareas. In the case of homogeneous areal probability at each site, Eq. (5.52) reduces to Eq. (5.3). To verify the validity of Eq. (5.52) a simple calculation procedure is shown in the following two examples.

Example 5.2

The composition of Eq. (5.52) for $m=4$ and $n_1=2$ reduces to

$$P_{\Delta t}(A_d = 2|4) = \frac{1}{2!} \sum_{\substack{j_1=1}}^{4} \sum_{\substack{j_2=1 \\ j_2 \neq j_1}}^{4} p_{r_{j_1}} p_{r_{j_2}} \prod_{\substack{j_3=1 \\ j_4 \neq j_1, j_2}}^{4} q_{r_{j_3}}$$

Table 5.4 presents in detail all the necessary steps for the calculation of the final probability from this last expression with similar calculations from Eq. (5.3) in a comparative manner.

TABLE 5.4 Station Characteristics

j_1	j_2	$\sum_{\substack{j_1=1}}^{4}\sum_{\substack{j_2=1 \\ j_2 \neq j_1}} p_{r_{j_1}} p_{r_{j_2}} \prod_{\substack{j_3=1 \\ j_3 \neq j_2}}^{4} q_{r_{j_3}}$	$P_{\Delta t(A_d=24)}$	$P_{\Delta t(A_d=24)}$
		Heterogeneous Probabilities ($p_{r1} \neq p_{r2} \neq p_{r3} \neq p_{r4}$)		**Homogeneous ($p = p_{r1} = p_{r2} = p_{r3} = p_{r4}$)**
I	**II**	**III**	**IV**	**V**	**VI**
1	2	$p_{r1}p_{r2}$	$q_{r3}q_{r4}$	$p_{r1}p_{r2}q_{r3}q_{r4}$	p^2q^2
	3	$p_{r1}p_{r3}$	$q_{r2}q_{r4}$	$p_{r1}p_{r3}q_{r2}q_{r4}$	p^2q^2
	4	$p_{r1}p_{r4}$	$q_{r2}q_{r3}$	$p_{r1}p_{r4}q_{r2}q_{r3}$	p^2q^2
2	1	$p_{r2}p_{r1}$	$q_{r3}q_{r4}$	$p_{r2}p_{r1}q_{r3}q_{r4}$	p^2q^2
	3	$p_{r2}p_{r3}$	$q_{r1}q_{r4}$	$p_{r2}p_{r3}q_{r1}q_{r4}$	p^2q^2
	4	$p_{r2}p_{r4}$	$q_{r1}q_{r3}$	$p_{r2}p_{r4}q_{r1}q_{r3}$	p^2q^2
3	1	$p_{r3}p_{r1}$	$q_{r2}q_{r4}$	$p_{r3}p_{r1}q_{r2}q_{r4}$	p^2q^2
	2	$p_{r3}p_{r2}$	$q_{r1}q_{r4}$	$p_{r3}p_{r2}q_{r1}q_{r4}$	p^2q^2
	4	$p_{r3}p_{r4}$	$q_{r1}q_{r3}$	$p_{r3}p_{r4}q_{r1}q_{r3}$	p^2q^2
4	1	$p_{r4}p_{r1}$	$q_{r2}q_{r3}$	$p_{r4}p_{r1}q_{r2}q_{r3}$	p^2q^2
	2	$p_{r4}p_{r1}$	$q_{r1}q_{r3}$	$p_{r4}p_{r2}q_{r1}q_{r3}$	p^2q^2
	3	$p_{r4}p_{r3}$	$q_{r1}q_{r2}$	$p_{r4}p_{r3}q_{r1}q_{r2}$	p^2q^2

The summation of the terms in column 6 gives the right-hand side of the previous expression except $1/2!$ factor. Hence, the formulation can be written explicitly as

$$P_{\Delta t}(A_d = 2|4) = \frac{1}{2!}(p_{r_1}p_{r_2}q_{r_3}q_{r_4} + p_{r_1}p_{r_3}q_{r_2}q_{r_4} + p_{r_1}p_{r_4}q_{r_2}q_{r_3} + p_{r_2}p_{r_1}q_{r_3}q_{r_4}$$
$$+ p_{r_2}p_{r_3}q_{r_1}q_{r_4} + p_{r_2}p_{r_4}q_{r_1}q_{r_3} + p_{r_3}p_{r_1}q_{r_2}q_{r_4} + p_{r_3}p_{r_2}q_{r_1}q_{r_4} + p_{r_3}p_{r_4}q_{r_1}q_{r_3}$$
$$+ p_{r_4}p_{r_1}q_{r_2}q_{r_3} + p_{r_4}p_{r_2}q_{r_1}q_{r_3} + p_{r_4}p_{r_3}q_{r_1}q_{r_2})$$

In case of homogeneous point probabilities, this detailed expression reduces to the following simple form:

$$P_{\Delta t}(A_d = 2|4) = \frac{1}{1.2}(4 \times 3 \times p^2 q^2) = \frac{12}{2}p^2 q^2$$

On the other hand, Eq. (5.3) yields

$$P_{\Delta t}(A_d = 2|4) = \binom{4}{2}p^2 q^{4-2} = 6p^2 q^2$$

Hence, Eqs. (5.3) and (5.52) yield the same result in both cases of homogeneous and heterogeneous subareal probabilities, respectively.

Example 5.3

Now let us consider the case of $m = 6$ and $n_1 = 4$. According to Eq. (5.52) the corresponding formulation becomes

$$P_{\Delta t}(A_d = 4|6) = \frac{1}{4!}\sum_{j_1=1}^{6}\sum_{\substack{j_2=1 \\ j_2 \neq j_1}}^{6}\sum_{\substack{j_3=1 \\ j_3 \neq j_1, j_2}}^{6}\sum_{\substack{j_4=1 \\ j_4 \neq j_1, j_2, j_3}}^{6} p_{r_{j_1}} p_{r_{j_2}} p_{r_{j_3}} p_{r_{j_4}} \prod_{\substack{j_5=1 \\ j_5 \neq j_1, j_2, j_3, j_4}}^{6} q_{r_{j_5}}$$

which has all the terms explicitly as

$$P_{\Delta t}(A_d = 4|6) = \frac{1}{1.2.3.4}(6 \times 5 \times 4 \times 3 \times p^4 q^2) = 15p^4 q^2$$

On the other hand, Eq. (5.3) yields

$$P_{\Delta t}(A_d = 4|6) = \binom{6}{4}p^4 q^2 = 15p^4 q^2$$

It is possible to try similar calculations with different sets of p, m, and n values, and the two formulations yield exactly the same results. This way is referred to as induction methodology in the mathematical literature; hence, the validity of Eq. (5.52) is confirmed.

5.4.3 Regional Persistence Model

As explained in Section 5.2.1, out of m possible drought-prone subareas, n_1 have dry spells; hence, the areal coverage of drought is equal to n_1 or in areal percentages, $100(n_1/m)$. For the subsequent time interval, Δt, there are $(m - n_1)$ drought-prone possibility subareas or sites. Assuming that the evolution of possible dry and wet spells along the time axis is independent over mutually exclusive subareas, similar to the concept in Eq. (5.3), it is possible to write the following conditional probability as

$$P_{2\Delta t}(A_d = n_2 | m - n_1) = P_{\Delta t}(A_d = n_1 | m) P(A_d = n_2 - n_1 | m - n_1)$$

or unconditional probability can be written by making use of the mutual exclusiveness property of the probability theory, which implies the addition of different alternatives as

$$P_{2\Delta t}(A_d = n_2) = \sum_{n_1=0}^{n_2} P_{\Delta t}(A_d = n_1 | m) P(A_d = n_2 - n_1 | m - n_1) \qquad (5.53)$$

where $P(A_d = n_2 - n_1 | m - n_1)$ is the probability of additional $(n_2 - n_1)$ subareas to be affected by dry spell during the second time interval out of the remaining $(m - n_1)$ potential subareas from the previous time interval. This expression reduces to its simplest case, which does not consider time variability of drought occurrences theoretically as presented by Şen (1980b) and experimentally on digital computers by Tase (1976). The right-hand side of Eq. (5.53) can be rearranged explicitly by consideration of Eq. (5.1) and, finally, leads to Eq. (5.5) where $n_2 - n_1$ is the total number of drought affected subareas during the second time interval.

One can write Eq. (5.53) by induction methodology in the form of the recurrence relationship as

$$P_{i\Delta t}(A_d = n_i | m) = \sum_{n_{i-1}=0}^{n_i} P_{(i-1)\Delta t}(A_d = n_{i-1} | m) P(A_d = n_i - n_{i-1} | m - n_i) \quad (5.54)$$

Substitution of Eq. (5.52) into Eq. (5.53) leads to areal drought probability values when the point probabilities are not equal to each other (heterogeneous case).

$$P_{2\Delta t}\left(A_d = n_2{}^2\right) = \sum_{n_1=0}^{n_2} \left[\frac{1}{n_1!} \underbrace{\sum_{\substack{j_1=1}}^{m} \underbrace{\sum_{\substack{j_2=1 \\ j_2 \neq j_1}}^{m} \cdots \sum_{\substack{j_{n_1}=1 \\ j_{n_1} \neq j_k \\ k=1,2,\ldots,i-1}}^{m}}_{n_1-\text{fold}} pr_{j_1} pr_{j_2} \cdots pr_{j_k} \underbrace{\prod_{\substack{j_k=1 \\ j_k \neq j_l \\ l=1,2,\ldots,i}}^{m} qr_{j_k}}_{(m-n_1)-\text{fold}} } \right.$$

$$\left. \frac{1}{(n_2-n_1)!} \underbrace{\sum_{j_1=1}^{m-n_1} \sum_{\substack{j_2=1 \\ j_2 \neq j_1}}^{m-n_1} \cdots \sum_{\substack{j_{(n_2-n_1)}=1 \\ j_{\left(n_1-n_1'\right)} \neq j_k \\ k=1,2,\ldots,i-1}}^{m-n_1}}_{(n_2-n_1)-\text{fold}} pt_{j_1} pt_{j_2} \cdots pt_{j_k} \underbrace{\prod_{\substack{j_k=1 \\ j_k \neq j_l \\ l=1,2,\ldots,i}}^{m-n_1} qt_{j_k}}_{(m-n_1)-\text{fold}} \right] \quad (5.55)$$

Similarly, according to Eq. (5.54) one can find the successive evolution of the drought probabilities up to the required time instant.

The probability of drought-stricken area to be equal to or less than a specific number of subareas, say j, can be evaluated from Eq. (5.54) according to the mutual exclusiveness property of the probability theory.

$$P_{i\Delta t}(A_d \leq j) = \sum_{k=0}^{j} P_{i\Delta t}(A_d = k) \quad (5.56)$$

5.4.4 Multiseasonal Model

In the first time interval, the same probability pattern is valid according to Eq. (5.3) for homogeneous and Eq. (5.52) for heterogeneous subareal probabilities. The probability of having n_1' dry subareas in the second time interval, given that there are n_1 dry subareas at the end of the previous time interval within the whole region, can be expressed at two stages for $n_1' < n_2$ and $n_1' > n_2$ as

$$P_{2\Delta t}\left(A_d = n_1' | A_d = n_1\right) = \begin{cases} \binom{m}{n_1}\binom{n_1}{n_1-n_1'} p_r^{n_1} q_r^{m-n} p_t^{n_1-n_1'} q_t^{n_1'} & \text{for } n_1' < n_1 \\[3mm] \binom{m}{n_1}\binom{m-n_1}{n_1'-n_1} p_r^{n_1} q_r^{m-n} p_t^{n_1'-n_1} q_t^{m-n_1'} & \text{for } n_1' > n_1 \end{cases}$$

$$(5.57)$$

or, shortly, as

$$P_{\Delta t}(A_d = n_1' \mid A_d = n_1) = \begin{cases} P_{\Delta t}(A_d = n_1)\dbinom{n_1}{n_1 - n_1'} p_t^{m-n_1'} q_t^{n_1'} & \text{for } n_1' < n_1 \\[2mm] P_{\Delta t}(A_d = n_1)\dbinom{m - n_1}{n_1' - n_1} p_t^{n_1' - n_1} q_t^{m-n_1'} & \text{for } n_1' > n_1 \end{cases} \tag{5.58}$$

or the heterogeneous equivalent can be written by considerations from Eq. (5.52) as follows:

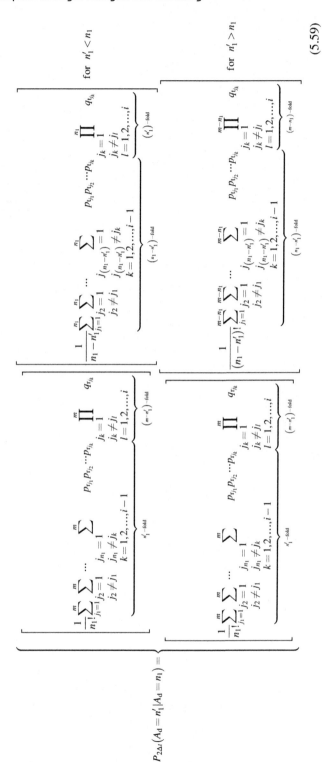

$$\tag{5.59}$$

or succinctly,

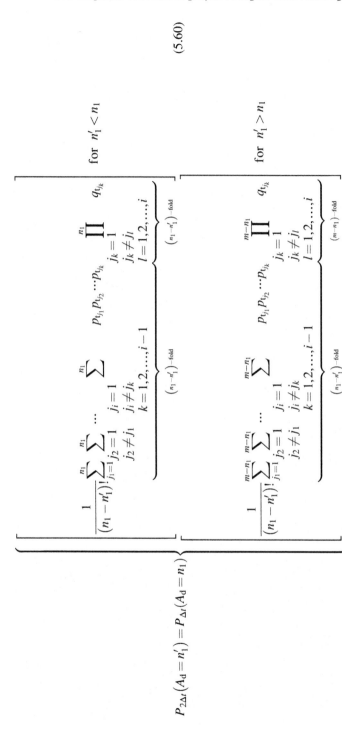

$$P_{2\Delta t}(A_d = n_1') = P_{\Delta t}(A_d = n_1)$$

$$
\begin{cases}
\dfrac{1}{(n_1 - n_1')!} \underbrace{\sum_{j_1=1}^{n_1} \sum_{\substack{j_2=1 \\ j_2 \neq j_1}}^{n_1} \cdots \sum_{\substack{j_i=1 \\ j_i \neq j_k \\ k=1,2,\dots,i-1}}^{n_1}}_{(n_1 - n_1')-\text{fold}} p_{t_{j_1}} p_{t_{j_2}} \cdots p_{t_{j_k}} \underbrace{\prod_{\substack{j_k=1 \\ j_k \neq j_l \\ l=1,2,\dots,i}}^{n_1} q_{t_{j_k}}}_{(n_1 - n_1')-\text{fold}} & \text{for } n_1' < n_1 \quad (5.60) \\[4ex]
\dfrac{1}{(n_1 - n_1')!} \underbrace{\sum_{j_1=1}^{m-n_1} \sum_{\substack{j_2=1 \\ j_2 \neq j_1}}^{m-n_1} \cdots \sum_{\substack{j_i=1 \\ j_i \neq j_k \\ k=1,2,\dots,i-1}}^{m-n_1}}_{(n_1 - n_1')-\text{fold}} p_{t_{j_1}} p_{t_{j_2}} \cdots p_{t_{j_k}} \underbrace{\prod_{\substack{j_k=1 \\ j_k \neq j_l \\ l=1,2,\dots,i}}^{m-n_1} q_{t_{j_k}}}_{(n_1 - n_1')-\text{fold}} & \text{for } n_1' > n_1
\end{cases}
$$

On the basis of Eq. (5.54), a general equation for the marginal probability of observing n_i' dry spells at the end of the same time interval, after simple algebra, becomes

$$P_{i\Delta t}\left(A_d = n_i'\right) = \begin{cases} \displaystyle\sum_{k=0}^{n_i'} P_{i\Delta t}(A_d = k+n_i') \binom{k+n_i'}{k} p_t^k q_t^{n_i'} & \text{for } n_i' < n_{i-1}' \\[2em] \displaystyle\sum_{k=n_{i-1}'}^{n_i'} P_{i\Delta t}(A_d = k - n_i') \binom{k+n_i'}{k} p_t^k q_t^{n_i'} & \text{for } n_i' > n_{i-1}' \end{cases} \tag{5.61}$$

Its corresponding expression for the heterogeneous probability distributions at a set of sites can be written by considerations from Eq. (5.58) as

$$P_{i\Delta t}\left(A_d = n_i'\right) = \begin{cases} \displaystyle\sum_{k=0}^{n_i'} P_{i\Delta t}(A_d = k+n_i') \left[\frac{1}{k!} \underbrace{\sum_{\substack{j_1=1}}^{k+n_i'}\sum_{\substack{j_2=1 \\ j_2 \neq j_1}}^{k+n_i'}\cdots\sum_{\substack{j_i=1 \\ j_i \neq j_k \\ k=1,2,\dots,i-1}}^{k+n_i'}}_{k-\text{fold}} p_{t_{j_1}}p_{t_{j_2}}\cdots p_{t_{j_k}} \underbrace{\prod_{\substack{j_k=1 \\ j_k \neq j_l \\ l=1,2,\dots,i}}^{k+n_i'}}_{n_i'-\text{fold}} q_{t_{j_k}} \right] & \text{for } n_i' < n_{i-1}' \\[4em] \displaystyle\sum_{k=n_{i-1}'}^{n_i'} P_{i\Delta t}(A_d = k+n_i') \left[\frac{1}{k!} \underbrace{\sum_{\substack{j_1=1}}^{k+n_i'}\sum_{\substack{j_2=1 \\ j_2 \neq j_1}}^{k+n_i'}\cdots\sum_{\substack{j_i=1 \\ j_i \neq j_k \\ k=1,2,\dots,i-1}}^{k+n_i'}}_{k-\text{fold}} p_{t_{j_1}}p_{t_{j_2}}\cdots p_{t_{j_k}} \underbrace{\prod_{\substack{j_k=1 \\ j_k \neq j_l \\ l=1,2,\dots,i}}^{k+n_i'}}_{n_i'-\text{fold}} q_{t_{j_k}} \right] & \text{for } n_i' > n_{i-1}' \end{cases} \tag{5.62}$$

Another version of the multiseasonal model is interesting when the number of continuously dry subareas appear along the whole observation period. In such a case the probability of dry spell in the first time interval can be calculated from Eq. (5.3). At the end of the second time interval the probability of j subareas with two successive dry spells, given that already n_1 subareas had deficit in the previous interval, can be expressed as

$$P_{2\Delta t}(A_d = j|A_d = n_1) = P\Delta t(A_d = n_1)\binom{n_1}{j}p_t^j q_t^{n_1-j} \qquad (5.63)$$

Eq. (5.3) provides a way of obtaining the general formulation for the heterogeneous probabilities by considering Eq. (5.52) so the following expression can be written:

$$P_{2\Delta t}(A_d = j|A_d = n_1)$$

$$= P_{\Delta t}(A_d = n_1)\left[\frac{1}{j!}\underbrace{\sum_{\substack{j_1=1 \\ }}^{n_1}\sum_{\substack{j_2=1 \\ j_2 \neq j_1}}^{n_1}\cdots\sum_{\substack{j_i=1 \\ j_i \neq j_k \\ k=1,2,\ldots,i-1}}^{n_1}p_{t_{j_1}}p_{t_{j_2}}\cdots p_{t_{j_k}}}_{j\text{-fold}}\underbrace{\prod_{\substack{j_k=1 \\ j_k \neq j_l \\ l=1,2,\ldots,i}}^{n_1}q_{t_{j_k}}}_{(n_1-j)\text{-fold}}\right]$$

$$(5.64)$$

This expression yields the probability of having n_1 subareas to have dry out of which j subareas are hit by two deficits; that is, there are $(n_1 - j)$ subareas with one deficit. Hence, the marginal probability of continuous dry subarea numbers can be written as

$$P_{2\Delta t}(A_d = j) = \sum_{k=0}^{m-j}P_{\Delta t}(A_d = k+j)\binom{k+j}{j}p_t^j q_t^k \qquad (5.65)$$

However, the heterogeneous correspondence to this formulation may be obtained by considering Eq. (5.52) and its convenient substitution in the previous expression.

$$P_{2\Delta t}(A_d = j) = \sum_{k=0}^{m-j} P_{\Delta t}(A_d = k + j)$$

$$\left[\frac{1}{j!} \underbrace{\sum_{j_1=1}^{k+j} \sum_{\substack{j_2=1 \\ j_2 \neq j_1}}^{k+j} \cdots \sum_{\substack{j_i=1 \\ j_i \neq j_k \\ k=1,2,\ldots,i-1}}^{k+j}}_{j-\text{fold}} p_{t_{j_1}} p_{t_{j_2}} \cdots p_{t_{j_k}} \underbrace{\prod_{\substack{j_k=1 \\ j_k \neq j_l \\ l=1,2,\ldots,i}}^{k+j}}_{k-\text{fold}} q_{t_{j_k}} \right]$$

$$(5.66)$$

In general, for ith time interval, it is possible to write

$$P_{i\Delta t}(A_d = j) = \sum_{k=0}^{m-j} P_{(i-1)\Delta t}(A_d = k + j) \binom{k+j}{j} p_t^j q_t^k \qquad (5.67)$$

Under the light of the aforementioned explanations, the general heterogeneous formulation becomes

$$P_{i\Delta t}(A_d = j) = \sum_{k=0}^{m-j} P_{(i-1)\Delta t}(A_d = k + j)$$

$$\left[\frac{1}{j!} \underbrace{\sum_{j_1=1}^{k+j} \sum_{\substack{j_2=1 \\ j_2 \neq j_1}}^{k+j} \cdots \sum_{\substack{j_i=1 \\ j_i \neq j_k \\ k=1,2,\ldots,i-1}}^{k+j}}_{j-\text{fold}} p_{t_{j_1}} p_{t_{j_2}} \cdots p_{t_{j_k}} \underbrace{\prod_{\substack{j_k=1 \\ j_k \neq j_l \\ l=1,2,\ldots,i}}^{k+j}}_{k-\text{fold}} q_{t_{j_k}} \right]$$

$$(5.68)$$

Example 5.4

The application of the presented methodology is provided for six precipitation stations in the northwestern part of Turkey, within the metropolitan boundaries of Istanbul City, as shown in Fig. 5.10. These stations are distributed as three and three in the European and Asian

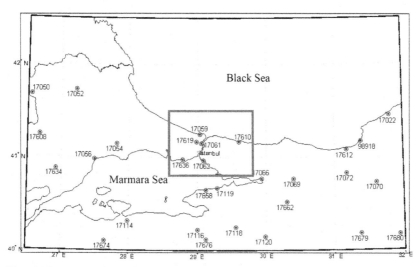

FIG. 5.10 Istanbul area meteorology station.

continental parts of the city, respectively. The records at each one of these stations are available from 1930 to 2010. In the area, there are two different types of climate patterns. In the northern parts, near the Black Sea, where the mountainous areas extend, the direct effect of the North Atlantic polar and maritime air movements are effective. In the southern part, along the Marmara Sea, a rather temperate climate prevails with intrusions from the Mediterranean type of modified climate.

Table 5.5 indicates the location, averages, standard deviations, and the regional as well as temporal dry and wet spell probabilities.

TABLE 5.5 Probability Calculations

Station No (1)	Longitude (E) (2)	Latitude (N) (3)	Precipitation (mm)		Regional Probability		Temporal Probability	
			Average (4)	Std. Dev. (5)	Wet, p_r (6)	Dry, q_r (7)	Wet, p_t (8)	Dry, q_t (9)
17062 (18)	40.97	29.08	680.76	118.00	0.5584	0.4416	0.1711	0.8289
17636 (128)	40.98	28.78	596.06	199.85	0.3766	0.6234	0.4079	0.5921
17061 (17)	41.15	29.05	616.68	363.45	0.2987	0.7013	0.6318	0.3682
17619 (119)	41.17	28.98	854.84	490.94	0.3117	0.6883	0.6316	0.3684
17610 (116)	41.17	28.60	719.81	323.35	0.4545	0.5455	0.3816	0.6184
17059 (16)	41.25	29.03	572.62	323.35	0.3247	0.6753	0.6184	0.3816
Average					0.3874			

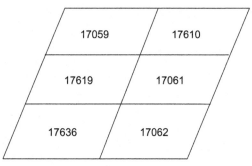

FIG. 5.11 Istanbul six-station representative subareas.

Next to the international station numbers in the first column, the local station numbers are given in parentheses. In the dry and wet spell probability calculations, the annual averages are taken respective to threshold levels.

Fig. 5.11 is a template similar to the one in Fig. 5.9, where each subarea (site) is shown representatively by a square.

The regional and temporal dry spell probabilities are given in Fig. 5.12 for Δt time period, and the same repetition may continue for subsequent time periods into the future.

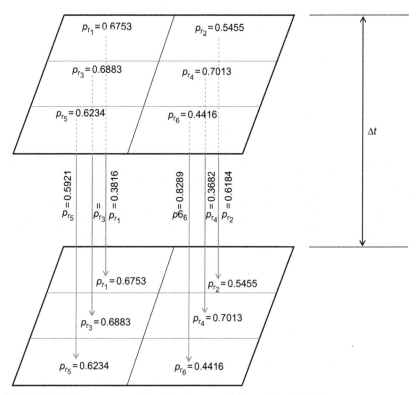

FIG. 5.12 Heterogeneous regional and temporal dry spell probabilities.

TABLE 5.6 Subarea Probabilities Between Δt Apart Two Successive Time Intervals

| | Subareas at 1Δt Time İnstant | | | | | | | |
	0	1	2	3	4	5	6	Marginal Probability
Subareas at 2Δt time instant								
0	0	0.0002	0.0006	0.001	0.0008	0.0003	0	0.0029
1	0.0002	0.0019	0.0063	0.0099	0.0079	0.0031	0.0005	0.0298
2	0.0008	0.0079	0.0265	0.0415	0.0329	0.0128	0.002	0.1244
3	0.0018	0.0173	0.0578	0.0905	0.0718	0.028	0.0043	0.2715
4	0.0021	0.0206	0.0689	0.1079	0.0856	0.0334	0.0051	0.3236
5	0.0013	0.0127	0.0423	0.0663	0.0526	0.0205	0.0031	0.1988
6	0.0003	0.0031	0.0104	0.0163	0.0129	0.005	0.0008	0.0488
Marginal probability	**0.0065**	**0.0637**	**0.2128**	**0.3334**	**0.2645**	**0.1031**	**0.0158**	**1.0000**

Input values are p_r, p_t and number of subareas is $n = 6$ in this case. Table 5.6 provides p_r and p_t values for these six stations in Istanbul.

Fig. 5.13 is the result of the program, which shows the change of areal drought coverage versus the cumulative probability of coverage area. It is observed that after the third year the cumulative probability curve becomes steady state, indicating that in the Istanbul region the average drought duration is about 3 years. This is in concurrence with the historical drought duration

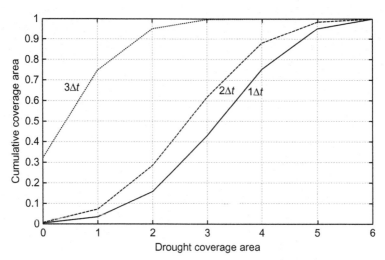

FIG. 5.13 Areal coverage versus cumulative probability of drought relationship.

average, which is referred to as 2.5–3 years, similar to the results in chapter: Temporal Drought Analysis and Modeling (Şen et. al., 2010).

Detailed calculations are shown in Table 5.6 for the cases between 1Δ*t* and 2Δ*t* instances with marginal probabilities in the last column and row. Herein, subarea numbers 0 represents that none of the subareas is under dry condition (ie, the whole region is covered by wet spell), whereas column 6 indicates that the whole region is under dry conditions. The values in the last row indicate the drought hit area probability after one time interval, which is 1 year in this study. One can calculate the cumulative probability for any given condition. Each cell value in this table shows conditional probability that if any subarea at the first instance is dry then it is also dry in the next time instant. For instance, the probability that subarea 5 is dry, given that subarea 4 was dry initially is equal to 0.0526. If one is interested in the probability that subarea 1 is dry irrespective of initial subareas, she or he will end up with 0.0065.

The entrance of the last row probabilities into the same computer program provides the next successive probability expositions as in Table 5.7. The last row in this table yields the curve in Fig. 5.13 that corresponds to 3Δ*t*

Table 5.6 can also provide information to questions such as "If a given number of subareas are under dry conditions initially, what is the probability that more than this number will be under dry conditions?" Say, if three subareas are dry initially, what is the probability that five subareas will be under dry condition after 1 time interval? The answer can be found after fixing subarea in the second column in this table and by adding all the conditional probabilities in this column up to 5, inclusive. Hence, the answer is 0.2024. The reader can enumerate similar questions and find the corresponding answers from the same table.

TABLE 5.7 Subarea Probabilities Between 2Δt Apart Two Successive Time Intervals

			Subareas at 2Δt time instant					Marginal probability
	0	1	2	3	4	5	6	
Subareas at 2Δt time instant								
0	0.0021	0.0205	0.0686	0.1073	0.0852	0.0332	0.0051	0.322
1	0.0028	0.0271	0.0905	0.1417	0.1124	0.0438	0.0067	0.425
2	0.0013	0.013	0.0436	0.0682	0.0541	0.0211	0.0032	0.2045
3	0.0003	0.0028	0.0094	0.0147	0.0117	0.0045	0.0007	0.0441
4	0	0.0003	0.0009	0.0014	0.0011	0.0004	0.0001	0.0042
5	0	0	0	0	0	0	0	0
6	0	0	0	0	0	0	0	0
Marginal probability	**0.0065**	**0.0637**	**0.213**	**0.3333**	**0.2645**	**0.103**	**0.0158**	**1.0000**

5.5 AREAL PRECIPITATION COVERAGE PROBABILITY FROM A SET OF HETEROGENEOUS POINT PROBABILITY

Various meteorological studies require probabilistic modeling such as flood, drought, and wet and dry spell occurrences. The most commonly employed approaches are based on the binomial PDF, which is valid for homogeneous temporal and spatial probabilities. However, in practice, heterogeneous probabilities are very common and in this case the classical binomial PDF cannot be employed. This section presents a rigorous probabilistic methodology for heterogeneous probabilities by considerations from the random field (RF) concept. The general form of the binomial PDF is derived and it is shown that for heterogeneous probabilities, it yields the classical binomial PDF exactly. The necessary analytical formulations are derived with the assumption that the meteorological occurrences at a set of sites or at a set of time periods (day, month, and year) are independent of each other. The probability statements derived in this section are helpful in predicting the spatiotemporal probable potential future meteorological occurrences.

5.5.1 Definitions of Basic Probabilities

Precipitation evolves temporally as well as spatially. Under given specific circumstances, precipitation has only a single value at each instant of time. Such a space–time distribution is referred to as a field quantity. Because meteorological quantities cannot be predicted with certainty, they are assumed to be random; hence, it is necessary to study a new class of fields [ie, RFs]. Definition and a detailed account of RFs are presented in chapter: Regional Drought Analysis and Modeling. Here, a precipitation phenomenon is modeled as a RF where time-wise probabilities at a fixed site are referred to as the precipitation occurrence probability (PoP) and spatial probabilities for a fixed time instant are areal coverage probabilities (ACP). The areal probability is the fraction of the area hit by rainfall. It does not provide a means for depicting which of the subareas is going to be hit with a precipitation event (see Fig. 4.1). Instead, it simply represents the estimate of what the fractional coverage will be. Furthermore, it is a conditional probability being conditioned by whether at least one precipitation event actually occurs in the subarea during a certain time interval.

The PoP at any desired threshold value, x_0, such as standard 0.01 in. (0.25 mm), is equivalent to the exceedance probability of this value. If the PDF of precipitation at site i is denoted by $f_i(X)$, then the PoP, p_i, at this site is

$$p_i = \int_{X_0}^{\infty} f_i(X) \mathrm{d}X \quad (i = 1, 2, 3, ..., n) \tag{5.69}$$

where n is the number of sites and p_is are the probabilities of one or more events over the study area during time T within the ith subarea, A_i. The average areal probability $\bar{p}$ can be defined formally in terms of areal weights as a weighted average:

$$\bar{p} = \frac{\sum_{i=1}^{n} A_i p_i}{\sum_{i=1}^{n} A_i} \tag{5.70}$$

If each subarea is equal in size, then this expression becomes equivalent to simple arithmetic average.

$$\bar{p} = \frac{1}{n} \sum_{i=1}^{n} p_i \tag{5.71}$$

Note that since $\bar{p}$ is a lumped value, it does not provide any information about the areal distribution of precipitation occurrences. It should be noted that associated subareas need to be small enough; hence, a single probability value can be applied to each one.

5.5.2 Theoretical Treatment

In the most general case, none of the sites have equal PoPs, which implies that the RF is heterogeneous. Let the wet and dry spell probabilities in a subarea be denoted by p_j and q_j, $(j=1, 2, ..., n)$, respectively, where n is the total number of sites and $p_j + q_j = 1$. The areal coverage of probability, $P(A=i|n)$, including i dry sites, can be evaluated through enumeration technique by the application of the mutual exclusiveness and exhaustiveness rules of probability theory. For the necessary derivations the conceptual model can be visualized as in Fig. 5.14, where there are n mutually exclusive sites represented by in-square boxes.

If out of n sites, i sites are wet and the remaining $(n-i)$ sites are dry, then the whole spectrum of possible cases become to have two mutually exclusive groups. The first group includes i cases with n possibility in each site, and

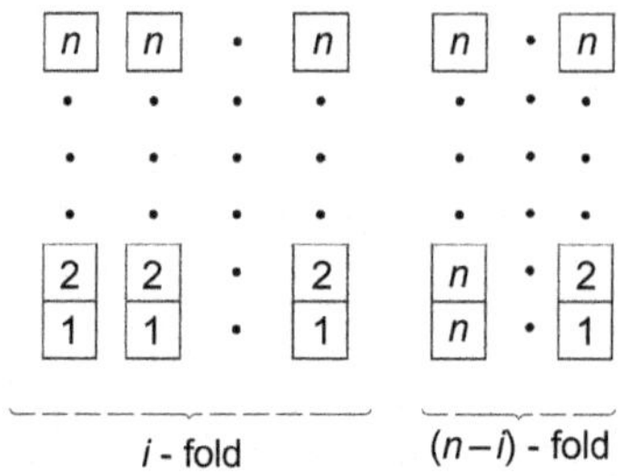

FIG. 5.14 Probability derivation conceptual model.

the second $(n-i)$ sites again each with n possibilities. Because the wetness on each site is independent from the others, then the probability of wet spell occurrence collectively can be found from the multiplication rule of independent events as $p_1.p_2.p_3...p_i$ for one pattern. In the case of i wet spell sites, there are $i(i-1)(n-2)...1=i!$, mutually exclusive patterns each with probability of wetness. Mutually exclusive events imply summation in the probability theory; therefore, the successive summation terms express all the possible, collective, and exhaustive joint probabilities at i sites. The second part on the right-hand side of Fig. 5.14 corresponds to the joint probability of the remaining $(n-i)$ sites each belonging to respective pattern of joint wet occurrence at i precedent patterns. Again, the independence principle of the probability theory provides the multiplication of the remaining dry occurrence sites. These explanations can be translated into a mathematical formulation as in Eq. (5.52).

$$P(A=i|n) = \frac{1}{i!}\underbrace{\sum_{j_1=1}^{n}\sum_{\substack{j_2=1\\j_2\neq j_1}}^{n}\cdots\sum_{\substack{j_i=1\\j_i\neq j_k\\k=1,2,...,i-1}}^{n} p_{j_1}p_{j_2}\cdots p_{j_i}}_{i-\text{fold}}\underbrace{\prod_{\substack{m=1\\m\neq j_l\\l=1,2,...,i}}^{n} q_m}_{(n-i)-\text{fold}} \quad (5.72)$$

where the i summation terms in the first horizontal big bracket includes all the possible combinations of i precipitation occurrences at n sites, whereas the second horizontal big bracket implies the multiplication term corresponding to possible dry combinations.

For heterogeneous wet spells the term in the first part in Eq. (5.72) simplifies to $n(n-1)...(n-i+1)p^i$ and the next part multiplication yields to q^{n-i}; hence, it reduces similar to Eq. (5.3),

$$P(A=i) = \binom{n}{i}p^i q^{n-i} \quad (5.73)$$

This is the well-known binomial distribution with two-stage Bernoulli trials (Feller, 1967).

The probability, p_A, that all the sites, hence the whole area, are covered by wet spell at any instant can be found from Eq. (5.72) as

$$p_A = \prod_{i=1}^{n} p_i \quad (5.74)$$

In practical applications, a group of randomly or systematically selected k wet or dry spell occurrences is of great interest. The joint probability that any k_1 of n

available sites has significant wet spell occurrences has already been obtained for homogeneous PoPs in chapter: Temporal Drought Analysis and Modeling, Chapter 4, Section 4.4. The corresponding general expression similar to Eq. (4.18) for the areal heterogeneous probabilities can be obtained by the substitution of Eq. (5.72) instead of the binomial distribution yields

$$P(A = i | X_1 > x_0, X_2 > x_0, \ldots, X_{k_1} > x_0)$$

$$= \left(\frac{i}{np}\right)^{k_1} \frac{1}{i!} \underbrace{\sum_{\substack{j_1 = 1}}^{n} \sum_{\substack{j_2 = 1 \\ j_2 \neq j_1}}^{n} \cdots \sum_{\substack{j_i = 1 \\ j_i \neq j_k \\ k = 1,2,\ldots,i-1}}^{n}}_{i-\text{fold}} p_{j_1} p_{j_2} \cdots p_{j_k} \underbrace{\prod_{\substack{m = 1 \\ m \neq j_l \\ l = 1,2,\ldots,i}}^{n} q_m}_{(n-i)-\text{fold}}$$

$$(5.75)$$

Similarly, the conditional ACP of wet spell, given that a group of k_2 sites have dry spell, can be found for homogeneous PoPs as

$$P(A = i | X_1 < x_0, X_2 < x_0, \ldots, X_{k_1} < x_0) = \left(\frac{n-i}{n} q\right)^{k_2} \binom{n}{i} q^i p^{n-i} \qquad (5.76)$$

and

$$P(A = i | X_1 < x_0, X_2 < x_0, \ldots, X_{k_1} < x_0)$$

$$= \left(\frac{n-i}{n} q\right)^{k_2} \frac{1}{i!} \underbrace{\sum_{\substack{j_1 = 1}}^{n} \sum_{\substack{j_2 = 1 \\ j_2 \neq j_1}}^{n} \cdots \sum_{\substack{j_i = 1 \\ j_i \neq j_k \\ k = 1,2,\ldots,i-1}}^{n}}_{i-\text{fold}} p_{j_1} p_{j_2} \cdots p_{j_k} \underbrace{\prod_{\substack{m = 1 \\ m \neq j_l \\ l = 1,2,\ldots,i}}^{n} q_m}_{(n-i)-\text{fold}}$$

$$(5.77)$$

Finally, the conditional ACP of wet spell, given that a group of k_1 sites has wet spell and another group of k_2 sites has dry spell, is obtained as

$$P(A = i | X_1 > x_0, X_2 > x_0, \ldots, X_{k_1} > x_0, X_{k_1+1} < x_0, X_{k_1+2} < x_0, \ldots, X_{k_2} < x_0)$$

$$= \left(\frac{i}{nq}\right)^{k_1} \left(\frac{n-i}{n} q\right)^{k_2} \binom{n}{i} q^i p^{n-i} \qquad (5.78)$$

With its heterogeneous counterpart as

$$P(A = i | X_1 > x_0, X_2 > x_0, \ldots, X_{k_1} > x_0, X_{k_1+1} < x_0, X_{k_1+2} < x_0, \ldots, X_{k2} < x_0)$$

$$= \left(\frac{i}{nq}\right)^{k_1} \left(\frac{n-i}{n}q\right)^{k_2} \frac{1}{i!} \underbrace{\sum_{j_1=1}^{n} \sum_{\substack{j_2=1 \\ j_2 \neq j_1}}^{n} \cdots \sum_{\substack{j_i=1 \\ j_i \neq j_k \\ k=1,2,\ldots,i-1}}^{n} p_{j_1} p_{j_2} \cdots p_{j_k}}_{i-\text{fold}} \underbrace{\prod_{\substack{m=1 \\ m \neq j_l \\ l=1,2,\ldots,i}}^{n} q_m}_{(n-i)-\text{fold}}$$

$$(5.79)$$

The probability expressions in Eqs. (5.76)–(5.79) can be effectively employed to study regional wet (dry) spell spatiotemporal occurrence patterns.

5.5.3 Numerical Applications

The numerical application will be the confirmation of the basic expression (Eq. 5.72) with the well-known binomial PDF. For this purpose, a MATLAB program given in Appendix 5.1 can be used, which has been developed for heterogeneous point probabilities of different sizes. In Fig. 5.15, Eqs. (5.72) and (5.73) are plotted versus each other. The scatter of points for different sample sizes fall exactly on the (1:1) 45 degree line, confirming the validity of heterogeneous probability areal formulation.

Figs. 5.16 and 5.17 are valid for the areal heterogeneous probabilities given as 0.1, 0.2, 0.3, 0,4, 0.5, 0.6, 0.7, 0.8, and 0.9 with 9 and 12 subareas, respectively.

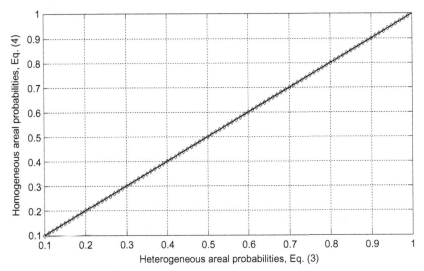

FIG. 5.15 Comparisons of Eqs. (5.3) and (5.4).

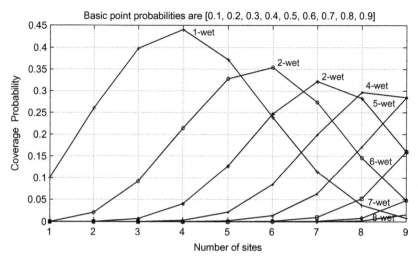

FIG. 5.16 Some heterogeneous areal coverage probabilities (ACP) variation for 9 sites.

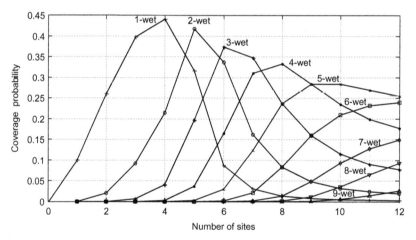

FIG. 5.17 Some heterogeneous ACP variation for 12 sites.

The application of the methodology is presented for factual data from the six meteorology stations around Istanbul City, Turkey, as shown in Fig. 5.10 within the rectangle. These are along the Bosporus, representing both the European and Asian sites of the city.

The station numbers, statistical parameters, mean and variance, locations, and the wet spells are given in Table 5.5 for each site, based on the average threshold levels. In the application these probabilities are the wet spell at each site. Application of Eq. (5.77) with $n=6$ and the probability set in Table 5.5 yields the coverage probabilities with 1, 2, 3, 4, 5, and 6 stations as shown in Table 5.8, which indicates that the maximum coverage probability in the vicinity of Istanbul appears at two stations (17636 and 17061), whereas other combinations have less probability of occurrence.

TABLE 5.8 Coverage Probability Values

Station No	Site No	Coverage Probability
17062	1	0.1988
17636	2	0.3235
17061	3	0.2715
17619	4	0.1245
17610	5	0.0297
17059	6	0.0029

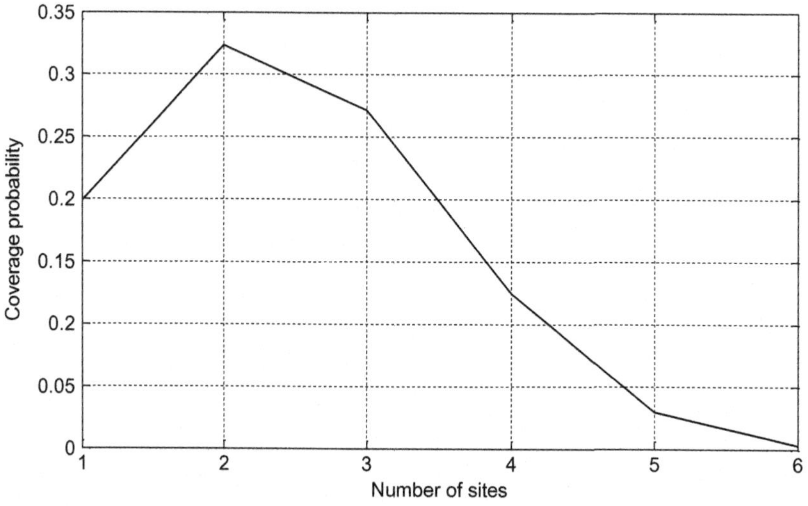

FIG. 5.18 Istanbul stations wet coverage probabilities.

The graphical variation of the coverage probabilities are presented in Fig. 5.18. If requested, any set of probabilities at these six sites can be calculated from Eq. (5.72) for finding the coverage area.

Analytical derivations of various point wet spell, including probability of wetness concept and the ACP, are presented for heterogeneous and homogeneous precipitation evaluations over a region. The basis of the derivations is the concept of RFs of independent processes. To derive the extreme value probabilities of the wet spell areal coverage, it is assumed that the precipitation phenomenon in the region evolves independently along the time axis. The basic requirements of the theory developed herein are the probability of wet spell at different sites. By using the probability statements in this chapter, it is possible to make risk evaluation of areal precipitation occurrences.

APPENDIX 5.1 HETEROGENEOUS REGIONAL BINOMIAL PDF CALCULATION SOFTWARE

```
function [Pr] = PoP6(P,n)
% Written by Zekâi Şen on 27 October 2011
% This program calculates probability of precipitation and areal
% coverage probability (ACP)
% P      : Probability vector including probability of exceedence in each
%           site of n site; this vector should come from the main program
% n      : Number of sites
% np     : Number of wet sites n > np
sum=0;
for i1=1:n
    for i2=1:n
        if (i2~=i1)
            for i3=1:n
                if (i3~=i1) && (i3~=i2)
                    for i4=1:n
                        if (i4~=i1) && (i4~=i2) && (i4~=i3)
                            for i5=1:n
                                if (i5~=i1) && (i5~=i2) && (i5~=i3) && (i5~=i4)
                                    for i6=1:n
                                        if (i6~=i1) && (i6~=i2) && (i6~=i3) &&
                                           (i6~=i4) && (i6~=i5)
                                        ppp=P(i1)*P(i2)*P(i3)*P(i4)*P(i5)*P(i6);
                                        Pro=1;
                                        for l=1:n
                                            if (l~=i1) && (l~=i2) &&(l~=i3) &&
                                               (l~=i4) && (l~=i5) && (l~=i6)
                                                Pro=Pro*(1-P(l));
                                            end
                                        end
                                        sum=sum+ppp*Pro;
                                        end
                                    end
                                end
                            end
                        end
                    end
                end
            end
        end
    end
end
Pr=sum/factorial(6);
end
```

REFERENCES

Anis, A.A., Lloyd, E.H., 1953. On the range of partial sums of a finite number of independent normal variates. Biometrika 40, 35–42.

Chung, C.H., Salas, J.D., 1999. Drought occurrence probabilities and risks of dependent hydrologic processes. ASCE, J. Hydrol. Eng. 5 (3), 259–268.

Downer, R., Siddiqui, M.M., Yevjevich, V., 1967. Applications of runs to hydrologic droughts. Hydrology paper 23, Colo. State Univ., Fort Collins, CO.

Feller, W., 1951. The asymptotic distribution of the range of sums for independent random variables. Ann. Math. Stat. 22, 427–432.

Feller, W., 1967. An Introduction to Probability Theory and its Application. John Wiley and Sons, New York, 509 pp.

Hurst, H.E., 1951. Long range storage capacity of reservoirs. Trans. ASCE 116, 770–808.

Lee, K.S., Sadeghipour, J., Dramp, J.A., 1986. An approach for frequency analysis of multi-year drought durations. Water Resour. Res. 22 (5), 655–662.

Lee, T., Modarres, R., Ouarda, T.B.M.J., 2012. Data-based analysis of bivariate copula tail dependence for drought duration and severity. Hydrol. Process. 27 (10), 1454–1463. http://dx.doi.org/10.1002/hyp.9233.

Llamas, J., Siddiqui, M.M., 1969. Runs of precipitation series. Hydrology paper 33, Colo. State Univ, Fort Collins, CO.

Mishra, A.K., Singh, V.P., 2010. A review of drought concepts. J. Hydrol. 391, 202–216.

Mishra, A.K., Singh, V.P., Desai, V.R., 2009. Drought characterization: a probabilistic approach. Stoch. Environ. Res. Risk Assess. 23, 41–55.

Mood, A.M., 1949a. The distribution theory of runs. Ann. Math. Stat. 98–119. http://dx.doi.org/10.1090/S0002-9947-0032114-7.

Panu, U.S., Sharma, T.C., 2002. Challenges in drought research: some perspectives and future directions. Hydrol. Sci. J. 47, S19–S30.

Salazar, P.G., Yevjevich, V., 1975. Analysis of drought characteristics by the theory of runs. Hydrology paper no. 80, Colorado State University, Fort Collins, pp. 1–44.

Saldarriaga, J., Yevjevich, V., 1970. Application of run-lengths to hydrologic time series. Hydrology paper 40, Colo. State Univ, Fort Collins, CO.

Şen, Z., 1976. Wet and dry periods of annual flow series. J. Hydraul. Div., ASCE 102 (HY10), 1503–1514, Proc. Pap. 12497.

Şen, Z., 1977. Run-sums of annual flow series. J. Hydrol. 35, 311–324.

Şen, Z., 1978. Wet and dry periods in regional water exchange. In: Symposium on Inputs for Risk Analysis in Water Systems, Canada.

Şen, Z., 1980a. Statistical analysis of hydrologic critical droughts. J. Hydraul. Div., ASCE 106 (HY1), 99–115, Proc. Pap. 14 134.

Şen, Z., 1980b. Regional drought and flood frequency analysis: theoretical considerations. J. Hydrol. 46, 265–279.

Şen, Z., 1998. Average areal precipitation by percentage weighted polygon method. J. Hydrol. Eng. ASCE 3 (1), 67–72.

Şen, Z., Uyumaz, A., Cebeci, M., Öztopal, A., Küçükmehmetoğlu, M., Özger, M., Erdik, T., Sırdaş, S., Şahin, A.D., Geymen, A., Oğuz, S., Karsavran, Y., 2010. The İmpacts of Climate Change on Istanbul and Turkey Water Resources. Istanbul Metropolitan Municipality, Istanbul Water and Sewerage Administration, 1500 pp.

Şen, Z., 2014. Regional wet and dry spell analysis with heterogeneous probability occurrences. J. Hydrol. Eng. http://dx.doi.org/10.1061/(ASCE)HE.1943-5584.0001144, 04014094.

Smith, K., 2001. Environmental Hazards. Assessing Risk and Reducing Disaster. Routledge, New York, 392 pp.

Takeuchi, K., 1974. Regional water exchange for drought alleviation Hydrology paper 70, Colorado State University, Colorado.

Tase, N., 1976. Area-deficit-intensity characteristics of droughts. Hydrology paper 87, Colo. State Univ., Fort Collins, CO.

Tase, N., Yevjevich, Y., 1978. Effects of size and shape of a region on drought coverage. Hydrol. Sci. Bull. 23 (2), 203–213.

Thiessen, M., 1911. Precipitation average for large areas. Mon. Weather Rev. 39, 1082–1084.

Yevjevich, V., 1967. An objective approach to definitions and investigations of continental hydrologic drought. Hydrology paper no. 23, Colorado State University, Fort Collins, CO.

Chapter SIX

Climate Change, Droughts, and Water Resources

CHAPTER OUTLINE

Applied Drought Modeling, Prediction, and Mitigation

6.1 GENERAL

Climate change and its impacts are among the most significant national, international, and global issues of the current century. Natural climate changes have forced many societies to migrate from arid and drought (water stricken) locations for better living conditions because every culture and civilization has to depend on and adapt to environmental conditions, among which the most significant factors are pleasant effects of weather and atmospheric conditions in terms of climate, hydrological cycle, and water resources. Global climate change is evident in many parts of the world, which affects droughts, floods (especially flash floods), agricultural products, and many human activities in an unusual manner.

The average of weather events over long time spans and large regions defines climate change due to changes in ocean currents, atmospheric chemistry composition, and circulation, in addition to weather variables such as sunshine radiation, precipitation, pressure, and temperature. Temperature and precipitation records are the two most important factors that should be taken into consideration for any climate change impact study. It is not effective only on human life, culture, and civilization but also on animals, plants, and materials. For instance, the surface features of any area are formed due to geological climate change effects. In contrast, humans and animals affect the weather as a result of atmospheric pollution and, consequently, present climate change with its direct and indirect effects on atmospheric composition due to anthropogenic (human) activities. Life is sustainable with appropriate climate only, the change of which returns to the living creatures directly or indirectly. Suitable hydrologic, lithological (geologic), biologic, and weather environments in balance with each other temporally, and spatially are the basic requirements for all life on Earth. Adverse situations begin to appear with the disturbance of the balance between these different sectors. All this information reminds us that it is necessary to care for the weather and especially climate by trying not to disturb the balance limits concerning, in particular, the atmosphere, where the meteorological and climatic events take place.

During the geological past of the Precambrian the climate situation was not able to support living organisms; therefore, there is no fossil signature in these geological layers. Later on, through millions of years the climate started to improve and allow for the activities of living organisms and it continued improving for the appearances of other living features such as plants, animals, and finally, human beings (very recently according to the geological time scale). During the geological epochs climate has changed several times and, accordingly, some creatures were extinguished and their remnants in the form of fossils are available in different sedimentary rocks. The most affected material related to living organisms and creatures is water, and this is the reason wherever life is searched for, the first thing to look for is suitable climate, but this is not sufficient if water resources do not exist. Even today, in search for life on other planets the first question is whether the climate is suitable for the survival of any organism or creature. Water is the major agent in development and sustenance. Although water has gas, liquid, and solid forms, for life to exist its liquid form must be available and this is the main reason why early civilizations developed along the Nile, Euphrates, Tigris, Indus, and Ganj Rivers and similar watercourses.

Climate change is also related to slight deviations from regular sun irradiation, which affects not only water resources but also the environment as a whole. The scarcity or abundance of sunspots is a simple indicator of the sun's radiation, which is the major agent in weather and climate affairs. Depending on the environmental conditions, ice caps advance or retreat throughout the geological time scales and dependently climate change effects take place.

Atmosphere and climate are available everywhere with equal share to all creatures but, unfortunately, the same availability is not possible for water resources due to their haphazard spatial distribution on the Earth. Human activities are the most active agents that cause pollution and contamination of water chemical outputs into the atmosphere, which lead to slow but accumulative changes in the atmospheric chemical composition toward climate change. The chemical composition prior to the Industrial Revolution (about the 1850s) has evolved during the past life of the universe and the only interactions with atmospheric composition were natural events such as solar irradiation of extraterrestrial and terrestrial types, consequent temperature and pressure differences, and natural surface features. Although human interferences were also continuous contributors, natural energy and animal force were used mainly for agricultural activities so their consecutive accumulations in the atmosphere were at a negligible scale with very short distance influences only. The most significant effects are weather dependence on small scales (hours, days, months, local, regional) and atmosphere dependence on large scales (annual, decade, continental, synoptic, meso, global) including the following natural events:

1. Atmospheric circulations
2. Cyclones
3. Precipitation
4. Droughts
5. Floods
6. Heat waves
7. Hydrological cycle
8. Ocean circulations
9. Solar transmission, emission, and absorption
10. Typhoons

11. Volcanic eruptions

12. Wind storms

There are also many artificial and anthropogenic effects that are reflected on top of the aforementioned natural events:

1. Contamination

2. Energy exploitation, especially fossil types

3. Industrial production

4. Land use

5. Pollution

6. Urbanization

7. Population growth

Human beings, animals, and plants alike are dependent for their sustainable survival on some gases, nutrients, and solids that are provided rather abundantly and almost freely in nature. The most precious ones are the air that living organisms breathe and water that is available in the form of hydrosphere either in the troposphere or in the lithosphere (especially groundwater). The atmosphere has evolved over geological time, and the development of life on Earth is closely related to the composition of the atmosphere. From the geological records, it seems that it was about 1.5 billion years ago that free oxygen first appeared in the atmosphere in appreciable quantities (Harvey, 1982). The appearance of life is dependent on the availability of oxygen and once a sufficient amount was accumulated for green plants, photosynthesis was able to liberate more into the atmosphere.

Even though the natural circulations in the atmosphere provide scavenging effects, continuous and long-term loading of atmosphere might lead to undesirable situations in the future. A close inspection and control should be directed toward various phenomena in the atmosphere. It is necessary that more applied and detailed research must be carried out to appreciate the meteorological events in the troposphere, ozone depletion in the stratosphere, pollution in the lower troposphere, and transboundary between the troposphere and hydro-lithosphere; energy, transport, and industrial pollutants generation and movement; effects of acid rain; and wastewater leakage into the surface and, especially, into the groundwater resources (Şen, 2008a). In order to achieve success in these areas, it is necessary to perform sound scientific basic research by taking climate change impacts into consideration. To this end, more extensive climatic, meteorological, hydrologic, and hydrogeological observation networks should be established with spatial and temporal monitoring. Any undesirable change in the atmospheric conditions may endanger forests, hydrosphere ecosystems, and economic activities such as agriculture mostly as a result of water stress and drought.

One of the greatest and most famous scientists of all time, Ibn Sina (Avicenna) (970–1030), recommended some 1000 years ago seven points for humanity to lead a healthy life in this world.

1. Cleanness of the inhaling atmospheric air

2. Choice of food and drinking water quality

3. Healthiness of the body

4. Getting rid of extra weight to feel fitness

5. Comfortable dressing

6. Healthiness in thinking and pondering

Two of these points, namely, choice of water and air clarity, are among the modern topics of this chapter.

In the past, water shortages and poor harvests were among the actual drought indicators because of the vulnerability of a region to climatic extremes. Recently, the prospect of a major climate change due to anthropogenic activities with impacts on extreme values and especially on droughts pose growing concern about the sustainability of any region.

Even though temperature increase can be predicted with confidence, the same does not apply for precipitation, and its projections are spatially and temporally much less certain. Globally, the impact of temperature is more certain, but even areas with precipitation increase projections are subject to increasing evaporation and, accordingly, temporal rainfall distribution and intensity are expected to change. It is recommended that the present intensity–duration–frequency curves must be updated following every 10-year period.

One of the consequences of global warming is the expectation that sea levels are bound to increase due to ocean expansion in addition to glacier melts. The rise in the sea level on the average is expected to be 1 m by 2100 (IPCC, 2007). Additionally, the sea level rise pushes saltwater intrusion in the coastal aquifers (Şen, 2014). It is stated in the literature that according to the Ghyben-Herzberg relation each 1 m rise in the sea level causes about 40 m of saltwater intrusion into the coastal aquifers.

Especially in the subtropical areas such as the Mediterranean region, where climate has high spatial and temporal variability, detection of climate change impacts is difficult to predict. If preliminary precautions are not taken, poor land use practices and hotter and drier spells are bound to occur more frequently than ever.

This chapter provides information about the potential climate change impacts on various human activities with sustainable development implications. Any sign of deficiency in food safety, human health, the local economy, and the ecosystem may exacerbate existing problems. It is possible to setup for long-term sustainability through effective adaptation studies, which must also include urgent action to cut climate emissions related to greenhouse gases. Otherwise, hotter and, consequently, drier times are ahead in many areas of the world.

6.2 BASIC DEFINITIONS AND CONCEPTS

Adaptation and disaster management should also include vulnerability information and exposure and climate impacts. Risk management is successful if based on climate change extreme features of the region concerned. Concerning this and the next chapter the following terminological definitions gain significance:

Climate change: It corresponds to a long-term (at least 30 years) change in the climate state, which can be depicted by effective statistical tests. Climate change can be noticed on the average by changes in the mean of the hydrometeorological variability (especially temperature and precipitation) that may persist for an extended period of a few decades or longer. Apart from the natural internal climate change, external effects (anthropogenic) also contribute to atmospheric composition impacts. Climate change may be due to natural internal processes or external forcing and land use. There are three descriptive statistical cases that indicate the climate change effects in any region or at global level. Fig. 6.1 indicates the shift in the mean value, gradual change on the average as a trend, and change in the standard deviation of the hydrometeorological variable. Of course, their combination may also appear in a sophisticated manner.

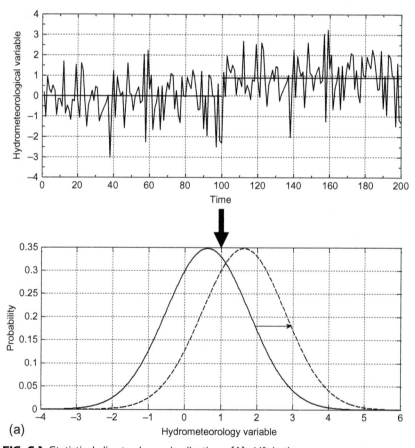

(a)

FIG. 6.1 Statistical climate change implications: (A) shift in the mean;

(Continued)

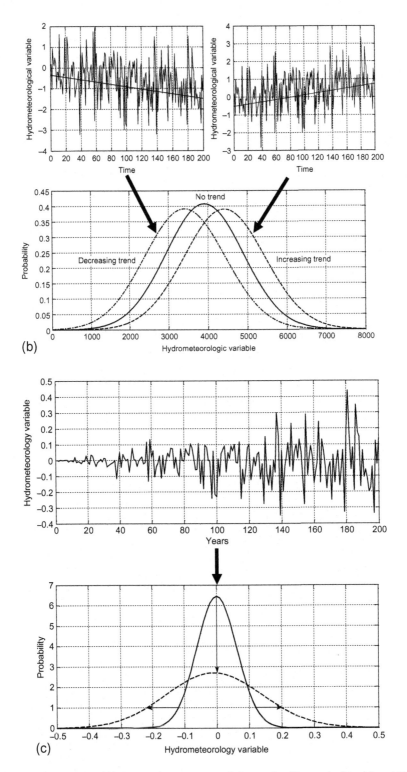

FIG. 6.1—CONT'D (B) change along a monotonic increasing (decreasing) trend; and (C) change in the standard deviation.

Climate extreme (extreme weather or climate event): These are the values that are significant in any extreme event (droughts, floods) assessment and they are expressible in terms of weather, climate, or hydrological variables (such as rainfall or runoff) with their above- (or below-) threshold value near the upper (or lower) ends of the variability domain ("see chapter: Temporal Drought Analysis and Modeling"). They are referred to as climate or hydrological extremes according to the context of the study.

Exposure: These are places that may be adversely affected by climate in terms of livelihoods; environmental services and resources; infrastructure; and economic, social, or cultural assets.

Transformation: To reduce climate change effects, various fundamental attributes can be altered toward better protection of different systems, including value attachment; regulatory, legislative, or bureaucratic regimes; financial institutions; and technological systems.

Disaster risk management and adaptation to climate change: Disaster risk management measures and systems should go hand in hand toward decreasing uncertainty by integrating all available knowledge and information at the exposure, vulnerability, and mitigation stages.

Socioeconomic differences: These are bound to influence local coping and adaptation capacity, which may be exacerbated during a drought event. Each society may have different challenges in drought assessment and response depending on the available infrastructure, institutions, instruments, and early warning facilities.

Resilience to climate change: This may be increased depending on risk assessments at global, regional, national, and local scales.

Weather, climate change and variability, past experience from drought periods, and future predictions reflect the capacity of a society or region to deal with long-duration vulnerability. Adaptation to climate change must be harmonious with drought risk management and involve the cooperation of local and central governors, in addition to alerting people. Climate change risks can be minimized by the consideration of past experiences and the understanding of extreme event behaviors with the incorporation of disaster risk management and adaptation approaches ("see chapter: Drought Hazard Mitigation and Risk").

6.3 ATMOSPHERIC COMPOSITION AND POLLUTION

The atmosphere is an almost infinite air mass composed of various gases that surround the Earth in every direction. It has several layers, depending on temperature profiles and gas compositions. The part of the atmosphere that is vital for all living organisms is adjacent to the Earth's surface and named the troposphere, constituting about 87% of the atmosphere by weight. It extends vertically up to about 8 km over the polar regions and 17 km over the equator. There are major circulation cells within this layer, the sun's radiation being the sole driving force (Held and Suarez, 1994). These major circulation units coupled with the atmospheric turbulent effects, especially within the planetary boundary layer, cause completely rapid and thorough mixing during less than a year in the atmosphere. Especially, the troposphere is constantly being mixed and stirred by turbulent winds as a result of solar energy interaction with dense gases. However, across the equator relatively complete mixing takes place over a few years. Consequently, whatever the pollutant released from the Earth's surface it will soon be mixed throughout the troposphere. A major agent for such a mixing is the vertical temperature gradient (ie, lapse rate), which supports large-scale vertical mixtures (Meyer, 1992).

Natural tendencies, and over the last several decades human activities, especially, started to influence and alter climate behavior, which led to climate change and its implications on the environment and society at large. Among the negative changes related to climate change are rises in sea levels, increases or decreases in precipitation amounts depending on geographical location, and changes in the ecosystem affecting snow amount. However, there are also expected positive changes such as longer growing seasons, especially in tropical and subtropical climate belt regions.

There have been several natural climate change effects during the past four billion years of Earth's existence, but they were impacted significantly by human activities. Use of fossil fuels in different parts of the industrialization process led to anthropogenic activities and greenhouse gases accumulation and, consequently, effects on climate change started to add extra trends on the possible natural variability. Various climate change models indicate that the global average temperature may increase up to 1.4–4.0°C until 2100 (IPCC, 2007).

One of the most significant consequences of global warming is an increase in the frequency and magnitude of droughts with increasing total deficit amounts in rainfall, and subsequently, in the runoff and soil-moisture content. The changes in the climate of a region impact on general and local hydrological cycles, which imply changes in regional water resources balances. For instance, water vapor buildup in a region as a result of atmospheric heating, and consequent changes in climate, may lead to more frequent drought occurrences than normal situations. Extensive drought periods may cause water stress and food supply decrease, which may end up with famine, as it already has in different parts of Africa.

Weather and climate impacts depend on their nature and severity, vulnerability, and exposure. In general, socioeconomic development interacts with natural climate variations, but anthropogenic climate changes cause an increase in disaster risk. Spatial and temporal climate changes including interannual variations lead to significant drought disasters on large spatial and interannual scales. Fig. 6.2 indicates that the economic losses incurred has increased during recent years much more than in previous decades.

Since 1980, extreme hot days and heavy precipitation became more common and the main cause of this event is understood as anthropogenic influences, including increasing atmospheric greenhouse gases, which must be reduced by

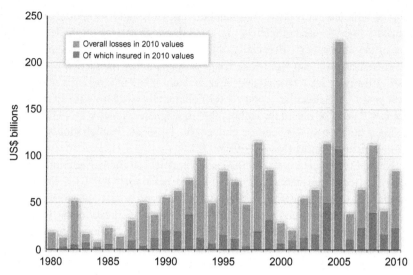

FIG. 6.2 Economic loss comparisons. *Data from Munich Re, 2011. Natural Catastrophe Year in Review. Munich Reinsurance America, Inc.*

any means. Many national and international climate models indicate expectations about an increase of heavy rainfall events throughout the 21st century, such as for the Arabian Peninsula (AP) (Şen et al., 2011). On the other hand, in some other areas, including the southern part of Turkey, more unusual rainfall reductions are expected (Şen et al., 2010). Weather and climate event risks can be reduced to the possible minimum level provided that there are improvements in forecasting through early warning systems, which require adaptation and disaster risk management for drought disaster reductions ("see chapter: Drought Hazard Mitigation and Risk").

Anthropogenic climate change impacts lead to depletion of the vegetation cover and loss of biophysical and economic productivity. As a result, the soil surface may be exposed to wind and water erosion and shifting sands, salinization of land and waterlogging, which are processes against groundwater recharge. These problems are especially severe in the dry areas such as the AP. In arid and semiarid regions, rainfall occurrences and distribution are sporadic with randomly variable low-rainfall quantities.

6.3.1 Natural Chemical and Hydrological Cycles

In pollutant intact atmosphere, naturally available gases, namely, nitrogen (N), oxygen (O_2), and carbon dioxide (CO_2), are replenished through many-year duration cycles due to the natural phenomena that take place between various spheres. Fig. 6.3 shows the interaction between the atmosphere, biosphere, lithosphere, and hydrosphere for the nitrogen cycle that is the main agent in the atmosphere, and it completes the renewal process about once every 100 million years (Galloway et al., 2004).

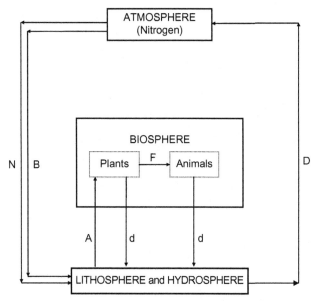

FIG. 6.3 Natural nitrogen cycle. *N*, Nitrogen compounds produced by atmosphere electrical storms; *B*, bacteria; *A*, assimilation; *d*, death; *F*, feeding; *D*, decomposition of organic matter.

Nitrogen is a major constituent not only of the atmosphere, but also of animals and plants or it is the principal element in proteins, the basic structural compounds of all living organisms. Certain microscopic bacteria converts the tremendous nitrogen supply of the atmospheric nitrogen into water soluble nitrate atom groups, which can then be used by plants and animals for protein manufacturing. The nitrogen reenters the atmosphere as dead animals, and plants are decomposed by other nitrogen-releasing bacteria.

The second major constituent of the lower atmosphere is oxygen, which is also the most abundant element in the hydrosphere and lithosphere. Most of the uncombined gaseous oxygen in the atmosphere is neither the hydrosphere nor the lithosphere, but photosynthesis by green plants. In the photosynthesis process, sunlight breaks down water into hydrogen and oxygen. The free oxygen is utilized by animals as an energy source being ultimately released into the atmosphere combined with carbon as CO_2, which is taken up by plants to begin the cycle shown in Fig. 6.4. Such a cycle recycles the whole oxygen available in the atmosphere in only 3000 years. Thus the free oxygen, like nitrogen, is closely interrelated with the life processes of organisms (Keeling, 1995).

Although CO_2 is a minor constituent of the troposphere, it plays a fundamental role in the atmospheric heat balance like ozone within the stratosphere, and it is a major controlling factor in the Earth's weather and climate change patterns.

The atmospheric CO_2 is very much dependent on the vegetative cover and an increase in its concentration affects moisture and rainfall areal distribution all over the world with acceleration or deceleration of the local hydrological cycle.

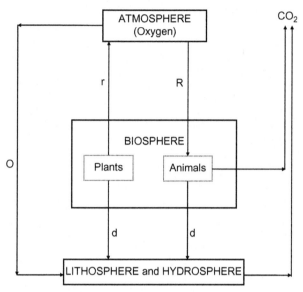

FIG. 6.4 Natural oxygen cycle. *d*, Death; *O*, oxidation of dead organic matter; *r*, released by photosynthesis; *R*, respiration.

Different experiments and experiences indicate that a doubling of CO_2 would increase stomatal resistance and reduce the rate of transpiration (the passage of water vapor from plants) by about 50% on average (Şaylan, 2000). Accordingly, any decrease in transpiration tends to increase runoff. On the other hand, CO_2 also causes an increase in plant growth, leading to a larger area of transpiration. Additionally, potential increase in leaf temperatures is caused by reduced transpiration rates, and species changes in vegetation communities offset increases in plant water-use efficiency associated within a CO_2-enriched atmosphere.

The CO_2 cycle is shown schematically in Fig. 6.5. Green plants directly use atmospheric CO_2 to synthesize more complex carbon compounds, which, in turn, are the basic food for animals. The carbon is ultimately returned into the atmosphere as a waste product of animal and plant respiration or decomposition just as free oxygen is contributed by green plant photosynthesis. The cycling of CO_2 through living organisms takes only 35 years for the relatively small quantity in the atmosphere to pass once through this cycle.

In the last 100 years, burning fossil fuels such as coal and oil has caused a significant increase in the level of CO_2. Because CO_2 has one of the large gas molecules that traps long-wave radiation to warm the lower atmosphere by the so-called "greenhouse effect," atmospheric scientists and meteorologists alike suggested that an increase in the CO_2 level might cause a general warming in the Earth's climate. Recent studies have indicated that CO_2 climate change does not yet pose a serious threat to the balance of atmospheric processes. However, worries about the effect of CO_2 on the climate gave rise to further detailed studies and investigations to focus attention on the complex interactions

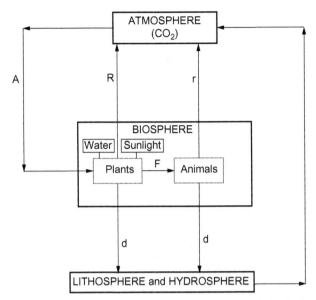

FIG. 6.5 Natural CO_2 cycle. *A*, Assimilation by photosynthesis; *d*, death; *O*, oxidation of dead organic matter; *r*, released by photosynthesis; *F*, feeding; *R*, respiration.

between human activities and the atmosphere, and thus may prevent still more serious problems from arising in the future.

Almost 70% of the Earth's surface is covered by bodies of water, which are referred to collectively as the hydrosphere. Although there is not extensive human activity in the hydrosphere itself, the intensive activities on land threatens the biological richness of oceans, and especially, in the beginning, the coastal areas along which about 60% of the world's population lives.

Although legislative measures can be enacted, their application often cannot be made due to the lack of reliable data, planning, management, international coordination, technology transfer, and adequate funds. The hydrosphere is polluted by sewage, agricultural chemicals, litter, plastics, radioactive substances, fertilizers, oil spills, and hydrocarbons. Landborne pollution gets into the major hydrosphere through rivers and the atmosphere. Dumping directly into the oceans and seas originate from ships and ocean liners. The hydrosphere is vulnerable to climate and atmospheric changes, including ozone depletion. The general form of the hydrological cycle under the effects of climate change-triggering events such as demography, extravagant style of life, fossil fuels, and land use, is shown in Fig. 6.6, with the resultant effects of food resources reduction, water stresses and shortages, droughts, floods, and pollution.

Foreign materials that humans release into the atmosphere at least temporarily and locally change its composition. The most significant man-made atmospheric additions (carbon monoxide, sulfur oxides, hydrocarbons, and liquid and solid particles) are gases and aerosol particles that are toxic to animal and plant life when concentrated by local weather conditions such as inversion

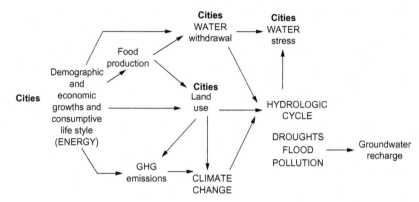

FIG. 6.6 Climate change, hydrological cycle, and environment (Oki et al., 2001).

layer development, orographic boundaries, and low-pressure areas. The principal pollution sources of toxic materials are automotive exhausts and sulfur-rich coal and petroleum burned for power generation and heating. Fortunately, most toxic pollutants are rather quickly removed from the atmosphere by natural weather processes, depending on the meteorological conditions.

Although pollutants may originate from natural or man-made activities, the term pollution is often restricted to considerations of air quality as modified by human actions, particularly when pollutants are emitted from industrial, urban, commercial, and nucleated areas at rates in excess of the natural dilution and self-purification processes prevailing in the lower atmosphere. Air pollution occurs as a local problem with three distinctive geographical factors. First of all, the wealth of human beings defines the distribution of housing, industry, commercial centers, and motor vehicle transportation among different centers. Such a system forms the major source of man-made pollution (see Fig. 6.6). The second agent that plays a significant role in the formation of air pollution is a natural phenomenon in the atmosphere, which controls the local and temporal climatic weather variations; hence, the pollutants introduced into the atmosphere are either scattered in various directions or carried away by air movements (winds). Finally, the interaction between pollution emissions and the atmosphere may well be modified by local relief factors, especially when pollution is trapped by relatively stagnant air within a valley.

6.4 CLIMATE BELT SHIFTS SIMPLE MODEL

It is possible to think about a simple model for the appreciation of climate change effect on the troposphere (Şen et al., 2010). The troposphere thickness varies from about 7 to 8 km at the polar regions and about 16 to 18 km at the equator. The thickness varies also according to season, being thinner in winter when the air is densest, and the seasonal effect is strongest at the mid-latitudes, where it varies around 11 km. This is the average thickness that corresponds to 15°C of global warming (Trenberth, 1992).

The simple question is, "What would be the tropospheric thickness increase if the global temperature increases by Δt °C?" This is tantamount to asking, "What will be the tropospheric thickness if the global temperature becomes $(15+\Delta t)$ °C?" Logically, the model is very simple because there is a directly proportional relationship between the temperature, T, and the tropospheric thickness, m. The question of this simple model is, "What will be the tropospheric thickness, m_x, at $(15+\Delta t)$ °C, if at 15°C the tropospheric thickness is m?" The solution appears in the form of the following expression:

$$m_x = \frac{T+\Delta t}{T} m = \left(1 + \frac{\Delta t}{T}\right) m \tag{6.1}$$

In the derivation of this simple model, the Earth is considered as having a smooth surface without any continental and oceanographic features.

Example 6.1 What Is the Tropospheric Thickness on the Equator at 16°C?

It is known that, on average, the tropospheric thickness at the equator is about $m=17$ km. Increase of the average global temperature to 16°C means a $\Delta t=1$°C increment in global warming; hence, according to Eq. (6.1) the tropospheric average thickness becomes

$$m_x = \left(1 + \frac{1}{15}\right) \times 17 = 18.13 \, \text{km}$$

Hence, the thickness increase in the troposphere at the equator is $18.13 - 17.0 = 1.13$ km $= 1130$ m.

Example 6.2 What Is the Tropospheric Thickness at the Polar Region at 16°C?

The solution is similar to the previous example. The only difference is that the tropospheric average thickness at this region can be taken as 7.5 km. The substitution of this value into the previous expression yields

$$m_x = \left(1 + \frac{1}{15}\right) \times 7.5 = 8 \, \text{km}$$

This is equivalent to saying that the tropospheric thickness increase at the polar regions is about $8.0 - 7.5 = 0.5$ km $= 500$ m.

The above calculations are achieved by consideration of vertical air expansion; however, there is also horizontal expansion, in fact, expansion in any direction. As far as the climate belts are concerned, horizontal air expansions have more significance than their vertical counterparts, because they cause climate belt shifts.

Let us consider the world as a sphere with its longest perimeter equal to 40,000 km. If only 30 degree partition of the latitudes is considered, then there are 12 portions, each with a length equal to 40,000/12 = 3333.33 km.

Example 6.3 What Is the 30 Degree Latitudinal Expansion Around the Equator at 16 °C?

The same simple logical model in Eq. (6.1) is valid, but this time as $T=15°C$, the thickness, $m=3333.33$ km, and if the average global warming is $\Delta t=1°C$, then what will be the horizontal expansion?

The substitution of the relevant quantities into Eq. (6.1) yields the horizontal expansion around the equator, say, in the Northern Hemisphere, as follows:

$$m_x = \left(1 + \frac{1}{15}\right) \times 3333.33 = 3555.55 \text{ km}$$

The horizontal air expansion can be calculated as the difference before and after the global warming, which yields $3555.55 - 3333.33 = 222.222$ km. This implies that the tropical belt around the equator will advance toward the polar regions by approximately 225 km.

All of these calculations are shown in Fig. 6.7, collectively, so that one can appreciate the approximate effect of the global warming.

Such horizontal expansions will affect especially the climate pattern of subtropical areas, which are bound to experience drier spells with extensive drought periods. As will be explained later in this chapter, especially in summer seasons, the AP desert and desertification conditions will shift toward the northern pole.

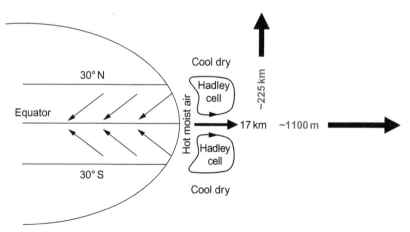

FIG. 6.7 Shifts of dry climate belts toward polar regions.

The most significant difference between the present impact of climate change and the impact of geological ones in the past is the anthropogenic effects and extra emissions into the atmosphere, which cause change in the chemical constituents. In other words today, apart from natural climate change, there is artificial climate change. Soon, the similarity principle of future events to past records will no longer be effective for water resources planning and management based on past statistical behaviors of the hydrometeorological records. However, today it is not more scientific to look for such planning procedures without taking into consideration the expected climate change impact role. Any drought parameters that have been effective in the past are no longer valid for future planning but depending on the location, either an increase or decrease in the parameters must be included in prediction and management calculations.

Among the types of droughts are, first, continuous dryness (deserts, arid, and semiarid regions); the second drought type includes astronomical effects; the third one includes regular droughts (natural effects or climate change); and finally, the forth one is the most random and it has irregular drought occurrences.

The IPCC (2007) fourth Assessment Report, second Working Group scrutinized thousands of pages of scientific and gray literature through a committee of experts and suggested climate change impacts, adaptation, and vulnerability strategies for freshwater resources (Kundzewitz et al., 2007). The effects of climate change on the hydrological processes are examined and future rules are suggested for better management of water resources (Rosenzweig et al., 2007). These variations may be regional and seasonal. Fig. 6.8 indicates the position of water-dependent regions by taking into consideration the climate change effects. This figure is prepared based on the results of a set of scenarios as their ensemble averages (Nohara et al., 2006).

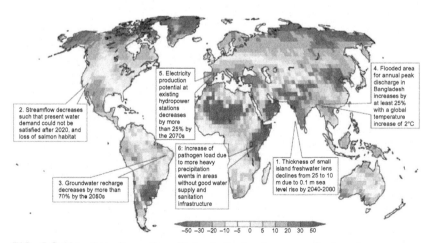

FIG. 6.8 Map that shows the effect of climate change on fresh water resources (Kundzewitz et al., 2007).

According to this map, the runoff will be improving in some watersheds, but the opposite will occur in others. Hence, in some regions there are beneficial impacts of the climate change, whereas in other regions drawbacks are expected. Toward the middle of the 21st century, in some tropical and subtropical regions, in dry areas there should be 10–30% water reductions, whereas in water-richer higher latitudes toward the polar regions, the reduction should remain between 10% and 40% (Milly et al., 2005). This situation can be related to the climate shift effects as explained in Fig. 6.7.

In winter seasons, the shift of precipitation from snow to rainfall and as a result of warming in many continental and mountainous regions, the snow melt process is expected to take place earlier, giving rise to additional early runoffs. This situation also brings the case of drought period increments. In spring seasons, early snow melts will bring the runoff periods closer to the winter season; consequently, summer seasons are bound to experience longer and intensified droughts. On the other hand, as a result of global warming the retreat of glaciers will increase the surface runoff in short durations but, with a retreat of the glaciers in the long run, the runoff volumes are expected to decrease. Hence, in summer and autumn seasons, water volumes are bound to decrease. Almost one-sixth of the world's population (about 1 billion people) live in places where the water resources are dependent on the mountain ice caps and snow melts and, therefore, due to the shifts of water availability by time, there may appear mismanagement problems in these areas. For instance, such effects are expected for the Euphrates and Tigris Rivers in the Middle East and their peak discharges as well as water amounts may show decreases (Şen et al., 2010).

The frequency and intensity of drought and flood occurrences are expected to change due to the climate impact on water resources. After intensive rainfall events, water volume augmentations are expected. Especially, the frequency and intensity of rainfalls in the equatorial and polar regional areas may increase and, in contrast in other regions just the opposite situations are expected to take place. In general, toward the end of the 21st century, the areal extent of droughts are expected to increase and the percentage of the areas that will be exposed to more extensive temperatures will also increase at least a few-fold. In some of the catchment areas, subsequent drought and flood intensities and frequencies may possibly increase.

6.5 ADAPTING TO CLIMATE CHANGE

In the last decade or so, water managers started to think about and take into consideration the impact of climate change on their water resources systems' management and adaptation. It is advised that preliminary adaptation studies should first begin with the existing water infrastructural capabilities; hence, one can identify the missing components for efficient climate change impacts, the necessary adaptation, and mitigation work. Although scientific methodological improvements and their applications are necessary, they cannot alone be sufficient for a successful solution. In addition to a numerical database, it is necessary to take into account the local features and even available verbal knowledge and information. Of course, a logical rule base is the most essential part for an effective solution. Even though there may be numerical climate change predictions, it is not logical to base the future climate change impact solutions for water resources systems on these scenarios only. The scenario projections must not be for more than 5 or 10 years; however, in many scientific studies the projection is taken up to 2100. But the greater the projection time duration, the less the reliability of the model predictions. It is advised to deal with a range of scenarios to identify the worst and best cases, with alternative suggestions in between. For average behavior description, one can take an ensemble average of the scenario results. These alternatives help to achieve efficient climate change adaptation work, which may fall into several of the following main subject areas.

1. *Data collection:* Most often this implies numerical data monitoring and records, but in this book, additionally verbal (linguistic) data are also recommended that can be collected from experts in the drought subject area or from experienced local people. Verbal knowledge, information, and numerical data help to make efficient decisions in model calibration, adaptation, verification, and prediction of the future. The data must not be confined to meteorological and climatological records only, but water supply and demand patterns also must be taken into consideration.

2. *Data treatment:* The collected data must be scrutinized visually and graphically with comparison to simple but effectively known similar studies or methods so as to appreciate data structure related to uncertainty ingredients, possible trend features, and extreme values (low and high values). It must not be forgotten that, especially in climate change studies, the past record features are not bound to happen in a stationary manner in the future. Hence, the recent past may not be a reliable guide to the hydrological resource base of the near future.

3. *Methodology:* There are many types of analytical tools and models for digestion of existing knowledge and, especially, numerical data and models are extensively available in the literature. In the predictions, especially for climate change studies, scenario and risk analyses play significant roles and they must be used for an efficient water management study.

4. *Decisions:* The final decisions must take into consideration scenario alternatives with their risk assessments for reaching a rational and reliable decision about the future assessments and their consequences. It is recommended than uncertainty techniques be used in the decision-making processes. Although there are sophisticated methodologies, for preliminary decisions and their guidance simple and effective decision rules may be employed.

5. *Data management:* After all the previous steps, it is necessary to use this information in an efficient management program, so as to provide rational and sustainable water resources distribution. For example, a new dam or reservoir may

be built or the demand management might be brought into the picture and reliable model results and recommendations may be distributed among the end users so as to conserve water and to diminish water consumption. Even though proper decisions and management studies may be affected in the area, the scientific research must be cared for continuously to control the present water resources infrastructure management and the climate change impacts.

After the completion of these points, it is possible to assemble an efficient and optimum management system for water resources systems in the area, including possible drought mitigation studies. Goals must be met in interactive cooperation among the hydrological science experts, the local managers, and water resources system operators. Even without climate change impact, the water managers are confronted with uncertain meteorological and hydrological consequences; accordingly, they are forced to adapt to the present situation and in the case of climate change, their usual reactions to similar changes may need more care and refinement.

Much of technological development may affect the environment in a way that causes various human health problems. Accordingly, measures should be taken for reducing health risks from environmental degradations including atmospheric pollution. Governments should build basic health infrastructures, paying attention to the provision of safe water and food supplies. Especially in the areas of water supply and quality, necessary monitoring work must be planned in a systematic manner.

6.6 DROUGHT DISASTERS

Unprecedented scientific and technological developments during the last century have brought new problems concerning the survival of living organisms on Earth. This century is expected to support and sustain living organisms, especially including the human race with its ever-increasing population, improved health services, and increased wealth, all of which can contribute to climate change. Recently, the public is becoming more aware of especially drought and flood natural hazards that impact many human societies in different parts of the world. The parlance of "global change" is advocated in every aspect of life, equally, in the social, economic, political, and many other domains, but physical global changes such as changes in the climate and the modern way of life bring extra burdens to the human population. Among the natural hazards, perhaps those related to water surplus or deficit (drought) are more risky to human societies than any of the geological hazards such as earthquakes, volcanic eruptions, and landslides. The forces of nature become more threatening and bring the human population under more severe risk dangers as the world population grows with more materialistic ambitions, extravagant wastages, and as the existing and expanding environments try to accommodate with their limited sources these societies. Especially, land and more precious water sources are under heavy burdens due to demographic and social trends that show continuous increase with time.

On the other hand, modern times have brought another sort of environmental risk to human lives and property due to "man-made" threats. Among these are industrial explosions, transportation accidents, and many other technological threats. Last but not the least, natural hazards are swiftly communicated to the whole world via high-speed media, especially on television screens. Presently, everywhere in the world natural hazards may appear unexpectedly; consequently, the following questions are of interest to scientists, administrators, and citizens. By considering the increase in natural hazards in recent decades, the following questions are among the most basic ones to ask:

1. Is the world as a whole or in many regions becoming more dangerous than before?
2. Is there an increase in the risk of a human society being more vulnerable to certain processes such as droughts?
3. Are these events genuinely natural or are there anthropogenic triggers (climate change) for the initiation of these so-called natural drought events?
4. Do innovative technologies increase or decrease the appearance, magnitude, and frequency of natural hazards?
5. Is it possible to eliminate completely the hazardous effects of the natural extreme (drought and flood) phenomena?
6. As it is not possible to avoid completely the appearances of natural hazards, is it possible to take necessary measures against them with certain risk expectations?
7. What are the best means of disaster mitigation ("see chapter: Drought Hazard Mitigation and Risk")?
8. Although sizable investments are made to reduce the destructive effects of these hazards, why do losses continue to rise?
9. How can we manage the temporal and spatial availabilities of water resources to meet the most important survival needs of the human population?
10. Are there enough monitoring investments and, thereafter, models to control the possible destructive occurrences of natural hazards?
11. What are the missing pieces or defects that prevent creating the most successful models and management programs for combating the negative effects of natural events?
12. Are the natural events that are communicated by the media really natural or natural coupled with human inducements such as climate change?
13. Is it better to take the necessary precautions against the impact of natural hazards on existing settlements of human population or to change the settlement locations to relatively less hazard-prone locations?
14. Are the hazards to be tamed by engineering and administrative calculations and regulations only or should the larger society become more involved?

The sequence of similar questions can be lengthened unendingly, but it is important to identify the most significant ones and, accordingly, try to reduce the hazards. Unfortunately, clear answers to these questions remain rather fuzzy and elusive, perhaps at least partially because broad-based natural hazard research and investigation did not begin until almost the middle of the 20th century.

Climate change is among the triggering agents of unusual droughts as natural hazards. It causes changes in timing, regional patterns, and intensity of precipitation events, and in particular in the number of the days with low and insufficient precipitation occurrences. This is mainly due to global climate change. Recent droughts seem to have some effects of global climate change, although they cannot be taken as proof that it is already taking place in other parts. Hence, global climate change should be taken into consideration in any drought and even flood study.

Regardless of whatever verbal information or numerical data is available, there remains uncertainty about rainfall and related aspects' future behaviors such as droughts. There have been tremendous advancements in prediction modeling during the last four to five decades, but their improvement rates decrease by time and recent improvements are marginal, although significant additions and improvements help to refine prediction reliability. The primary indicator of climate change is temperature records, and any increase in their values implies impact on evaporation and runoff amounts; especially, on extreme events. Climate change research and applicable studies are instructive as to the possible magnitude of uncertainty surrounding the hydrological implications.

Response of a region to climate impact on water resources and, especially, on any drought duration, is rather enormous as regards its consequences, including meteorological, hydrological, social, ecological, environmental, and economic activities. Any warming in any part of the world with occasional or continuous snow cover may delay the start of snow melting in addition to the amount of snow melt areal extent and time-wise occurrences.

Extreme weather events such as droughts and floods shape the sustenance of daily life in a society. Floods are initially more conspicuous than droughts because they can occur over periods of time, whereas droughts happen with a more persistent weather pattern and conditions before they are recognizable by people. Climate is not a consequence of steady meteorological happenings but may have haphazard variations over time and area that may not be expected with certainty. Climate has both regional and global ingredients even though one may talk about the climate of a locality. Additionally, climate trends cover long periods of time at different scales. Unfortunately, droughts are neither periodic nor have temporal and spatial regularities. The geographic location and meteorological events in an area are the preliminary knowledge and information sources for better understanding possible climate change at the location. Regional weather patterns should be considered for better drought identification in an area or drainage basin; hence, nearby drainage basin drought features and weather patterns should also be considered for an effective climate change assessment against possible hazards.

6.7 DESERTIFICATION AND CLIMATE CHANGE

Desertification might be due to different effects but the most significant indicators are temporal temperature increase, precipitation decrease, and vegetation cover area reduction. Over the last 30 years, climate change has had a very significant impact on the whole world in general, and on the arid and semiarid regions in particular. Climate change impacts on desertification process are among the acceleration factors that must be reduced by convenient adaptation procedures. This section concentrates on the possible relationships among climate change impact factors on the desertification process through temperature and evaporation increase, rainfall decrease or increase, as well as air movements as a result of global warming.

One of the dramatic long-term impacts of droughts, combined with human activities, is the degeneration of productive ecosystems into desert during the process called desertification, which is not exclusively a consequence of drought, but can be accelerated by climate change. Additionally, wind action in dry years, soil erosion in drought and postdrought periods, and particularly human activities are responsible for poor management of land, soil, crops, and herds. The desertification areas reflect more of the solar radiation than the original land by causing changes in the thermal regime of the atmosphere, which may tend to extend or intensify droughts.

Recent global warming, greenhouse effect, climate change, environmental degradation, water resources depletion, and demand on more agricultural products put extra pressure on some regions, and it is necessary to measure the scale of desertification for future developments and to preserve at least the present situation and, even better, to improve the local conditions with the purpose of combating against the droughts and desertification. Although there are different studies concerning social, economic, and water resources evaluation in some regions, unfortunately an extensive desertification study is not available with climate change impacts.

The process of desertification is a slowly creeping phenomenon similar to droughts, which may take place in any area during long time durations for different reasons. The first indications are due to variations in weather or meteorological parameters. In general, desertification implies decrease in some meteorological and agricultural quantities such as the rainfall amounts, vegetation coverage, surface water extensions, groundwater level drops, and crop yields (Şen, 2008b). On the other hand, increases in the temperature, sand coverage, areal drought coverage, and urban area expansion and sedimentation support desertification. In simple terms the historical records of these variables, either as time series measurements through local ground surface instruments or satellite images, help to identify increasing or decreasing trends. Any desertification study based on satellite images without the evaluation of available of necessary ground data is incomplete and cannot serve a purpose.

Desertification is associated with biodiversity loss and contributes to global climate change through loss of carbon sequestration capacity and an increase in land surface albedo. Additionally, global warming (greenhouse effect, climate change) increases evapotranspiration (ET), thus adversely affecting biodiversity; changes in community structure and diversity are also expected because different species might react differently to the elevated CO_2 concentrations. Climate change effects on desertification are complex and are neither understood nor examined sufficiently for the identification of main causes and their collective effects on the desertification process. Climate change may increase aridity and desertification; hence, many areas have great uncertainty in the prediction of consequences (Şen, 2008b, 2013a).

Different studies have shown conflicting results regarding climate change impact consequences. For instance, according to Tao et al. (2003) water shortages are expected to worsen in the AP. On the other hand, Dai et al. (2001) states large precipitation increases up to 50% are to be expected over the AP. As stated by Kotwicki and Al-Sulaimani (2009), hydrological considerations indicate that increased rain intensity combined with a reduction of annual rainfall is likely to reduce vegetation cover and increase surface runoff, leading to increased desertification. The resulting soil erosion, salinization, and loss of vegetation may further increase surface runoff. Agricultural fields are bound to become more saline from increased ET and saline intrusion into coastal aquifers.

Although one may think that desertification is a natural phenomenon only, unsustainable exploitation of the available local scarce natural sources by human beings is another significant factor in such a process. In general, there are three factors that may accelerate desertification in any region.
1. Local environmental situations that may prepare the basic features.
2. Human activities that may damage the natural balance in the area.
3. Recently, exacerbation of the climate change impact.
The first two factors are well known and climate change may trigger these two factors in addition to its genuine effect on vegetation, water resources, humidity, temperature, rainfall occurrences, and groundwater recharge. Water scarcity is one of the dominant factors in the desertification process. The intensive use of available natural resources might be more extensive under the impact of climate change; hence, arid and especially semiarid regions might come under greater threat than humid regions. Among the desertification implications are dust storms, flooding, and especially flash floods, regional, and global climate change, which may cause human migration of economic refugees because of poverty and political instability. In order to avoid these activities in a region with its socioeconomic effects in other regions, it is necessary to combat desertification, and hence, to provide local and global benefits under human-induced global climate change conditions.

One of the major logical conclusions is that climate change is linked to desertification and its impacts depend on regional climate change, future temperature, and rainfall trends, which can be predicted according to various scenarios as will be explained in Section 6.8. Additionally, the local hydrological cycle is also expected to be impacted by climate change (Section 6.3.1). Among the hydrological phenomena, extreme events might further intensify leading to unexpected drought and flood occurrences. As a result of these factors, fertile soil erosion may take place with desertification that affects global climate change through soil and vegetation losses. As desertification is a rather local phenomenon, the local manifestations of global climate change must be examined for future planning and management decisions.

To reduce expected desertification rates, climate change adaptation strategies and, accordingly, management of irrigation practices are among the strategic measures employed. Otherwise, overgrazing and extensive irrigation may be exacerbated by the climate change impacts and the ecosystem in the area may incur additional unrecoverable damages.

Understanding desertification is constrained by a set of uncertainties in meteorological, hydrological, geological, environmental, irrigational, and climate change impacts. To reduce such uncertainties, it is necessary to attach great significance to long-term remote sensing prospecting, hydrometeorological data, socioeconomic data, and local linguistic information from the settlers and refined future prediction methodologies with the consideration of climate change impacts under various scenarios (Section 6.8).

6.7.1 General Approach to Desertification Problem

The following points are among the initial phases for the confirmation of even the initiation or existence of any desertification signal in any part of the world (Şen, 2013a):

1. The temperature records (monthly, seasonal or annual) should be processed for any significantly increasing trend in the area regionally.
2. The rainfall records should be examined for any obvious or hidden decreasing trends and cyclic phases.
3. The wind direction and speed variations in the study area must be evaluated again with the possibility of discovering increasing or decreasing trends.
4. The dry and wet period durations can be identified from the monthly rainfall records with the possible identification of maximum drought duration. It is necessary to make a distinction in an area between lengthy drought durations and desertification.
5. Regional (maps) and directional variations (change with distance) in rainfall, temperature, soil moisture, sedimentation, and other variables should be sought; variations from coastal areas toward the inlands or from the highlands toward different directions are especially important.
6. Identification of urbanization effects concerning water supply and groundwater level depletions coupled with the desertification effects.
7. Consideration of global warming, greenhouse effect, and climate change on the region.

8. Crop yield amounts from season (year) to season (year), vegetation coverage, and changes in agricultural product types.
9. Sand dune area boundaries (satellite) and depths at a set of irregular points (ground) should be measured and mapped.
10. Correlations between the aforementioned variables should be established for the general determination of desertification initiation.

After the completion of the aforementioned points and close to the final stages of the study, satellite imagery and GIS techniques with pattern recognition can be used for the final assessment and delimitation of the desertification effects with the index development and classification.

6.7.2 Drought and Climate Change

Although drought is thought of initially as a lack of rainfall, more than this it is a persistent moisture deficiency below the average that is expected to balance precipitation and ET in an area ("see chapters: Basic Drought Indicators; Temporal Drought Analysis and Modeling"). Even though the same meteorological and climatological conditions may prevail in different areas, this does not mean that their impacts as drought occurrence will be the same, because there are other events that may trigger drought appearance such as soil-moisture content, water demand, and so forth.

Extreme events such as droughts and floods are dependent on climate change and variability. Climate change will be effective in some areas to increase the droughts, but in others the flood events will increase. All these facts are related to water resources. For instance, as drought demand on groundwater resources increases, drops in groundwater levels start to appear. On the other hand, floods assist groundwater recharges. It is the duty of administrators, planners, and decision makers to satisfy long-term sustainability by balancing these two events (groundwater abstraction and recharge rates) by taking into account the climate change impacts. In the meantime, possible groundwater pollution and quality variations must also be considered.

Those who are acquainted with climate science and hydrology must investigate past drought and flood events in a qualitative and quantitative manner for better predictions and mitigation. Although there is greater understanding of temporal and small local drought effects and predictions ("see this chapter and chapters: Temporal Drought Analysis and Modeling; Regional Drought Analysis and Modeling; Spatiotemporal Drought Analysis and Modeling"), in addition to this knowledge it is important to consider international research effort and their integration with local studies. In this manner, one can achieve better drought prediction, early warning possibilities, and mitigation preparations.

Prior to the appearance of drought, it is necessary to become acquainted with related weather patterns. It is well known that the subsequent drought events have some similarities to meteorological conditions that follow each other. Both droughts and floods appear apart from the normal weather patterns. In fact, the climate does not appear as a result of continuous weather patterns, but instead as a result of rather sophisticated global scale variations ("see chapter: Introduction"). The main reasons for drought and flood occurrences are the short-term climate variation in local, regional, and global scales and the effect of long-term temporal (tens of years) climate change impact. Before studying the past and future behaviors of these events, it is necessary to collect knowledge and information about the geographic location of the area, surface features (geomorphology) and, especially, about the meteorological conditions. This information must be confined to the drainage basin with adjacent meteorological conditions.

6.7.3 Water Resources, Climate Change, and Drought

Depending on the available information, present water resources must be operated accordingly on a regional scale through convenient manipulations for the common satisfaction of end users, administrators, and politicians. As stated by Redmond (2002), human decisions can affect water supply and demand, so that drought status can represent a highly nonlinear and even occasionally counterintuitive or paradoxical response to the climate drivers and their history.

One of the most significant social factors in any drought is the number of people that will need water supply for different purposes such as household, agriculture, and industrial uses. Water demand features including future planning, operation, and maintenance of water resources systems in an area requires reliable drought duration, magnitude, and intensity predictions by considering the possible climate change impacts. Without such planning and prediction studies the local people will be under the threat of possible water shortages during dry periods and there may also be appearances of haphazard water demands, which may affect the present water resources distribution system. The population growth rate and its increase are formidable challenges when coupled with climate change impacts on water resources systems. With the beginning of the 21st century more than half of the world's population started to live in cities, where extravagant water consumption takes place. Half of the world's 6 billion people now live in urban environments (projected to increase to 60% by 2030), and the majority of the globe's megacities (10 million or more residents) are in regions confronting mild to severe water stress, according to the United Nations (2003). Between 1950 and 2000, the world's population doubled (United Nations, 2002), and water demand roughly tripled (Postel and Vickers, 2004). From 2000 to 2050, global population is projected to grow 45%, reaching nearly 9 billion people (United Nations, 2002). Clearly, the world's water demands are increasing, but nature's present and future water budget remains largely fixed at the limits of its primordial creation.

Climate change cannot be appreciated during short time periods, but needs long-term climate records, preferably of 30-year duration or for preliminary studies at least for the last decade. Trend features in the temperature and rainfall records are preliminary indicators for droughts that may be experienced in short-term durations or expected to occur in the long-range future. To collect preliminary scientific information about future possible undesirable occurrences such as droughts, it is necessary to consider models as explained in the previous chapters to have reliable prediction capability on an interannual, seasonal, or monthly bases.

It is now very well appreciated by many researchers and administrators that the present anthropogenic climate change affects the hydrological cycle's natural behavior in any region and, consequently, the duration, severity, intensity, and frequency of rainfall occurrences take place more often than usual. The most important impact of climate change on drought features may be to cause accelerated ecological, social, and economic problems as a result of water supply shortages.

To understand about drought possibility in any region, it is first necessary to survey not only the weather and climate records but also the social and, especially, economic developments and their impacts on land use, vegetation cover, and so on. First the records must be examined for possible verbal relationships, such as for the effects of economic and social developments. It is then necessary to make numerical predictions of future temperature and rainfall and, subsequently, identify the various drought features as explained in chapters: Temporal Drought Analysis and Modeling; Regional Drought Analysis and Modeling; Spatiotemporal Drought Analysis and Modeling. In the prediction studies, the stationary property (statistical reflection of the past in the future) must not be employed any more, but instead trend features depending on different scenarios must be considered. After all these activities the conclusions must be presented as executive summaries for implementation by the central or local administrators, if the drought predictions are ever be useful for the society at large. The main point at this stage is to use the conclusive knowledge and practical information for drought preparedness, and subsequent combat and mitigation work.

6.8 CLIMATE MODELS

In the last three decades, the related processes of land, ocean, and atmosphere, including global circulations, have been improved extensively; therefore, it is possible to make future predictions not only with local modeling approaches, but also by their coupling with general circulation model (GCM) and regional climate model (RegCM). Such an integrated model success rate is dependent on the numerical data and verbal information in terms of rule bases (Şen, 2010), dense measurement stations, and input of all this information into software with high-speed computation techniques, which provide extensive scenarios, alternatives, and their combined results concerning any natural event occurrence, extension, and effects such as droughts. Each of these components is bound to be developed further; therefore, one can expect better refined modeling and prediction results in the future.

The scientific developments for drought prediction models are on the way, but their conclusive results, applications, and implementations must be followed by responsible local and central authorities. Otherwise, efforts of modeling and consequent predictions may be left on the shelf without any benefit except for academic fantasies. Computer models based on the relevant input knowledge, information, and numerical data help to simulate the coupled systems. Among the replications one can derive the necessary information, say, for droughts in some way either after individual inspection of prediction alternatives or their combined ensemble features may lead to certain numerical conclusions under accepted confidence levels such as 5% or 10%. At this stage the adaptation, calibration, and verification of any prediction model can be adjusted to local and present circumstances by experienced operators, who gain practical knowledge and information for the phenomenon concerned. This is a reflection of operator predictive skill. It is advisable not to depend only on a single computer model, but on several available alternatives, and the output of each model provides a member from the relevant ensemble, and consequently, the combination of all these alternatives in a simple risk assessment yields the most reliable prediction. Any modeling prediction results have, in general, the tendency toward refinement of spatial and temporal resolution by downscaling procedures (Section 6.8.1).

Modeling is a process that subsumes the characteristic of every information source about the past performance of an event for the purpose of future behavior predictions. This is a sort of road map for future planning, project, operation, and maintenance of socioeconomic and environmental systems. So far as the uncertainty (statistical) behaviors of any atmospheric phenomena is concerned, in the long run it is a composition of systematic periodic fluctuations due to astronomic effects superimposed by stochastic uncertainty, which also has stationary characteristics. Even though there are trends in the records, they are accounted for as the embedding of various environmental effects (climate change) such as land use, measurement instrument depreciation, urbanization, and similar man-made changes on the Earth's surface without any thought of atmospheric composition contribution. At this point, it is meaningful to ask the question whether atmospheric composition also embeds some components and, especially, long-term trends on local, regional, or global scales. If such an element shows itself as a long-term trend due to genuine atmospheric composition effect, then it is possible to conclude that the climate is changing on the long-term average (ie, climatologically).

Living creatures cannot survive without proper weather and climate circumstances, so any climate change is expected to penetrate different aspects of life such as social, economic, industrial, urban, law, and other activities. Short-term weather changes may impact the following systems:

1. Water resources
2. Hydroelectric energy
3. Environmental conditions
4. Social sustainability
5. Agriculture

It is necessary to provide means for the future effects of climate change on these aspects through logical and rational bases with the availability of global, regional, and local data and information, which are input into convenient mathematical models that take into consideration atmospheric dynamics with embedded uncertainty components and they provide a reliable basis for the prediction of future behaviors, but not as stationary processes. The most advanced climate models are coupled with atmospheric, oceanic, land surface, and sea ice models, known as coupled GCMs (Trenberth, 1992), which are based upon the laws of physics including mass, energy, and momentum conservation principles in addition to the state equation of gases. These equations appear as a set of simultaneous differential equations, which can be solved numerically on spatiotemporal grid nodes at various scales (ie, at different sets of resolutions). GCM simulates the whole Earth's climate for decades, even for centuries. The finer (coarser) the node spacing, the higher (lower) the resolution; and the longer (shorter) the computation time, the larger (smaller) the computer memory requirements for the solution of simultaneous differential equations. Although all efforts in the world at different GCM solution centers are directed toward refining solutions, computational restrictions allow presently about 250 km resolution in the horizontal directions and 1 km in the vertical direction with higher resolution at the atmospheric boundary layer near the land surface and coarser resolutions in the upper troposphere and stratosphere. Such coarse resolutions help to model continuous physical processes better than discontinuous processes, among which are precipitation and clouds. These and similar discontinuous meteorological variables cannot be resolved properly and necessitate some average descriptions based upon physical relationships with large-scale variables through parameterization. The projections of anthropogenic climate change in the future depend on assumptions about the future emissions of greenhouse gases, aerosols, and the proportion of emissions remaining in the atmosphere.

The solution of GCM at any research center for the prediction of future conditions necessitates its calibration and performance based on the baseline statistical behavior of the historical hydrometeorological and climate data at least over a past 30-year normal period. Hence, the performance of the GCM can be compared with the statistical baseline data. This comparison is typically performed in terms of a comparison of the observed mean values of the hydrometeorological and climate data state variables against their observed counterparts in time and space. After the proper validation of GCM outputs, this model is then used for the projection of future hydrometeorological and climate data, which correspond to specified external forcing scenarios, such as the changing greenhouse gas concentrations in the atmosphere. The differences between the projected hydrometeorological and climate data conditions and their historical counterparts provide a measure of the climate change due to the specified external forcing scenario. As stated by IPCC (1990), early climate change studies utilized atmospheric GCMs, which were coupled to a simplified model of oceans. By means of these GCMs, equilibrium climate studies were performed, where an equilibrium climate response to the doubling of the equivalent CO_2 concentration in the atmosphere was computed, and then these results were compared against the current climatic conditions. In other words the current atmospheric CO_2 content was doubled as a step forcing, and the GCM then simulated the atmospheric conditions as they approached a new climatic equilibrium. The comparisons among the mean conditions of the current climate and the mean conditions of the projected equilibrium climate then provide a quantification of the climate change under the specified CO_2 doubling scenario. These studies yielded global warming results in the range of 1.5–4.5°C (IPCC, 1990, 1992). However, such equilibrium climate studies do not provide any information on the climatic response of the hydrometeorological and climate system to the gradually forcing conditions in terms of the gradually changing CO_2 and aerosols (airborne particles) in the atmosphere due to human activities.

The GCM solutions for practical applications pose different questions and difficulties, among which are the following points:

1. The differential equations are derived under rather restrictive assumptions of uniformity, homogeneity, isotropy, and so on.
2. Geometrical configuration is considered as a very small cube with linear changes of the variables along the sides. Consideration of a very small size of cube in the derivations implies the assumptions of homogeneity, uniformity, and isotropy automatically. Such a combination of assumptions requires the use of the arithmetic average and linearity concepts at any stage.
3. In the derivations, no boundary or initial conditions are defined and the dynamic equations obtained in this manner are all very general and condition-free. The final shapes of the differential equations for the dynamic system can be applicable at any time and space, theoretically. However, practical applications bring to mind the identification of the boundary and initial conditions.
4. No chance element is incorporated in the derivations; therefore, they describe average behaviors of the system.
5. They are unsolvable analytically but their solutions are possible only through numerical (finite difference, element, or boundary element) techniques. Therefore, rather powerful computers and reliable solution algorithms are necessary. Furthermore, in any numerical solution technique stability condition is necessary with a definite amount of error acceptance such as 5% or 10%.
6. Although the equations are expected mechanistically to satisfy the ideal conditions under the light of a set of simplifying assumptions, their practical solutions cannot be achieved easily. This is due to the fact that even though a natural phenomenon such as the climate and some meteorological events are continuous in the atmosphere, their measurements are possible only at a set of irregularly scattered station locations. It is unfortunate that these locations do not coincide with the nodal points of the GCM finite difference solution mesh. Prior to attempting to solve the dynamic equations, it is necessary to transfer record values measured at ground station sites to already decided mesh nodes. Such a process is referred to as upscaling procedure. As

the solutions of dynamic systems by numerical techniques provide chaotic behavior due to the initial values (Lorenz, 1963), what if different researchers employ different transfer methodology for the calculation of nodal values? Because differences in the techniques used will lead to different values at the nodal points, which will trigger the numerical solution of the dynamic system equations, then each researcher is expected to have rather different solutions as a result of chaotic effects.

7. Another practical problem is after the solution of the dynamic system numerically, the predictions or estimations are available at the nodal points. Hence, this time their transference to measurement or any desired points is necessary. Such a transference process is referred to as downscaling procedure. This will introduce another source of error into the procedure.

Various analyses related to the observations have proven that rapid climate changes are possible and they might have occurred in the past (Appenzeller et al., 1998). Observations, paleoclimate data, and models show that such changes exist and they also might have happened in the past, too. The occurrence of such transitions should be taken into consideration in a changing climate. Model simulations show that similar changes are expected more frequently in this century if greenhouse gas emissions continue to increase. According to IPCC (2001a,b, 2007) reports, there is a growing change in the global climate due to anthropogenic effects, which cause changes in the chemical composition of the atmosphere by greenhouse gas emissions. It is, therefore, necessary to improve understanding of the global climate system to assess necessary possible impact of a climate change on meteorological processes.

Determination of significant changes in the observed climate and referring those to the specific reasons are reviewed in the previous assessments of IPCC (2001a,b, 2007). Determining the climate change and making the adaptation works to the climate change are a matter of adaptation process. It is much more complicated to determine the changes that have been observed in the systems, which are highly intense and complex, especially in subtropical areas.

6.8.1 Downscaling Methodology

All regional studies for future climate change use outputs from GCMs of the atmosphere coupled with oceanic effects. In these models the physical laws and empirical relationships that describe atmospheric and oceanic systems are represented by mathematical equations. It is possible to assess greenhouse gases effects on the climate change system by changing input variables to GCMs. No climate change model is without a number of limitations. For instance, the models cannot adequately depict different geographical and spatial variability due to insensitive atmospheric, ocean, and land surface inferences. GCM outputs are available on coarse resolutions; they represent neither the local climate variability nor the variation properties, because natural variations are much greater than in finer resolution. Furthermore, land-use changes such as deforestation and desertification are among the major local climate affected factors, which are not taken into account in GCM, but they affect substantially local climates. Such local effects are embedded in the local meteorological records of temperature, precipitation, and so on. Despite these limitations, researchers depend rather confidently on GCM outputs in their local downscaling work for local predictions. Such confidences may cause rather questionable downscaling predictions on small local and regional scales. One may tend to suspect the representativeness of current climate model results on these scales, and consequent projections can vary widely.

Usage of computer models became a certain scientific requirement for relating past behavior to future behavior. Water resources are exposed to the effects of climate change, and the structure of the proposed model in this section is given in short steps as follows (Şen et al., 2010):

1. *Global circulation model (GCM):* Atmosphere circulation model, whose analytic terms are given by different researchers (Trenberth, 1992). It constitutes a dynamic base of the project work.
2. *Scenarios:* These are global climate change scenarios that take into consideration social and cultural features in addition to economy, technology, energy variety, demography, agriculture, and environmental issues on local and global scales (Special Report on Emissions Scenarios, SRES, IPCC, 2007).
3. *Downscaling model:* Global data are gained via high-capacity computers and high-technology and they are available at seven different centers (the United States, Canada, the United Kingdom, Germany, Italy, Japan, and Australia). These centers run global modeling based on the previous two steps, and the outputs are referred to as the GCM results. The relevant outputs from GCM for Turkey are then treated with the local statistical downscaling model (SDM) by considering point cumulative semivariogram principles for spatial resolution as developed by Şen (2009a), and the white Markov process (Şen, 1974) is used for refining the serial dependence function of the scenario data so as to match them with the historical data structure.
4. *Field models:* Spatial transfer models have been used to make predictions, which are valid for any point in the study area. These models have been generated with the use of spatial dependence function (SDF) developed by Şen (1989, 2009b).

The modeling method used here basically consists of GCM integration with a local (national) model, which includes downscaling procedure. With the GCM and SRES scenarios, it is possible to obtain the global values at a set of nodes for the meteorological variables such as temperature, rainfall, relative humidity, evaporation, wind speed, and sunlight. So, it is possible to downscale the global model results verbally and numerically as in the following modeling structure shown in Fig. 6.9 (Şen et al., 2010).

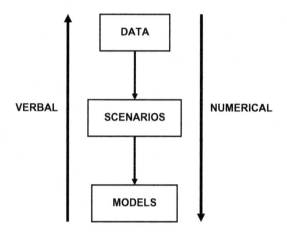

Data: Historic data, regional and local
Scenario: Verbal assumptions for now, GCM and SRES

Model: Numeric forecasting of the future, with economic and social extractions, regional downscaling and GCM

FIG. 6.9 Model developing structure.

These models give information about the average behavior of the activity with the simplifier and idealizer acceptances; however, natural activities are not such kind of values. They are below or above the mentioned values and may show specific cases that deviate unexpectedly from the averages. In addition, one can say that the acceptances, such as the activities that are being reviewed during the derivation of basic equations, are homogeneous and uniform, and they may be valid only in small scales. This assumption may not be valid for the long time periods and large space scales. In practice, it is possible to say that simple but effective methods should be used, which will facilitate the forecasting of results in a short and economically beneficial time duration. An example of local model development and its joint operation with the GCM is presented in Fig. 6.10, leading to various local outputs (Şen et al., 2010).

After serious feasibility studies, new infrastructural activities may prove demanding as convenient responses to climate-induced effects in hydrological cycle management and water demand regulations. Such approaches may be beneficial and effective provided that the necessary drought predictions are completed and future directions are planned in a proper manner. By all means for better predictions, the possible uncertainties in the models must be confined into narrow limits through validation studies. In any modeling of possible climate change impact, the following points must be taken into consideration (Şen, 2008a):

1. The global circular model (GCM) outputs must be considered with regional climate features and models that are properly downscaled in an appropriate manner (Section 6.8).

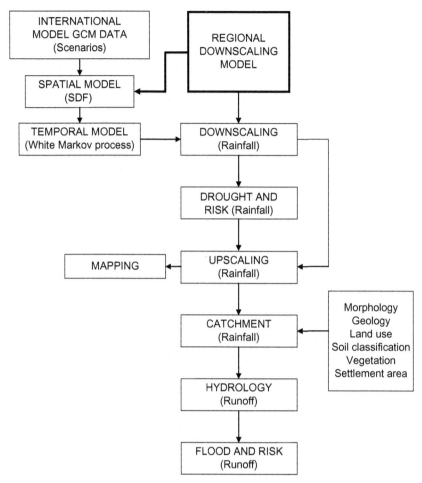

FIG. 6.10 National model compounds.

2. Hydrological features must be related to future drought prediction features and, accordingly, drought duration, magnitude, and intensity must be adjusted ("see chapters: Temporal Drought Analysis and Modeling; Regional Drought Analysis and Modeling; Spatiotemporal Drought Analysis and Modeling").
3. The ecosystem reaction and response to climate change must be examined both verbally according to local logical and rational knowledge and information, and also numerically according to model outputs.
4. Not only quantity but also quality of water resources must be taken into consideration in any climate change adaptation study.
5. Different scenarios with various CO_2 levels must be experienced after quantitative model runs and, accordingly, the most convenient adaptation strategy must be determined. It is also possible to consider quantitatively the possible risk, reliability, and safety of each water structure

individually and collectively so as to reduce the overall uncertainty in the system of water resources management and operation as well as preparation for future occurrences ("see chapter: Drought Hazard Mitigation and Risk").

In all these executions, first low-cost managerial or structural modifications must be given way, and if necessary, then high-cost solutions can be put into the climate change mitigation circuit.

6.8.2 Application for the AP

As mentioned in Section 6.7.3, temperature and rainfall records are the two major indicators for possible climate change coupled with the desertification assessments in any area. Herein, the Riyadh meteorology station in the AP with its surroundings is selected as the study area with total annual rainfall amount of about 130 mm. This station has longitude $24° 43' 00''$ E, and latitude $46° 44' 00''$ N at an elevation of 660 m above mean sea level. Fig. 6.11 indicates GCM nodal points together with the Riyadh meteorology station location.

Monthly rainfall records are available for 36 years from 1961 to 1996 (Fig. 6.12). Although more recent data could not be obtained, according to Arnell (2004) at least 30-year data as historical baseline climate horizontal (1960–90) is enough for climate studies.

Dry and arid conditions are the general features of Riyadh climate, with extreme heat and minimal rainfall amounts year round. For vulnerability assessment due to climate change on some activities such as the agriculture and water resources, it is necessary to construct models for prediction of what the future

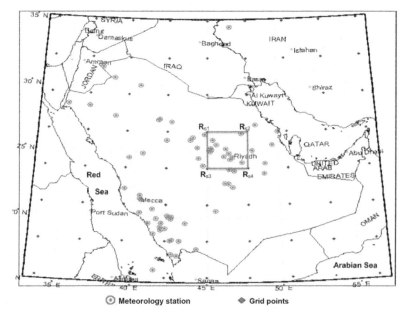

FIG. 6.11 Arabian Peninsula, KSA, and Riyadh City locations.

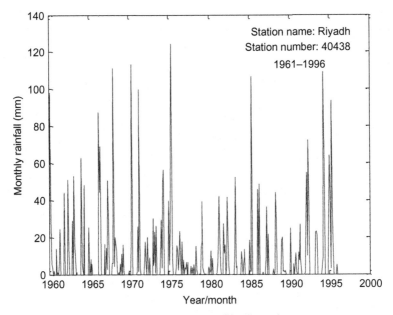

FIG. 6.12 Historical monthly rainfall amounts at Riyadh station.

climate would be under physically reasonable assumptions of greenhouse gas concentrations. The climate change–based scenarios coupled with the downscaling procedure lead to future projections that should have an acceptable comparative structure in accord with the historical baseline values (1960–90).

Annual average rainfall is about 15 mm and in Riyadh, the bulk of rainfall occurs in the months from January to May, averaging 100 mm/year. The climate at Riyadh also goes through a remarkably hot summer caused by the inland winds. During this season, temperatures can run as high as 50°C, with 45°C being the common temperature. In the winter months, daytime temperature averages are around 14°C. Nights on the AP are chilly and winter temperatures in Riyadh at night can plunge some points below freezing. Riyadh climate is characterized by average day temperatures of 8–21°C in January and 26–42°C in July (Al-Yamani and Şen, 1992).

6.8.2.1 QUADRANGLE DOWNSCALING METHODOLOGY

Although many climate change studies have been performed in various research centers worldwide, in the Kingdom of Saudi Arabia (KSA) such studies are still in their infancy, where few and limited investigations have been done. However, the Excellency Center for Climate Change Research, King Abdulaziz University, has developed the use of GCM and RegCM especially for climate change research in the AP and northern Africa (Almazroui, 2011a,b, 2012; Almazroui et al., 2012, 2014).

Many remarkable climate change features can be concluded from the GCM outputs by a simple quadrangle downscaling model (QDM) (Şen et al., 2012). In this section, the National Center for Atmospheric Research (NCAR) GCM outputs are used. They are available at a set of regular grid points (see Fig. 6.11) and it is necessary to bring them down to smaller and practically usable local scales through convenient downscaling methods. The grid interval to retrieve information is quite coarse as 5- by 5-angle quadrangles with about 250–300 km sides. The GCM scenario outputs provide global-level data for different climatic scenarios up to the year 2100. The main focus in this study is on A2 scenario from NCAR, United States, which suggests the future of the world as very heterogeneous, because the main emphasis is on family values with local traditions and cultural identity preservations. It implies a heterogeneous, market-led world, with less rapid economic growth, but more rapid population growth due to less convergence of fertility rates. Economic growth is regionally oriented; hence, both income growth and technological change are regionally diverse (IPCC, 2001a,b).

Although there are various existing models for climate change downscaling procedure from the GCM outputs to any local station position (such as Riyadh City location), herein QDM is proposed for monthly rainfall downscaling. An important point in any downscaling model is to consider a spatial estimation procedure, where there are different approaches as explained by Şen (2009b). For instance, in the well-known RegCM, spatial downscaling resolution is achieved by the Cressman (1959) radius of influence, which is a rather subjective approach because the definition of the radius is not determined objectively. It adapts geometrically without any physical or empirical base successive radius of influences as 5, 7.5, and 10 km around the prediction point, and all the sample (measurement) points that are within the radius affect the prediction value according to inverse distance square weighting. These subjective radiuses of influence are not derived from the available rainfall data at a set of meteorology stations. Besides, within such small radii there may not be any other meteorology station, which is the case in the AP, where there are sparse meteorology stations. However, in this section, SDF alleviates this drawback, because it helps to base the radius of influence on the available rainfall records around the prediction point. In any downscaling procedure, the spatial estimation method occupies the basic component, which should be obtained from a set of measurements (Şen, 1989, 2009a).

The SDF around Riyadh City shows the spatial dependence change of variable concerned (rainfall or temperature) with distance. Theoretically, if the distance between two stations is zero or close to zero, then the variable values at these stations are very similar (close) to each other; in fact the SDF value is equal to 1, whereas as the distance between the stations is very big, then the SDF value falls to zero. In between distances, SDF has values between 0 and 1. Practically, this leads to the empirical SDF for Riyadh City as shown in Fig. 6.13, where there are 12 SDFs, one for each month. This is a very typical pattern in arid regions, where the rainfall occurrences are rather scarce and haphazard.

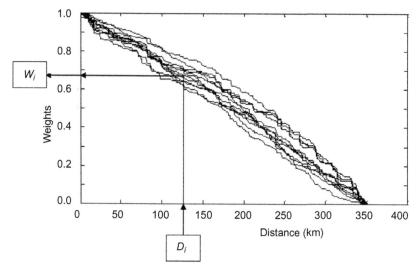

FIG. 6.13 Riyadh City SDF.

Fig. 6.13 indicates that the radius of influence around Riyadh City has the maximum effective distance up to 350 km. To calculate plausibly the rainfall estimation, R_E, at Riyadh location, the following expression is employed:

$$R_E = \frac{\sum_{i=1}^{4} R_{Si} f(D_i)}{\sum_{i=1}^{4} f(D_i)} = \frac{\sum_{i=1}^{4} R_{Si} w_i}{\sum_{i=1}^{4} w_i} \tag{6.2}$$

where $f(D)$ indicates the SDF weights, w_is on the vertical axis corresponding to distance, D_i on the horizontal axis as in Fig. 6.13.

It has been stated in the First National Climate Change Report of the Kingdom (2005) that on average, there is a general warming trend all over the country, but at different rates according to the topography and air mass movements on a synoptic scale. Fig. 6.14 indicates future minimum temperature predictions up to 2100 from NCAR-SRES-A2 scenario through QDM near Riyadh City.

This figure indicates the expectation of a patchy warming trend of about 4°C temperature increase during this century around Riyadh City. Especially after 2040, a steeper increase in temperature is predicted.

Finally, spatial estimations according to Eq. (7.2) for each month from 2000 to 2100 give rise to a series of scenario rainfalls in its unadjusted form to the actual monthly rainfall records at the stations. This unadjusted series exhibits future trends, wet and dry (drought) spells, relatively, but they must be adjusted with the available monthly rainfall amounts at the stations concerned. Such an adjustment is verified by considering rainfall baseline duration records from 1961 to 1996 and their discrepancy from the predictions, which led to $\pm 10\%$ relative error limits that are practically acceptable.

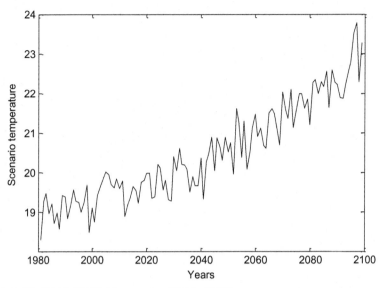

FIG. 6.14 NCAR-SRES-A2 scenarios (1980–2100).

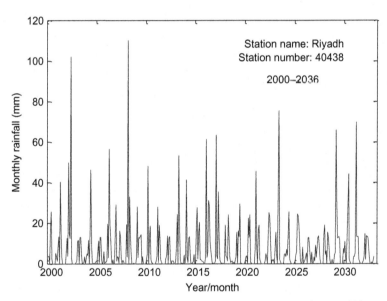

FIG. 6.15 Scenario monthly rainfall amounts at Riyadh City station (2000–36).

In order to depict the future features of the rainfall amounts and frequency at Riyadh City station, the scenario series are considered for the same duration as 36 years (432 months). Fig. 6.15 provides the monthly rainfall time series between 2000 and 2036. Its comparison with the baseline rainfall amounts in Fig. 6.12 indicates that the rainfall extremes (maxima) are expected to decrease.

In another Riyadh station scenario, monthly rainfall amounts between 2064 and 2100 are presented in Fig. 6.16, which has extremely high values compared with the two previous periods, 1961–96 and 2000–36 (Figs. 6.12 and 6.15).

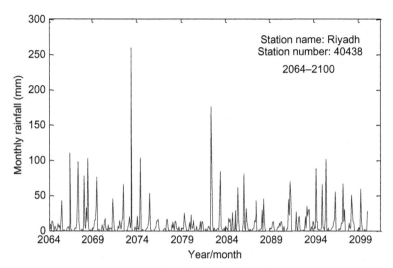

FIG. 6.16 Scenario monthly rainfall amounts at Riyadh station (2064–2100).

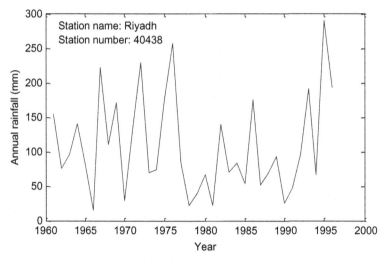

FIG. 6.17 Riyadh City observed annual rainfall amounts.

Monthly rainfall frequency and amount increases are compared with the 1961–96 and 2000–36 periods. Increase in the low-rainfall frequency is about 30%, whereas monthly rainfall amount increment is almost 67%.

On the basis of the previously mentioned explanations, it is possible to conclude that the monthly rainfall frequency and amount increments are expected to occur more severely in the second half of the 21st century even though there are more rainfall occurrences in the first half of the century than the historical baseline. To further elaborate on this point, observed annual and scenario monthly rainfall amounts are presented in Figs. 6.17 and 6.18, respectively. Projections indicate that there will be less intense rainfalls in the short run and more intense rainfall in the long run around Riyadh City.

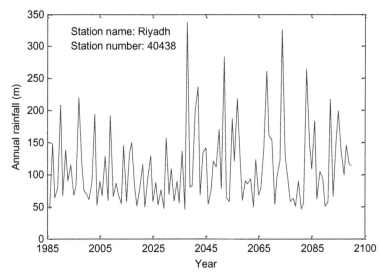

FIG. 6.18 Riyadh City scenario annual rainfall amounts.

During baseline period features in Fig. 6.17, it is obvious that there are two distinctive, wet (1966–76) and dry (1987–93), spells at Riyadh City location. The overall scenario annual rainfall amounts in Fig. 6.18 show explicitly that after around 2035, wetter spells are expected at the same station and its surrounding area. Comparison of Figs. 6.17 and 6.18 reveals the fact that in the future the number of low-rainfall amounts is expected to decrease.

The outcomes of this study are among the preliminary requirements for climate adaptation possibilities in the AP. Future predictions indicate drought and flood possibilities, but for arid regions more than droughts, flood and flash floods provide potential groundwater recharge possibilities. It is, therefore, possible to render the floodwater volumes during wet periods through water harvesting (WH) to recharge groundwater aquifers ("see chapter: Drought Hazard Mitigation and Risk") so as to use during later dry period occurrences. WH in central Saudi Arabia near Riyadh City is reported by the Prince Sultan Center for Water and Environment, which aims at the identification of potential WH areas by using GIS mapping as well as climate change models ("see chapter: Drought Hazard Mitigation and Risk"). Such maps can be used by decision makers for planning WH activity within the KSA, which are proven to be very effective.

6.9 CLIMATE CHANGE AND MAJOR CITIES

World population is in steady increase almost exponentially as if there is no limitation to natural resources for survival on the Earth. In the meantime, the population shifts are from rural areas to major cities, and currently half of the world's population lives in major cities (Vörösmarty, 2000). These centers consume about two-thirds of the world's energy and account for more than 70% of global CO_2 emissions. This brings extra burdens on the environment and socioeconomic structure of the cities and the generation of additional intensified greenhouse gas emissions. Extravagant lifestyle standards are observable generally everywhere, and they trigger consumption severely leading to additional chlorofluorocarbon emissions, which artificially affect climate causing an average temperature increase in the troposphere (Crutzen and Zimmermann, 1991). Even 1°C increase in the global temperature leads to about 250–300 km climate belt shifts from the equator toward the polar regions (Section 6.4).

Istanbul City, Turkey, and mid-latitude countries come into the effect of such shifts with hotter weather conditions on average in addition to an extraordinary increase in the frequency, occurrence, and amount of extreme events, droughts and floods, which are observable in many parts of the world. According to an IPCC (2007) report, until the end of the 21st century, global temperature increase is expected to reach at about 3.5–5°C, which implies additional increases in the frequency and amount of different weather events leading to extraordinary hydrological phenomena such as droughts, floods, water shortages, or stresses. It is, therefore, necessary to plan for the future through a reliable model for estimating possible scenarios of future meteorological events for the short-range of about 10–20 years, middle-range of 40–50 years, and long-range durations of 80–100 years. Such plans are necessary for adaptation and mitigation purposes (Kundzewitz et al., 2007).

To alleviate climate change impacts on major cities, three main activities must be taken into consideration. The first is the reduction of consumptive energy use by implementing energy efficiency improvements in buildings, outdoor lighting, and transportation systems. Among the most important resources are water impoundment reservoirs; hence, their exploration must not be considered according to supply levels only, but also jointly with climate change impact effects and water demand. In the meantime, clean energy usage facilities must be intensified within the city at district levels (Hill and Goldberg, 2001).

Climate change impacts are bound to be felt, especially in major cities of the world. Many major cities are at risk of flooding from rising sea levels due to climate change (see Figs. 6.6 and 6.8). Heat trapping in the form of heat islands in major urban landscapes (roofs, asphaltic roads, pedestrians, etc.) raises not only temperatures, but also causes air quality deteriorations. In some cities, a significant part of the population lives in squatter houses, where the people are vulnerable particularly to the environmental risk and health problems posed by climate change. Istanbul City is a major center of culture, history, politics, economy, and interactive social activities and the Istanbul Metropolitan Administration started a series of extensive climate research studies by taking into consideration the climate change report recommendations (IPCC, 2007). This section is the outcome this extensive climate change study.

In big cities there are certain leakage percentages from the main water distribution network. Therefore, among the water conservation approaches are leakage detection, remedial solutions, water utility consolidation, demand management, and especially integrated water resources management, drought prediction and management. On the other hand, alerting people through communication and the educational system is among the most effective conservation methods; after all, the end users are the people and they can help significantly in demand management. It is well known that reduction in water consumption is an obvious response to drought danger.

6.9.1 Istanbul Water Consensus

The Istanbul Water Consensus Local and Regional Authorities Declaration on the occasion of the fifth World Water Forum in Istanbul, the Local Government Declaration on Water of Mar. 21, 2008, expressed the awareness and responsibility of local and regional leaders concerning climate change impacts on major city environments. The consensus for local authorities was the primary output of the Local Authorities' Process for the fifth World Water Forum, which had the aim of promoting action concerning water resources and services at the local level around the world. Recognizing the urgent need to develop effective strategies, cities and regions depend on appropriate legal, institutional, and financial frameworks from local and central governments. Among the primary tasks are mainly climate change, population growth, urbanization, rapid economic development, and other pressures on the environment, water resources and systems' development, and to show their impacts more rapidly than political and social systems on scientific bases. Among this consensus are some of the following items that mention the global warming effect, climate change and, accordingly, water resources management procedures.

1. Water management should be approached at the regional level by taking into consideration climate change, to secure water supply in rural areas where marginalized people live to fight against rural depopulation.
2. Water resources and their management are also a tool to facilitate adaptation to global changes including climate impacts.
3. Climate change will impact every aspect of the water cycle affecting citizens and biodiversity, increasing water scarcity, extreme events such as droughts and floods, decreasing groundwater recharge, leading to a rise in sea levels, increasing temperatures, changing rainfall patterns, and streamflow regimes (Fig. 6.6).
4. To share with and associate local and regional authorities in the definition and implementation of political strategies taken at the regional and national level to improve access to water and to prepare for climate change and other global changes, which require new environmental and infrastructure projects to anticipate climate change-related effects in the design of water structures, sanitation, storm water, and other urban infrastructure.
5. To put the highest attention on forecasting and understanding future climate, demographic, and other developments affecting the water cycle and management systems at national and regional levels, and to share the knowledge gained with local governments and help to interpret these developments for their relevance at the local level.

6. Assemble the best available climate forecasts applicable for the hydrological factors that impact the city/local authority—from water source to sea.

7. Assess the city's water resources capacity to deliver services under major scenarios of climate and global changes. Determine other climate-related risks, potential benefits, and uncertainties with respect to water management.

8. Protect the environment, especially important aquatic habitats, against cumulative impacts of urban development and climate change.

Hence, the main purpose is to emphasize the preliminary scientific studies that have been carried out for the assessment of water resources planning, management, operation, and maintenance under the climate change effects (Şen, 2013a, b). For this purpose, future scenario projections for some meteorology station precipitation and storage reservoir catchment area rainfall projections can be generated and the necessary interpretations can be deduced for the next several decades.

6.9.2 Methodology

In the case of future possible climate change effects estimations on rainfall–runoff patterns, it is first necessary to consider the GCM output rainfall scenario data over the study area at a set of nodes. For Istanbul City, this has been achieved by downloading such data from the Max Plank Institute, Germany, on a monthly rainfall basis up to 2100. On the other hand, there is a set of meteorology stations, where monthly rainfall records are available for at least 40 past years. The GCM output data are available only at a set of regular grids, which are about 250–300 km away from each other. Hence, a downscaling procedure helps to combine GCM output with finer-resolution observation data at the meteorology stations. In general, there are two downscaling ways as either a dynamic or a statistical downscaling procedure (Wilby and Wigley, 1997; Wilby and Harris, 2006; Kilsby et al., 2007; Jones et al., 2009; Cavazos and Hewitson, 2005; Zorita and von Storch, 1999; Wilks, 1992; Şen, 2009b; Willems and Vrac, 2011). In this section, a SDM is adopted according to Şen et al. (2010) and its content is given below. GCM output resolution is not suitable for hydrometeorology application and, therefore, the following steps are executed by taking into account the meteorology stations for resolution reduction. Details of the proposed SDM are presented for the Istanbul Water and Sewerage Directory (Şen et al., 2010; see Figs. 6.9 and 6.10).

1. Monthly precipitation amounts and other meteorological variables (temperature, humidity, irradiation, wind speed) from GCM at the Max Plank Institute are downloaded up to 2100.

2. Historical monthly rainfall time series are checked for reliability, and simple methods are used for their reliability analysis.

3. Spatial downscaling of the GCM monthly data is achieved through the SDF concept, which provides the spatial relationship between any station and the others; hence, a function that shows the regional correlation with distance is obtained (Şen, 2009a). A sample for the Florya/Istanbul meteorology station

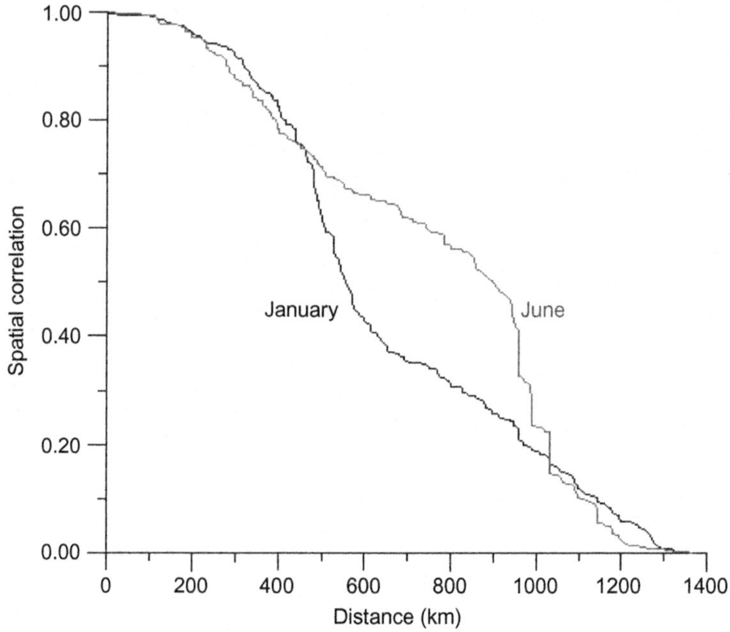

FIG. 6.19 Istanbul (Florya) station SDF.

is given in Fig. 6.19 for January and June months only. Such SDFs are available for each month at each station and they are based on the whole available meteorology stations in Turkey.

4. SDFs help to transfer the GCM outputs to meteorology stations and this is achieved through the weighted average procedure (Eq. 6.2) and updated until the maximum relative error between the downscaled and historical data becomes less than a certain limit, which is taken in the studies as ±10%. The weights are the spatial correlation values that are taken from the concerned SDF based on the distance between the GCM nodes and the station location, similar to Fig. 6.11.

5. Spatially downscaled series do not match the observation series temporally; hence, the final step is to adjust them to available monthly observation series. This is achieved through a white Markov stochastic process (Şen, 2009b).

Convenient software is written for the execution of all these steps with required downscaling, optimization, and estimation procedures (Şen et al., 2010).

6.9.3 Applications

Two different sets but consecutive applications are presented starting from the precipitation future patterns at two meteorology stations, Florya (Europe) and Göztepe (Asia) sides (see Fig. 1.10). However, these are the two examples only, but similar procedures can be applied to all the meteorology stations and drainage basins that serve for water supply in any other city.

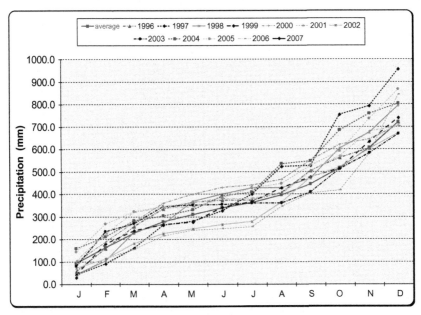

FIG. 6.20 Istanbul precipitation past CMP graphs.

The 10-year consecutive totals of the past monthly precipitation that Istanbul City received are given in Fig. 6.20. It reveals that for the initial months, the worst (best) year is 2001 (2005). However, in terms of annual totals, the worst (best) year is 2007 (1997). In planning endeavors, the total precipitation (annual) holds a special significance. In this section, such graphs are referred to as cumulative monthly precipitation (CMP) graphs, which provide many meteorological features for future planning and management of water resources.

The CMP graphs show the trace of the precipitation within each year, whereas they also serve as a basis of information in terms of annual rainfall amounts at the end of each year. Fig. 6.21 presents CMP results for two stations after the SDM application on the GCM results with A2 scenario up to the year 2050.

At Florya station, the heaviest rains within the next 10 years are expected to occur in 2016. The CMP graph of this year has no horizontal part, which means that each month has more or less precipitation. Especially in spring and winter seasons, the rate of precipitation (slope of the CMP graph) is steeper than any other year within this decade. Other years of the decade have more or less the same trends, and the annual precipitation range is about 500–700 mm. This station is at the verge of subbasin, Ayamama within the Istanbul metropolitan area, where a flash flood in 2009 caused 20 deaths. The CMP graphs indicate that a graver flood than the one observed in 2016 may arise in the same subcatchment. For this reason, proper measures must be taken beforehand.

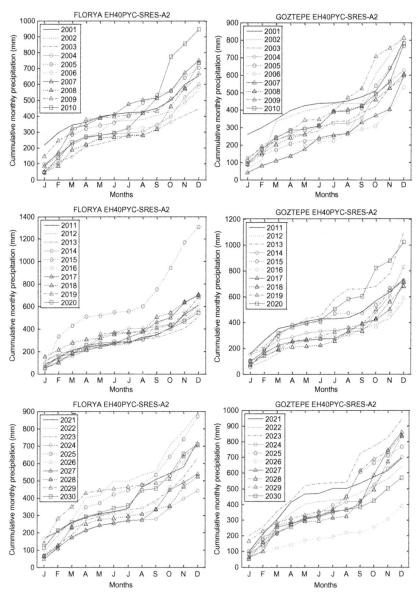

FIG. 6.21 EH40PYC-A2 scenario CMP decadal graphs for 2001–50 period: (A) Florya and

(Continued)

On the Asian side, Göztepe, in 2013 had an extreme event and is predicted to have another extreme event in 2020. Monthly precipitation variability at this station is more than Florya. Apart from the extreme years, the precipitation varies between 600 and 825 mm, which is about 15% more than the European side.

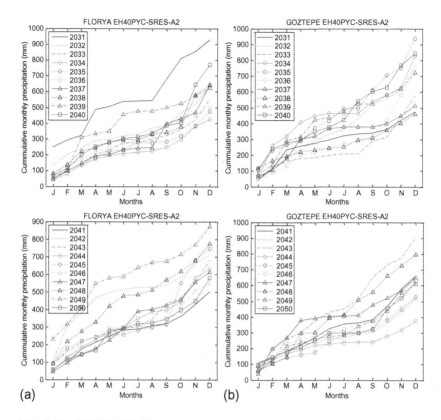

FIG. 6.21—CONT'D (B) Göztepe.

6.10 CLIMATE CHANGE AND WATER RESOURCES

In general, especially water supply and demand is considered classically dependent on temperature and precipitation, but recently also the role of CO_2 in the climate change process implies extra impact on water resources.

The primary hardware for water supply and their unusual functioning during dry (drought) periods due to climate change are dams, levees, small reservoirs, weirs, wells, pumps, and canals that may help to adapt also to climate change, hydrological cycle intensification, and hence, due to steadily growing water demands. Unfortunately, though new supply sources are almost always sought to meet increasing water demand even from far distance water sources, the demand end is not bothered with at all. In any adaptation study to water stress, shortage. or dry period onset and continuity, demand management is of utmost necessity by conserving and avoiding extravagant style of consumption, which helps improve and combat against rising climate change impacts.

Water resources managements are dependent on the regular quantity, variability (drought and flood), temporal and spatial distribution, intensity and frequency of precipitation, which are all bound to be affected by climate change impacts. Global climate change has important implications for water resources as evaporation rates, precipitation as rainfall and snow, earlier and shorter runoff seasons, water temperature augmentations, and water quality deteriorations. Increases in the evaporation rates are expected to reduce the water supply system capacity in many regions. Especially during hot seasons, soil-moisture levels may decrease leading to more frequent and severe meteorological, hydrological, and especially agricultural droughts. Such serious variations as a result of climate change are expected to cause serious water resources management implications for water resource consumers. Agricultural producers and urban areas are particularly vulnerable and the necessary precautions must be taken into account according to a systematic plan ("see chapter: Drought Hazard Mitigation and Risk"). Due to climate change impacts there will be more frequent and severe drought expectations in the future. Accordingly, gradual changes in the frequency and severity of drought will be difficult to distinguish from normal interannual variations in precipitation.

6.10.1 Climate Impacts on Water Supplies

Global warming, greenhouse gases effect, and climate change are the three terminologies that are interrelated to each other and collectively affect extreme weather and climate events such as droughts as rainfall, runoff, and soil-moisture deficiencies and floods as rainfall, runoff, and soil-moisture surpluses, compared to normal natural capacities. The International Panel on Climate Change (IPCC, 2007) announced that climate change will have the following effects on water supplies:

1. The timing and regional patterns of precipitation will change and more intense precipitation days in the form of rainfall are likely.
2. GCMs used to predict climate change suggest that a 1.5–4.5°C rise in global mean temperature would increase global mean precipitation about 3–15%.
3. Although the regional distribution is uncertain, precipitation is expected to increase in higher latitudes, particularly in winter. This conclusion extends to the mid-latitudes in most GCM results.
4. Potential ET rate (water evaporation from the surface and transpiration from plants) rises with air temperature. Consequently, even in areas with precipitation increase, higher ET rates may lead to runoff reductions, implying a possible decrease in renewable water supplies.
5. More annual runoff caused by precipitation increase is likely in the high latitudes. In contrast, some lower latitude basins may experience large reductions in runoff and water shortage increments as a result of a combination of increased evaporation and decreased precipitation.
6. Flood frequencies are likely to increase in many areas, although the amount of increase for any given climate scenario is uncertain and impacts will vary among basins. Floods may become less frequent in some areas.

7. The frequency and severity of droughts could increase in some areas as a result of a decrease in total rainfall, more frequent dry spells, and higher ET.
8. The hydrology of arid and semiarid areas is particularly sensitive to climate variations. Relatively small changes in temperature and precipitation in these areas could result in large percentage changes in runoff, increasing the likelihood and severity of droughts and/or floods.
9. Seasonal disruptions might occur in the water supplies of mountainous areas, if more precipitation falls as rain than snow, and if the length of the snow storage season is reduced.
10. Water quality problems may increase where there is less flow to dilute contaminants originating from natural and human sources.

6.10.2 Climate Impacts on Water Demand

Among the most consumptive components of water demand are irrigation activities and, in general, more than 80% of all groundwater and surface water is consumed by such activities because of the demand for food for survival. As climate conditions become drier and hotter, dry-land farming is bound to increase, which requires increase in water supply per square of crop area. Water use efficiency must be improved to cope with higher atmospheric CO_2 levels, which would tend to counter the tendency to apply more water as temperature rises.

On the other hand, domestic water use for household activities accounts on the average for about 10% of overall water consumption in the world. Annual domestic water consumption is not very sensitive to climate change; after all, whatever the circumstances, the primary order of exploitation will be given to domestic water consumption, of course, with lowest possible level restrictions. For instance, in many parts of the world, in any planning and domestic water resources operation, per capita water consumption rate is adopted as 150 Liter per day (L/day); however, in cases of droughts, water stresses, or shortages, it can be reduced down to 50 L/day, without many problems considering the drought circumstances (Şen et al., 2013).

Industrial water consumption is, in general, 5–10% of the overall consumption rate. About 95% of the used industrial water is returned to groundwater or surface water. In case of hot industrial water release to the environment and especially to aquatic media, it might be necessary to impose necessary restrictions in order to protect the natural ecosystem in the area. Cooling towers or ponds might help to reduce the consumption and damage to the local ecosystems. However, they may lead to evaporative losses. Climate change might also have indirect effects on industrial activities. For example, summer energy consumption will increase and the same will occur during the winter for space heating. These activities may require cooling water systems.

In any environmental ecosystem if convenient navigation, hydroelectric energy generation, and recreation activities are dependent on streamflow, then its different quantities, qualities, low (drought), and high (flood) discharges may be affected further by climate change impacts. Changes in the quantity, quality, and timing of runoff stemming from greenhouse warming would affect instream water uses such as hydroelectric power generation, navigation, recreation, and maintenance of ecosystems. These changes might also affect instream water demands, directly or indirectly. For example, changes in streamflows would alter actual and potential hydroelectric power generation. Changes in water quantity affect hydroelectric power generation directly or indirectly. As a result, shifts in hydroelectric power production could have a significant impact on water demand within a watershed.

Primary factors in any navigation system are discharge of river confluences or lake water and also ice-free periods. Seasonal water demands are also associated with recreational activities such as fishing, swimming, and surfing. Climate change may also cause aquatic system behavioral changes, vulnerability to their benefits, and shifts in hydrological regimes and cycles. In order to maintain ecosystem sustainability, necessary minimum water levels must be maintained so as not to endanger the existing situation. The water balance between instream and water withdrawal may be endangered when water resources become scarce, especially during drought periods with the additional impact of climate change.

It is certain that climate change is considered as the main water resources system impact factor for water demand and supply balance in the future, and the more periods are expected to be dry spells leading to extensive drought. Even if the impact of climate change is not effective, social, economic, cultural, industrial, technological, agricultural, land use, and many other human activities will have effects on the maintenance of sustainable water supply in order to satisfy steadily increasing demands in any part of the world. All these activities induce impact on the hydrological cycle and, additionally, human activity influenced climate change will also have impact, which is not possible to assess without necessary predictions, in general, and drought predictions, in particular. Unfortunately, almost in any place in the world, the existing water supply and distribution systems are based on past weather and climate impacts that are extractable from the available meteorological and hydrological records. However, return periods are considered for future planning, design, management, and operation, but it provides a global approach without future predictions and a room for extreme value occurrences such as droughts and floods. In practical applications, the return periods and associated risk levels are considered for floods rather than droughts. Hence, it is necessary to establish some rules (this chapter) for future drought occurrences based on effective prediction models ("see chapters: Temporal Drought Analysis and Modeling; Regional Drought Analysis and Modeling; Spatiotemporal Drought Analysis and Modeling").

The climate change impacts will cause alteration in the traditional or usual ways of water resources at unexpected times under quite unknown and severe impacts. These impacts may cause sensible and feelable effects on present management and operations and the operator or investigator may put in operation such alterations and provide warnings to responsible authorities for necessary additional and innovative adaptation studies. Adaptation studies without accurate information may lead to worsening of the present situation. In order to achieve cost-effective adaptation, mitigation, and solutions, it is necessary to have an integrated management of existing supplies and infrastructure at drainage basin levels, which leads to an increase of reliable supplies and efforts to resolve water conflicts in many regions. Appropriate incentives can be provided to local people to conserve and protect resources and opportunities in response to changing conditions. Climate change adds an additional element of uncertainty to matching future supplies with demands; however, the logical steps of any management and operation thoughts are independent of such changes.

6.11 GLOBAL WARMING THREAT ON WATER RESOURCES

Global warming, greenhouse gases effect, and climate change problems are long-term anthropogenic consequences that are expected to threaten water-related supply and demand patterns in the near future. These problems may be identified linguistically on a logical basis by taking the necessary precautions, and implementations of mitigation strategies after vulnerability possibilities are assessed using fuzzy logic principles (Şen, 2010). Climate change effects are the focus of many scientific, engineering, economic, social, cultural, and global studies, and these effects await cost-effective remedial solutions. Extreme events such as droughts and floods and modified groundwater recharge may be influenced by climate change (Şen, 2008a).

A series of IPCC reports on climate change, including the third assessment in 2001 and the fourth assessment in 2007, noted that although few would argue that completely unambiguous detection and attribution of climate change had already occurred, the "balance of evidence suggests a discernible human influence on global climate" (Santer et al., 1996; Kundzewitz et al., 2007). Some evidence indicates that 20th century global mean temperature was at least as warm as any other century since 1400. Data prior to 1400 are too sparse, at the moment, to allow the reliable estimation of global mean temperature.

Climate change will affect the availability of water, its quality, distribution, the complex infrastructure and systems in place to manage it, and existing climate variability. Unfortunately, nowadays global climatic change is presented only as a series of disasters, but the positive aspects are neglected. For instance, it is true that some areas start to suffer from desertification, lack of precipitation, or flooding, but on the other hand, today some arid zones could change into more humid zones with significant groundwater recharges. There is very little literature about how different climate changes may affect the infrastructure and complex systems built to manage water resources. Significant knowledge gaps remain and far more research is needed. Priorities and directions for future studies should come from water managers and planners as well as from the more traditional academic and scientific research community.

The cycling of water in its various solid, liquid, and gaseous states is a primary process within the Earth's climate system. Information on variability of states and fluxes over time is crucial for the understanding of the sustainability of local, regional, national, and international economies and ecosystems. It is essential to establish rates of cycling, changes in these resulting from human intervention, and the consequences of those changes for regional freshwater availability. It is a priority scientific objective to establish how much of the variability in the water cycle is predictable over a range of time and space scales.

There are many opportunities to adapt to changing hydrologic conditions, and the net costs are sensitive to the institutions that determine how water is managed and allocated among users.

An increase in atmospheric moisture may lead to increased relative humidity and increased clouds, which could cut down on solar radiation and reduce the energy available at the surface for evaporation. Those feedbacks are included in the climate models and alter the magnitude of the surface heat available for evaporation in different models.

The rainfall rates and frequencies as well as accumulations are important in understanding what is going on with precipitation locally. The accumulations depend greatly on the frequency, size, and duration of individual storms, as well as the rate, and these depend on atmospheric static stability (vertical structure) and other factors as well. In particular, the need for vertical heat absorption transport at the surface is a factor in convection and extratropical weather systems, both of which act to stabilize the atmosphere. Increased greenhouse gases also stabilize the atmosphere. Those are additional considerations in interpreting model responses to increased greenhouse gas simulations.

Changes in runoff and the source of a region's renewable water supply are direct results of changes in precipitation and evaporation, which is strongly influenced by temperature.

Uncertainties about how the climate and hydrology of a region will change in response to a global greenhouse warming are enormous. However, one of the more likely impacts involves areas where precipitation currently comes largely in the form of winter snowfall and where streamflow comes largely from spring and summer snowmelt. A warming would likely result in a distinct shift in the relative amounts of snow and rainfall and in the timing of snowmelt and runoff. A shift from snow to rainfall could increase the likelihood of flooding early in the year and reduce the availability of water during periods of peak demand, especially for irrigation.

Detection of climate change requires demonstrating that the observed change is larger than would be expected to occur by naturally inherent fluctuations alone. An observed variation in climate is highly unusual in a statistical sense. Attribution of change to human activity requires showing that the observed change cannot be explained by natural forced or unforced causes. Climate change impacts are observable through numerical measurements and/or linguistic expressions in a fuzzy manner (Şen, 2010). Assessments of the statistical significance of the observed global mean temperature trend over the last century have detected a significant change and shown that the warming is unlikely to be entirely natural in its origin.

1. The two most key climate variables are temperature and precipitation and their numerical measurements are necessary in a systematic manner temporally.
2. Past climate variable records in terms of time series provide an objective basis for climate variability and change assessments.
3. Time series plots may reveal hidden or obvious trends, which may be observable by the naked eye or better by the application of competent statistical methodologies. Increasing or decreasing trends may indicate the climate impact provided that other side effects are isolated.
4. People make their decision about the climate change based on their daily experience with climate. They make statements such as "the climate is changing" and "it's raining more compared to the past."
5. Observations on snow occurrence indicate that in some parts of the world snow events have decreased.
6. Drought and flood frequency as well as magnitude variations.
7. Observations of the relevant human inputs to the atmosphere (greenhouse gases, sulfate aerosols).

6.11.1 Effect of Climate Change on Water Availability

Climate is only one of many factors influencing future supply and demand for water. Population, technology, economic conditions, social and political factors, and the values society places on alternative water uses are likely to have more of an impact on the future availability and use of water than changes in the climate. Demands are outpacing supplies, water costs are rising sharply, and current capacities are depleting or some valuable resources are subject to contamination.

Humans are influencing the global climate and are thereby altering the hydrological cycle inadvertently. The impact of climate change on water resources is dealt with in more detail in the Second Assessment Report of the IPCC issued in 1995 and in the Report of the Second International Conference on Climate and Water held in 1998, which made a number of recommendations:

1. Research priorities (data networks, problems of scale, need for interdisciplinary dialogue).
2. Research management (large-scale land surface experiments, advanced planning for remote sensing, communication with decision makers and the public).
3. Project design and management (effect of climate change, broad dialogue on practical operational problems, conflict resolution on water issues).
4. Policy formulation (national planning based on up-to-date information, respect for local culture and level of development, involvement of all stakeholders at an early stage).

While future water demand depends as much on consumer preference (or individual unit consumption) as it does on population, much less attention has been given to consumer preferences, which can be determined by market purchase analyses and varies from place to place depending on cultural, environmental, and other features. Future preference trends will result from intersection between the introduction of new goods and services and the changes in the means to purchase and time to enjoy them.

6.11.2 Effect of Climate Change on Extremes

Perhaps one of the greatest interests in weather and climate relates to extremes of climate. Due to inadequate monitoring as well as prohibitively expensive access to weather and limited data held by the world's national weather and environmental agencies, only limited reliable information is available about large-scale changes in extreme weather or climate variability. The time varying biases that affect climate means are even more difficult to effectively eliminate from the extremes of the distributions of various weather and climate elements. There are a few regions and climate variables, however, where regional and global changes in weather and climate extremes have been reasonably well documented.

There is a risk of increased flooding in many parts of the world that experience large increases in precipitation (medium confidence). The Intergovernmental Panel on Climate Change concluded in 1996, that "the flood-related consequences of climate change may be as serious and widely distributed as the adverse impacts of droughts" and "there is more evidence now that flooding is likely to become a larger problem in many temperate regions, requiring adaptations not only to droughts and chronic water shortages, but also to floods and associated damages, raising concerns about dam and levee failure."

By 2025, one-third of the population of the developing world is expected to face severe water shortages. Yet, even in many water-scarce regions, large amounts of water annually flood out to the sea. Some of this floodwater is committed flow to flush salt and other harmful products out of the system and maintain the ecological aspects of estuaries and coastal areas (Seckler et al., 1998). When drought comes everybody is concerned, if it lasts everybody tries to do his/her best for the duration, but when it passes away, everybody forgets except those who have been hurt (Yevjevich et al., 1983).

The crop yield in a year can be compared with its long-term average and drought intensity can be classified linguistically in a fuzzy manner as nil, mild, moderate, severe, or disastrous, based on the difference between the current yield and the average yield. Various researchers in different parts of the world have developed drought indices that can also be included along with weather and climate variables to estimate crop yield ("see chapter: Basic Drought Indicators"). For example, Boken and Shaykewich (2002) modified the western Canada wheat yield model (Walker, 1989) drought index using daily temperature and precipitation data and advanced very high-resolution radiometer satellite data. The modified model improved the predictive power of the wheat yield model significantly. The performance of the regression techniques have been found to have improved significantly by including satellite data-based variables.

Droughts represent temporary imbalance in the irradiative transfer with characteristics that it is persistently lower than average precipitation with uncertain frequency, duration, and severity; has unpredictable consequences; represents overall diminishing water resources; and has diminishing average carrying capacity of ecosystem.

6.11.3 Effect of Climate Change on Groundwater Recharge

Groundwater and climate are linked in many ways. Impacts of climate change on groundwater are poorly understood and the relationship between climate variables and groundwater is more complicated than that with surface water. Groundwater hydrologists need to be more attuned to the effects of climate. The links between groundwater and climate should receive greater attention as part of drought monitoring and assessments, studies of climate change and variability, analysis of water quality, and assessment of the availability and sustainability of groundwater resources. Groundwater level responses to climate variation have a time delay. Without proper consideration of variations in aquifer recharge and sound pumping strategies, the aquifer water resources could be severely impacted under a warmer climate. Many aquifer regions are very vulnerable to climate change impacts because of the following reasons:

1. The region is largely dependent on the aquifer to meet municipal, agricultural, industrial/military, and recreational water demands with limited large-scale alternative water supplies, which are subject to large climatic variability.
2. There is a strong linkage between climatic inputs (precipitation to be specific) and regional hydrology, through the conversion of rainfall to runoff and runoff to aquifer recharge by streambed seepage.
3. The historical climatic records show large variability in precipitation and the occurrence of occasional multiyear droughts (North et al., 1995), which can reduce natural aquifer recharge to a negligible level (Loaiciga et al., 2000).

Because groundwater resources are naturally replenished by infiltration after precipitation and subsequent percolation of water through geologic materials, a decline in precipitation or an increase in ET result in a decline in recharge, possibly resulting in decreased groundwater levels.

The resulting effects on mean annual groundwater recharge and streamflow are small, as increased atmospheric CO_2 levels reduce stomata conductance, thus counteracting increasing potential evapotranspiration induced by the temperature rise and decreasing precipitation (Eckhardt and Ulbrich, 2003). In spring, the decreased snowmelt, the temperature rise, and the earlier beginning of the growing season reduce groundwater recharge.

Both precipitation and annual mean temperature may display a strong correlation with annual groundwater levels in the aquifer. The correlation with temperature becomes stronger in the periods of higher annual mean temperatures. In areas where the aquifer is found at shallow depth, temperature has a greater influence than precipitation on groundwater levels. Results suggest that a trend of increasing temperatures predicted by global climate models may reduce net recharge and affect groundwater levels (Chen et al., 2001).

Studies suggest that climate change will lead to less surface water availability in the Canadian Prairies (Schindler, 2001), thus increasing need for groundwater development.

With increasing evapotranspiration and decreasing precipitation, the impact of climate change may result in declining groundwater levels, which may cause some wells to become dry while others may become less productive due to the loss of available drawdown.

Predicting the impacts of climate change and developing adaptation strategies are essential for ensuring a sustainable groundwater supply in any region. However, potential impacts are difficult to assess accurately, as the influence of climate change on groundwater levels cannot be detected immediately. The potential impacts of climate change on groundwater supplies in the upper carbonate aquifers could include the following points (Chen et al., 2001):

1. less groundwater recharge available due to more evaporative loss of surface water and less precipitation in the winter and spring,
2. longer residence time due to changes in hydraulic properties in the aquifer after prolonged droughts, and
3. groundwater quality could deteriorate as a result of saline water intrusion if in response to climate change fresh groundwater head drops relative to saline waters, which are buffered by a regional-scale flow system.

As an important component of the global hydrologic cycle, groundwater not only contains information about environmental changes, but it is also directly affected by these changes.

Consideration of climate variations can be a key factor in ensuring the proper management of groundwater resources (Şen, 2012; Şen, 2013b). Generally, the importance of climate impact on groundwater at seasonal to decadal scales is neglected by groundwater hydrologists. Groundwater systems tend to respond more slowly to variability in climate conditions than do surface water systems. Droughts are the most widely recognized climate perturbations relative to groundwater. However, high groundwater levels from extended wet periods also cause problems such as flooded basements and waterlogged lands. Water storage is critical in dealing with climate variability. As surface water storage becomes more limited, use of groundwater storage to modulate the effects of droughts increases in importance, as do potential enhancements by artificial recharge (Alley, 2001). Among the most important links between climate and groundwater are the effects of surface water and the land. For example, groundwater development may significantly affect low flows in streams, minimum levels in lakes, and hydroperiod of wetlands. Likewise, reduced heads in response to increases in pumping during droughts can cause land subsidence as a result of compaction of aquifer material.

The quality of groundwater is also affected in many ways by climate. Reduced freshwater discharges to coastal areas during droughts can cause seawater to intrude into aquifers beyond previous landward limits.

Human-induced climate change in the coming decades may further affect groundwater resources in several ways, including changes in groundwater recharge resulting from changes in annual and seasonal distribution of precipitation and temperature, more severe and longer-lasting droughts, changes in evapotranspiration resulting from changes in vegetation, and possibly increased demands for groundwater as a backup source of water supply. Surficial aquifers that supply much of the water in streams, lakes, and wetlands are likely to be the part of the groundwater system most sensitive to climate change. However, little attention has been directed at determining the possible effects of climate change on shallow aquifers and their interaction with surface water (Alley, 2001).

Rise in temperature and decrease in precipitation lead to a reduction of groundwater recharge and level (Eitzinger et al., 2003). Study in the Northrhine-Westfalia region of Germany indicate that in mountainous areas the groundwater recharge change would be smaller than in the plains, where a reduction of 30% is predicted (Krüger et al., 2002). The groundwater level is influenced by water withdrawals, too.

Groundwater is the major freshwater source especially for arid and semiarid regions, but unfortunately there has been very little attention or study of the potential climate change effects on these freshwater resources. Most of the studies are concentrated on humid regions. Aquifers in arid and semiarid regions are replenished by floods at possible recharge outcrop areas through fractured and fissured rocks, solution cavities in dolomite or limestone geological setups, as well as through main stream channels of quaternary alluvium deposits. At convenient places along the main channel engineering infrastructures such as levees, dikes, and successive small-scale groundwater recharge dams may be constructed for groundwater recharge augmentation (Şen et al., 2011). The groundwater recharge areas must be cared for isolation from fine silt accumulation after each flood occurrence or at periodical intervals. Furthermore, flood inundation areas are among the most significant groundwater recharge locations in arid and semiarid regions. Accordingly, their extents must be delimited by considering future climate change effects.

Understanding the relative importance of climate, vegetation, and soils in controlling groundwater recharge is critical for estimating recharge rates and for assessing the importance of these factors in controlling aquifer vulnerability to contamination. Also, understanding the role of climate and vegetation in controlling recharge will be valuable in determining impacts of climate change and land-use change on recharge.

The aquifers that are in contact with the present-day hydrological cycle will be affected by climate change. These are referred to as unconfined or shallow aquifers. On the other hand, deep and especially confined aquifers are not in contact with the present-day hydrological cycle and, consequently, their effect from climate change is virtually negligible. They include fossil groundwater storages (Şen, 2014).

Due to global warming, a smaller proportion of the winter precipitation falls as snow. The spring snowmelt peak, therefore, is reduced while the flood risk in winter is probably increased. In summer, mean monthly groundwater recharge and streamflow are reduced by 50%, potentially leading to problems concerning water quality, groundwater withdrawals, and hydropower generation (Eckhardt and Ulbrich, 2003). Some water supplies could become unusable due to the penetration of saltwater into rivers and coastal aquifers as the sea level rises. Changes at the surface water resources frequency and magnitude would influence the aquifer storage replenishment through natural recharge. Water quality may also respond to changes in the amount and timing of precipitation.

Coastal aquifers may be damaged by saline water intrusion into groundwater changes due to sea level rise (Şen, 2014). The movement of the salt front up in aquifers would affect freshwater pumping plants near the coastal line. Relative sea level rise adversely affects groundwater aquifers and freshwater coastal ecosystems. Rising sea level causes an increase in the intrusion of saltwater into coastal aquifers. Other impacts of sea level rise are likely to include changes in salinity distribution in estuaries, altered coastal circulation patterns, destruction of transportation infrastructure in low-lying areas, and increased pressure on coastal levee systems. Higher sea levels associated with thermal expansion of the oceans and increased melting of glaciers are expected to push saltwater further inland in rivers, deltas, and coastal aquifers. It is well understood that such advances would adversely affect the quality and quantity of freshwater supplies in many coastal areas (IPCC, 2007).

Unsaturated zone has a unique capability in helping to assess impacts of climate change on groundwater resources. The potential impacts of climate change can be assessed if focused on fractured carbonate aquifer systems (having variable degrees of fracture development, ie, karstification) because of the following points:

1. Fracture flow aquifers are the most responsive to changes in recharge as typically they have low specific yields (ie, they have drainable porosities) in comparison with intergranular flow systems.
2. Carbonate rocks are soluble and the aquifers might show exacerbated water table lowering, if predicted increases in atmospheric CO_2 contents along with temperature rise induce rapid enlargement of fracture apertures.
3. If dissolution of carbonate rocks does become more vigorous then potentially, the hardness of groundwater could be expected to increase leading to possibly unacceptable water quality.

Application of the climate change scenarios and software is necessary to forecast changes in aquifer geometries, hydraulic parameters (permeabilities, storage coefficients), flows, water balances, and water quality.

6.11.3.1 GROUNDWATER RECHARGE

Groundwater recharge depends on several factors such as infiltration capacity, stochastic characteristics of rainfall, and climate factors. The spatial and temporal distribution of the rainfall mainly controls the natural groundwater recharge. In arid regions, recharge occurs through the ephemeral streams, which flow through the wadi course but most of the water is absorbed in the unsaturated zone before reaching the aquifer. In semiarid regions, the recharge is irregular and occurs only in the periods of heavy rainfall. In humid regions, recharge is mainly in the winter period. In the summer period, most of the rainfall becomes soil moisture and evaporates. In cold areas the melting of ice suddenly recharges the groundwater.

The groundwater recharge in a region depends mainly on the precipitation change during the major recharge season. In temperate climates, an increase in precipitation is generally foreseen during the winter season where most recharge occurs. During the hotter summers, however, there might be increased ET in particular if the groundwater table is close to the land surface. Using a coupled groundwater and soil model for a groundwater basin in Belgium, Brouyère et al. (2004) projected a future decrease of groundwater recharge (being severe in dry years) for climate models predicting less summer and more winter precipitation.

Less groundwater recharge leads to a drop in the groundwater table, which can have a negative impact on vegetation. Indirect effects of climate change on groundwater quantity can result from climate-induced changes of groundwater withdrawals or land use. The former may increase due to the following reasons:

1. Irrigation water requirements increase.
2. River discharge decreases or its temporal variability increases, so that the reliance on surface water goes down.

If irrigated areas decrease due to less available surface waters, groundwater recharge via leached irrigation water decreases too. Climate change may lead to vegetation changes, affecting groundwater recharge.

With respect to groundwater quality, climate change is likely to have a strong impact on coastal saltwater intrusion as well as on salinization of groundwater. For two small and flat coral islands off the coast of India, Bobba et al. (2000) computed the impact of sea level rise on the thickness of the freshwater lenses. With a sea level rise of only 0.1 m, the thickness of the freshwater lens decreased from 25 to 10 m for the first island and from 36 to 28 m for the second island. In addition to the sea level rise, any change in groundwater recharge affects the location of the freshwater/saltwater interface, and saltwater intrusion is expected to increase if less groundwater recharge occurs. This can also happen inland, where saline water is located next to or below freshwater (Chen et al., 2001). For many semiarid areas, a decrease in precipitation is projected and enhanced ET in the warmer world might cause a salinization of groundwater.

Unsaturated zone has a unique capability in helping to assess impacts of climate change on groundwater resources. The potential impacts of climate change can be assessed by focusing on porous, fractured, and karstic (carbonate rock, dolomite, limestone) aquifer systems. Especially, fractured and karstic aquifers are the most responsive to changes in recharge as typically they have low specific yields (ie, they have drainable porosities) in comparison with porous flow systems. Karstic rocks are soluble and the aquifers might show exacerbated water table lowering if predicted increases in atmospheric CO_2 contents along with temperature rise induce rapid enlargement of fracture apertures and enlargement in the solution cavities. Dissolution of carbonate rocks (karstic media) might become more vigorous by time and, accordingly, the hardness of groundwater sources is expected to increase, leading to possibly unacceptable water quality.

As a result of climate change, in many aquifers of the world the spring recharge retreats toward winter with more or less the same rates, but summer recharge declines dramatically. Coupled with the changes in the hydrological cycle and probable inducement of climate change basic elements, the groundwater recharge is also interactively affected due to the following events:

1. Changes in precipitation, evapotranspiration, and runoff are expected to influence recharge. It is possible that increased rainfall intensity may lead to more runoff and less recharge.
2. Sea level rise may lead to increased saline intrusion of coastal and island aquifers, depending on the relative position of sea level to the groundwater table level.
3. Changes in precipitation imply changes in CO_2 concentrations, which may influence carbonate rocks dissolution and, hence, formation and development of karstic groundwater aquifers.
4. Natural vegetation and crop pattern changes are reflected in the climate change influencing recharge.
5. Increased flood events contribute through recharge to unconfined aquifers in arid and semiarid zones; hence, they affect groundwater quality in alluvial aquifers of wadis (Şen, 2008a).
6. Changes in soil organic carbon may affect the infiltration properties of shallow aquifers and, consequently, the groundwater recharge.

The abovementioned factors indicate that the groundwater-focused organizations should take interest in global climate change issues to protect the groundwater resources effect from the implications.

6.11.4 Climate Change and Engineering Systems

Engineering structures include dams, reservoirs, weirs, levees, aqueducts, dikes, wells, subsurface dams, qanats, culverts, canals, distribution pipe networks, treatment and desalination plants, and so on, which help humans to cope, adopt, operate, and manage inherent temporal and/or spatial variability in water resources planning, design, operation, management, and maintenance. To the average citizen, engineering structures are largely invisible and taken for granted; yet, they help insulate people from wet and dry years and moderate other aspects of our naturally variable climate. Indeed, the engineering structures help society to almost forget about the complex dependences on climate. Water resources engineering systems are major social engineering units that are essential for individuals, societies, countries, and humanity in general. The development of any country is measured with the water resources system availability and adaptation to natural (droughts and floods) and man-induced variations (including climate change).

Engineering systems help to manage water resources utilization according to supply and demand requirements in the best possible (optimum) manner. Climate change can be regarded as one of accumulating variability. Different management alternatives are developed and applied in the water sector over many years, but they do not take into account the climate change effects explicitly. However, some water resources managers care about the climate change effects, which will become rather significant in the coming decades, especially in the mid-latitudes and some subtropical climate belts of the world. If reservoirs are full after a wet period, then a short-lived summer flood may not end a water resources drought caused by prolonged lack of dam inflows. Hence, droughts are not dependent on possible climate changes only but critically on the water resources system characteristics and especially on their management.

Climate change will affect the complex infrastructure of engineering systems to manage society's water and existing climate variability. It is not that the construction of additional dams indicates the development level of a country, but rather an efficient management program of existing dams is the most important task to achieve.

The potential impact of climate change on the hydrologic regime is a crucial question for water resources engineering system management. Potential change in hydrologic regime resulting from climate change is an important topic in contemporary hydrology and water resources management.

An understanding of mechanisms linking large-scale climate variability with regional conditions also forms the basis for reducing the uncertainty associated with assessing regional impacts of climate change over decadal-to-centennial periods. A region-specific ability to project the consequences of global change is necessary, for example, by decision makers concerned with long-term fixed capital investments in infrastructures such as dams, water diversion systems, and flood damage mitigation systems that are vulnerable to shifts in hydroclimatic regime.

The IPCC (2001a,b) review of climate impact studies suggests large differences in the vulnerability of water resource systems to climate variables. Isolated single-reservoir systems in arid and semiarid areas are extremely sensitive. They lack the flexibility to adapt to climate impacts that could vary from decreases in reservoir yields in excess of even more than 50% at one extreme to increasing seasonal flooding at the other. In contrast, highly integrated regionally interconnected systems are inherently more robust.

Climate change has the potential to either aggravate or alleviate an area's water situation. On balance, however, the impacts are likely to be adverse because the existing water infrastructure and uses are based on an area's past climate and hydrology records. During most of this century, dams, reservoirs, pumps, canals, and levees have provided the primary means of adapting to climate and hydrological variability and meeting the growing demands for water. While the focus was on supply side solutions, institutions that establish opportunities as well as incentives to use, abuse, conserve, or protect water resources are slow to adapt to the challenges of growing scarcity, rising in-stream values, and the vulnerability and variability of supplies. In recent decades, however, the high financial and environmental costs of water projects, along with limited opportunities for building additional dams and reservoirs to develop new water supplies, have shifted the focus away from new construction to improved management of existing supplies and facilities and also toward demand management. New infrastructure may, in some instances, eventually prove to be an appropriate response to climate-induced shifts in hydrological regimes and water demands. However, it is difficult to plan for and justify expensive new projects when the magnitude, timing, and even the direction of the changes at the basin and regional levels are unknown. Narrowing the range of uncertainty for improved water planning, operation, and management depends on a better understanding of the following points.

1. The processes governing global and regional climates.
2. The links between climate and hydrology.
3. The impacts of the climate on unmanaged ecosystems.
4. The impacts of ecosystem change on the quantity and quality of water.
5. The impacts of increased atmospheric CO_2 on vegetation and runoff.

In the meantime, the possibility that a warming could result in greater hydrological variability and storm extremes should be considered in evaluating margins of safety of long-lived structures such as dams and levees that are under consideration. In particular, low-cost structural and managerial modifications that ensure against the possibility of a range of climate-induced impacts should be sought. Unlike the structural supply side approach, demand management that introduces additional incentives to conserve and opportunities to reallocate supplies as conditions change does not require long lead times, large financial commitments, or accurate information about the future climate. Integrated management of existing supplies and infrastructure at the river basin and watershed levels offer a potentially cost-effective means of increasing reliable supplies and resolving water conflicts in many regions. While the prospect of climate change adds another element of uncertainty to the challenge of matching future supplies with demands through different storages (dams, weirs, dikes, etc.), it does not alter what needs to be done to ensure that water is managed and distributed wisely.

In general, climate change is expected to lead to more precipitation coupled with more evaporation, for instance in the AP, but the important question is how much of this precipitation will end up at water deficit areas? If not, then regional management of water engineering infrastructures comes into view with sustainable water distribution programs. On the other hand, probable precipitation increase in some areas and decline in others is another indication for regional water resources distribution to needed areas through efficient management programs.

Certain aspects of water resources and engineering infrastructure are very sensitive to both climate and to how to manage complex water systems. It is, therefore, necessary to have mediators in the form of engineering structures (such as dams) in order to offset or diminish the sensitivity to various expected and unexpected changes in the future. Changes in management of the engineering systems requires understanding what changes would be the most effective and then applying the will and direction of those responsible. Water managers and policymakers must start considering climate change as a factor in all decisions about water investments and the operation of existing facilities and systems.

A continued reliance solely on current engineering practice may lead to making incorrect and potentially dangerous or expensive decisions. Conventionally, water resources system operation and distribution practices should be designed and for the most part are operated assuming that future climatic and hydrologic conditions will look like past conditions, which is no longer true. Accordingly, two of the most important coping strategies must be to try and understand what the consequences of climate change will be for water resources and to begin planning for and adaptation to those changes through real-time operation and management programs. Dynamic management strategies of dams can be effective in mitigating the adverse impacts of climate change, but such policies need to be implemented before such changes occur to maximize their effectiveness. Such management studies are available on a dynamical daily basis for Istanbul City including European and Asian sides' dams in an integrated manner (Şen and Kadioğlu, 2000).

6.12 SOME RECOMMENDATIONS

The present industrial activities are most likely to accelerate the natural phenomena but the share of human involvement in global change is difficult to establish precisely. Very dramatic changes in temperature and concentration of greenhouse gases occurred in the geological history of the Earth with no human intervention. For this reason, all activities on how to mitigate negative impacts of global climatic change are necessary.

Although previous studies in the literature are comprehensive as to the basic understanding of climate change and implications on hydrological cycle components and water resources, there are still many gaps, uncertainties, and unanswered questions, which can be summarized as follows:

1. Cost-effectiveness of various climate change impacts such as droughts, floods, and mismanagements.
2. There have been a few studies that have summarized potential response strategies and assessment of how water managers might respond in practical applications.

3. There are different trends in different parts of the world and even among subareas of the same region, but unfortunately such regional variations have not been taken into consideration through regional water resources management. Climate change is expected to affect regional integrated water resources planning, operation, and maintenance.

4. Although soil moisture is related to probable climate change, infiltration capacity as one of the significant hydrological components for groundwater resources storage is not addressed with the perspective of groundwater storage increase or decrease on decadal time scales.

5. The significance of infiltration and floods for groundwater recharge in arid and semiarid regions are not assessed by considering climate change.

6. Artificial groundwater recharge enhancement possibilities can be effective as preliminary solutions against climate change. Especially, fractured medium and karstic region (solution cavity) groundwater recharge can be related to climate change. For instance, the effect of warmer rainfall on solution cavities should play a role in the expansion of solution cavities and, consequently, groundwater recharge.

7. Drainage basin areas are considered as they appear in humid regions, but the arid wadi concept and freshwater resources are not covered properly (Şen, 2008b).

8. The rate of groundwater recharge estimated by using carbon-14 isotopes is mentioned, but the same estimation in arid regions by chloride concentration is rarely considered (Subyani and Şen, 2006).

9. Rather than the trend as the indication of climate change, it is preferable in future studies to consider the breaking of record.

10. Flash flood appearances in arid and semiarid regions are not covered from the climate change point of view.

11. Adaptive real-time reservoir and freshwater distribution must be presented through an effective management project.

12. Multipurpose and multidam operation and management studies must be considered with the future climate change effects.

It is expected that future studies will cover the following points toward more comprehensive understanding and application of climate impact:

1. New scenarios for future climate change effectiveness. These are gathered in the SRES.

2. Different application of scenarios on water resources availability and management assessments.

3. Climate change and variability increased effects including more illuminating and detailed knowledge on extreme events (droughts, floods), engineering aspects, management strategies.

4. Strategic planning of freshwater resources storages under the effect of different scenarios.

5. New techniques and suggestions for adaptation against climate change-related man-made influences.

6. In freshwater resources assessment, the role of technological developments and especially desalination plants in arid regions must be considered.

7. The effect of climate change on expected flood inundation maps must be documented.

There are increasing or decreasing trends in streamflow discharge, which may be partially due to climate change effect, but this effect cannot be quantified definitively. Climate change glacier retreats toward the winter season are observable and this is obvious in streamflow timing in many areas. Rather than streamflow variable only, an integrated assessment of different hydrological elements must be evaluated simultaneously in order to arrive at more sound evidence of climate impact through detectable trends.

A growing number of studies suggest that climate change will increases the frequency and intensity of the heaviest precipitation events, but there is little agreement on detailed regional changes in storminess that might occur in a warmed world. Contradictory results from models support the need for more research, especially to address the mismatch between the resolution of models and the scales at which extreme events can occur.

The current state-of-the-science suggests that plausible climate changes, projected by GCMs, raise a wide range of concerns that should be addressed by national and local water managers and planners, climatologists, hydrologists, policy makers, and the public.

Little work has been done on the impacts of climate change for specific groundwater basins, or for general groundwater recharge characteristics or water quality. Some studies suggest that some regional groundwater storage volumes are very sensitive to even modest changes in available recharge.

Variability in climate already causes fluctuations in hydroelectric power generation. Climate changes that reduce overall water availability will reduce the productivity of hydroelectric power facilities. Reliable increases in average flows would increase hydropower production. Changes in the timing of hydroelectric power generation can affect the value of the energy produced. Specific regional impacts are not well established.

Catchments with a substantial snow pack in winter are expected to experience major changes in the timing and intensity of runoff due to climate change. Reductions in spring and summer runoff, increases in winter runoff, and earlier peak runoff are all common responses to rising temperatures. The ability of existing systems and operating rules to manage these changes has not been assessed adequately.

Climate models are still unable to make precise regional predictions. In addition, the hydrological cycle is extremely complex. A change in precipitation may affect surface wetness, reflectivity, and vegetation, which then affect evapotranspiration and cloud formation that, in turn, affect precipitation. Meanwhile, the hydrological system is also responding to other human activities such as deforestation, afforestation, urbanization, and the overuse of water supplies.

Little information is available on how climate changes might affect ground-water aquifers, including quality, recharge rates, and flow dynamics. New studies on these issues are needed.

REFERENCES

Alley, W.M., 2001. Ground water and climate. Ground Water 39, 161.

Almazroui, M., 2011a. Calibration of TRMM rainfall climatology over Saudi Arabia during 1998–2009. Atmos. Res. 99 (3), 400–414.

Almazroui, M., 2011b. Sensitivity of a regional climate model on the simulation of high intensity rainfall events over the Arabian Peninsula and around Jeddah (Saudi Arabia). Theor. Appl. Climatol. 104 (1–2), 261–276.

Almazroui, M., 2012. Dynamical downscaling of rainfall and temperature over the Arabian Peninsula using RegCM4. Clim. Res. 2, 49.

Almazroui, M., Nazrul, I.M., Athar, H., Jones, P.D., Rahman, M.A., 2012. Recent climate change in the Arabian Peninsula: annual rainfall and temperature analysis of Saudi Arabia for 1978–2009. Int. J. Climatol. 32 (6), 953–966.

Almazroui, M., Awad, A.M., Nazrul, I.M., Al-Khalaf, A.K., 2014. A climatological study: wet season cyclone tracks in the East Mediterranean region. Theor. Appl. Climatol. http://dx.doi.org/10.1007/s00704-014-1178-z.

Al-Yamani, M.S., Şen, Z., 1992. Regional variations of monthly rainfall amounts in the Kingdom of Saudi Arabia. J. King Abdulaziz Univ., Earth Sci. 6, 113–133.

Appenzeller, C., Stocker, T.F., Anaklin, M., 1998. North Atlantic oscillation dynamics recorded in Greenland ice cores. Science 282, 446–449.

Arnell, N.W., 2004. Climate change and global water resources: SRES scenarios and socio-economic scenarios. Global Environ. Change 14, 31–52.

Bobba, A., Singh, V., Berndtsson, R., Bengtsson, L., 2000. Numerical simulation of saltwater intrusion into Laccadive Island aquifers due to climate change. J. Geol. Soc. India 55, 589–612.

Boken, V.K., Shaykewich, C.F., 2002. Improving an operational wheat yield model for the Canadian Prairies using phonological-stage-based normalized difference vegetation index. Int. J. Remote Sens. 23 (20), 4157–4170.

Brouyere, S., Carabin, G., Dassargues, A., 2004. Climate change impacts on groundwater resources: modelled deficits in a chalky aquifer, Geer basin, Belgium. Hydrogeol. J. 12, 123–134.

Cavazos, T., Hewitson, B.C., 2005. Performance of NCEP–NCAR reanalysis variables in statistical downscaling of daily precipitation. Clim. Res. 28, 95–107.

Chen, C.C., Gillig, D., McCarl, B.A., 2001. Effects of climatic change on a water dependent regional economy: a study of the Texas Edwards Aquifer. Clim. Change 49 (4), 397–409.

Cressman, G.D., 1959. An operational objective analysis system. Mon. Weather Rev. 87 (10), 367–374.

Crutzen, P.J., Zimmermann, P.H., 1991. The changing photochemistry of the troposphere. Tellus 43AB, 136–151.

Dai, A., Wigley, T.M.L., Boville, B.A., Kiehl, J.T., Buja, L.E., 2001. Climates of the twentieth and twenty-first centuries simulated by the NCAR climate system model. J. Clim. 14 (4), 485–519.

Eckhardt, K., Ulbrich, U., 2003. Potential impacts of climate change on groundwater recharge and streamflow in a central European low mountain range. J. Hydrol. 284, 244–252.

Eitzinger, J., Stastna, M., Zalud, Z., Dubrovsky, M., 2003. A simulation study of the effect of soil water balance and water stress in winter wheat production under different climate change scenarios. Agric. Water Manage 61, 195–217.7.

First National Climate Change Report of the Kingdom, 2005.

Galloway, J.N., Dentener, F.J., Capone, D.G., Boyer, E.W., Howarth, R.W., Seitzinger, S.P., Asner, G.P., Cleveland, C., Green, P.A., Holland, E., Karl, D.M., Michaels, A.,

Porter, J.H., Townsend, A., Vorosmarty, C., 2004. Nitrogen cycles: past, present, and future. Biogeochemistry 70, 153–226.

Harvey, J.G., 1982. Atmosphere and Ocean. Our Fluid Environment. The Vision Press, London, 143 pp.

Held, I.M., Suarez, M.J., 1994. A proposal for the intercomparison of the dynamic cores of atmospheric general circulation models. Bull. Am. Meteorol. Soc. 75 (10), 1825–1830.

Hill, D., Goldberg, R., 2001. Energy demand. In: Rosenzweig, C., Solecki, W.D. (Eds.), Climate Change and a Global City: An Assessment of the Metropolitan East Coast Region. Columbia Earth Institute, New York, pp. 121–147, 210 pp.

IPCC, 1990. Climate change: the IPCC scientific assessment. IPCC first assessment report (FAR).

IPCC, 1992. The supplementary report to the IPCC scientific assessment.

IPCC, 2001a. Climate change 2001: The scientific basis. In: Houghton, J.T., Ding, Y., Griggs, D.J., Noguer, M., van der Linden, P.J., Dai, X., Maskell, K., Johnson, C.A. (Eds.), Contribution of Working Group I to the Third Assessment Report of the Intergovernmental Panel on Climate Change. Cambridge University Press, Cambridge, p. 881.

IPCC, 2001b. Climate change 2001: impacts, adaptation and vulnerability – contribution of working group II to the third assessment report (TAR). Intergovernmental Panel on Climate Change.

IPCC, 2007. The physical science basis. IPCC fourth assessment report working group I report, Cambridge University Press, New York.

Jones, P.D., Kilsby, C.G., Harpham, C., Glenis, V., Burton, A., 2009. UK climate projections science report: projections of future daily climate for the UK from the weather generator. University of Newcastle, UK, ISBN 978-1-906360-06-1.

Keeling, R.F., 1995. The atmospheric oxygen cycle: the oxygen isotopes of atmospheric CO_2 and O_2 and O_2/N_2 ratio. Rev. Geophys. (Supplement), 1253–1262, U. S. national report to International Union of Geodesy and Geophysics.

Kilsby, C.G., Jones, P.D., Burton, A., Ford, A.C., Fowler, H.J., Harpham, C., James, P., Smith, A., Wilby, R.L., 2007. A daily weather generator for use in climate change studies. Environ. Model Softw. 22, 1705–1719.

Kotwicki, V., Al-Sulaimani, Z., 2009. Climates of the Arabian Peninsula – past, present, future. Int. J. Clim. Change Str. Manage. 1 (3), 297–310.

Krüger, A., Ulbrich, U., Speth, P., 2002. Groundwater recharge in Northrhine-Westfalia by a statistical model for greenhouse gas scenarios. Phys. Chem. Earth, Part B: Hydrol. Oceans Atmos. 26, 853–861.

Kundzewitz, Z.W., Mata, L.J., Arnell, N.W., Döll, P., Kabat, P., Jimenez, B., Miller, K.A., Oki, T., Şen, Z., 2007. Freshwater resources and their management. In: Parry, M.L., Canziani, O.F., Palutikof, J.P., van der Linden, P.L., Hanson, C.E. (Eds.), Climate Change 2007: Impacts, Adaptations and Vulnerability, Contribution of Working Group II to the Fourth Assessment Report and of the Intergovernmental Panel on Climate Change. Cambridge University Press, Cambridge.

Loaiciga, H.A., Maidment, D.R., Valdes, J.B., 2000. Climate-change impacts in a regional karst aquifer, Texas, USA. J. Hydrol. 227, 173–194.

Lorenz, E.N., 1963. Deterministic non-periodic flow. J. Atmos. Sci. 20, 130–141.

Meyer, H.W., 1992. Lapse rates and other variables applied to estimating paleo altitudes from fossil floras. Paleogeogr. Paleoclimatol. Paleoecol. 99 (Issues 1–2), 71–99.

Milly, P.C.D., Dunne, K.A., Vectis, A.V., 2005. Global pattern of trends in streamflow and water availability in a changing climate. Nature 438, 347–350.

Nohara, D., Kith, A., Osaka, M., Oki, T., 2006. Impact of climate change on river runoff. J. Hydromet. 7, 1076–1089.

North, G.R., Bomar, G., Griffiths, J., Norwine, J., Valdes, J.D., 1995. The changing climate of Texas. In: North, G.R., Schmandt, J., Clarkson, J. (Eds.), The Impact of Global Warming on Texas. The University of Texas Press, Austin, TX, pp. 24–49.

Oki, T., Agata, Y., Kanae, S., Saruhashi, T., Yang, D.W., Musiake, K., 2001. Global assessment of current water resources using total runoff integrating pathways. Hydrol. Sci. J. 46, 983–995.

Postel, S., Vickers, A., 2004. Bosting water productivity. In: Starke, L. (Ed.), State of the World 2004. Worldwatch Institute, Washington, D.C., pp. 46–65.

Redmond, K.T., 2002. The depiction of drought: a commentary. Bull. Am. Meteorol. Soc. 83, 1143–1147.

Rosenzweig, C., Casassa, G., Imeson, A., Karoly, D.J., Chunzhen, Liu, Menzel, A., Rawlins, S., Root, T.L., Seguin, B., Tryjanowski, P., 2007. Assessment of observed changes and responses in natural and managed systems. In: Parry, M.L., Canziani, O.F., Palutikof, J.P., van der Linden, P.J., Hanson, C.E. (Eds.), Climate Change 2007: Impacts, Adaptation and Vulnerability. Contribution of Working Group II to the Fourth Assessment Report of the Intergovernmental Panel on Climate Change. Cambridge University Press, UK, pp. 79–131.

Santer, B.D., Wigley, T.M.L., Barnett, T.P., Anyamba, E., 1996. Houghton, J.T., Meira Filho, L.G., Callander, B.A., Harris, N., Kattenberg, A., Maskell, M. (Eds.), Detection of Climate Change, Contribution of Working Group I to the Second Assessment Report on the Intergovernmental Panel on Climate Change. Cambridge University Press, New York, pp. 407–443.

Şaylan, L., 2000. The effect of the environmental factors on the evapotranspiration in different growth phases of soybean. Die Bodenkultur 51 (2), 127–134.

Schindler, D., 2001. The cumulative effects of climate warming and other human stresses on Canadian freshwaters in the new millennium. Can. J. Fish. Aquat. Sci. 58, 18–29.

Seckler, D., Amarasinghe, U., Molden, D., de Silva, R., Barker, R., 1998. World water demand and supply, 1990 to 2050, scenarios and issues. International Water Management Institute Research Report 19, Sri Lanka.

Şen, Z., 1974. Small sample properties of hydrologic processes and the Hurst phenomenon. Unpublished Ph.D. ThesisUniversity of London, Imperial College of Science and Technology, 256 pp.

Şen, Z., 1989. Cumulative semivariogram model of regionalized variables. Int. J. Math. Geol. 21, 891.

Şen, Z., 2008a. Global warming threat on water resources and environment: a review. Environ. Geol. 57 (2), 321–329.

Şen, Z., 2008b. Wadi Hydrology. CRC Press, Taylor and Francis Group, 347 pp.

Şen, Z., 2009a. Spatial Modeling Principles in Earth Sciences. Springer, New York, 351 pp.

Şen, Z., 2009b. Precipitations downscaling in climate modeling using a spatial dependence function. Int. J. Global Warm. 1 (1–3), 29–42.

Şen, Z., 2010. Fuzzy Logic and Hydrological Modeling. CRC Press, Taylor and Francis Group, 340 pp.

Şen, Z., 2012. Groundwater risk management assessment in arid regions. Water Resour. Manag. 26 (15), 4509–4524.

Şen, Z., 2013a. Desertification and climate change: Saudi Arabian case. Int. J. Global Warm. 5 (3), 270–281.

Şen, Z., 2013b. Urban climate change impact and Istanbul Water Consensus. Int. J. Global Warm. 5 (2), 210–229.

Şen, Z., 2014. Practical and Applied Hydrogeology. Elsevier, 486 pp.

Şen, Z., Kadioğlu, M., 2000. Simple daily dynamic adaptive operation rules for water resources optimization. Water Resour. Manag. 14 (5), 349–368.

Şen, Z., Uyumaz, A., Cebeci, M., Öztopal, A., Küçükmehmetoğlu, M., Özger, M., Erdik, T., Sırdaş, S., Şahin, A.D., Geymen, A., Oğuz, S., Karsavran, Y., 2010. The Impacts of Climate Change on Istanbul and Turkey Water Resources. Istanbul Metropolitan Municipality, Istanbul Water and Sewerage Administration, Istanbul, 1500 pp.

Şen, Z., Al Alsheikh, A., Al-Turbak, A.S., Al-Bassam, A.M., Al-Dakheel, A.M., 2011. Climate change impact and runoff harvesting in arid regions. Arab. J. Geosci. 6 (1), 287–295.

Şen, Z., Alsheikh, A.A., Alamoud, A.S.M.A., Al-Hamid, A.A., El-Sebaay, A.S., Abu-Risheh, A.W., 2012. Quadrangle downscaling of global climate models and application to Riyadh. J. Irrig. Drain. Eng. 138 (10), 1–7.

Şen, Z., Al Sefry, S.A., Al Ghamdi, S.A., Wahib, A.A., Bardi, W.A., 2013. Strategic groundwater resources planning in arid regions. Arab. J. Geosci. 6, 4363–4375.

Subyani, A., Şen, Z., 2006. Refined chlorine mass-balance method and its application in Saudi Arabia. Hydrol. Process. 20, 4373–4380.

Tao, F., Yokozawa, M., Hayashi, Y., Lin, E., 2003. Future climate change, the agricultural water resources, and agricultural production in China. Agric. Ecosyst. Environ. 95, 203–215.

Trenberth, K.A. (Ed.), 1992. Climate System Modeling. Cambridge University Press, Cambridge.

United Nations, 2002. International Migration Report. Department of Economic and Social Affairs Population Division, New York, 78 pages.

United Nations, 2003. Water for People, Water for Life. World Water Development Report, World Water Assessment Report, New York, 32 pages.

Vörösmarty, C.H., 2000. Change and population growth global water resources: vulnerability from climate. Science 289, 284.

Walker, G.K., 1989. Model for operational forecasting of western Canada wheat yield. Agric. For. Meteorol. 44, 339–351.

Wilby, R.L., Harris, I., 2006. A framework for assessing uncertainties in climate change impacts: low flow scenarios for the River Thames, UK. Water Resour. Res. 42, http://dx.doi.org/10.1029/2005WR004065, W02419.

Wilby, R.L., Wigley, T.M.L., 1997. Downscaling general circulation model output: a review of methods and limitations. Prog. Phys. Geogr. 21, 530–548.

Wilks, D.S., 1992. Adapting stochastic weather generation algorithms for climate change studies. Clim. Chang. 22, 67–84.

Willems, P., Vrac, M., 2011. Statistical precipitation downscaling for small-scale hydrological impact investigations of climate change. J. Hydrol. 402, 193–205.

Yevjevich, V., Cunha, L.D., Vlachos, E., 1983. Coping with Droughts. Water Resources Publications, Littleton, CO, 417 pp.

Zorita, E., von Storch, H., 1999. The analog method as a simple statistical downscaling technique: comparison with more complicated methods. J. Clim. 12 (8), 2474–2489.

Chapter SEVEN

Drought Hazard Mitigation and Risk

CHAPTER OUTLINE

7.1 GENERAL

Recently, drought hazard mitigation sides of scientific activities started to play a vital role through various prediction models, algorithms, and formulations more than ever before. These opportunities are empowered by weather forecasting procedures, agricultural practice, natural resources management, disaster

Applied Drought Modeling, Prediction, and Mitigation

prevention, and preparedness. Rational decisions can be made on the basis of proper modeling and forecasting work, which support disaster vulnerability reduction and drought management with insignificant hazard and disaster inflictions. Provided that the necessary scientific and social knowledge and information sources are properly prepared for a region, it is easy to combat against the hazardous future drought occurrences and to set into effect the mitigation directions.

Droughts cannot be easily identified, especially at their early development stages. The early drought warning comes from any unusual dry period that results in water stress and shortage. The first drought trigger is rainfall deficiency, which leads to water shortages in the soil, rivers, or reservoirs and the consequent impact may appear as a natural hazard. In practice, rather than via rainfall deficiency, drought disasters are better understood according to their impacts on human activities and resources, such as agricultural production, water supplies, and food availability. Areas in the world with scarce rainfall are in arid regions, but drought is different than aridity ("see chapter: Basic Drought Indicators"). In arid regions, the settlers adapt their daily lives according to the present circumstances without caring much about drought effects. These areas have rather low annual rainfall amounts (less than 200 mm), but droughts are not confined only to areas of low rainfall.

The expected climate change highlights the need to have a new basis and framework for planning water resources structures by taking into consideration innovative information and knowledge for necessary adaption and mitigation actions. Public awareness about local climate change impacts is of prime importance as individuals' responsibility, otherwise there might be crucial consequences in facing the impact of fierce weather. Public information can be enriched by experience exchange based on logical verbal data and a common base of climate change impacts about future rainfall and subsequent runoff occurrences. Top-down plans on vulnerability, sensitivity, and capacities should be shared and complemented by communities for groundwater augmentation, especially under climate change impacts.

Droughts are among slowly developing natural hazards that evolve over time and cover extensive areas with various damage potentials. The only way to effectively combat drought hazard impact is through preparedness, assessment of vulnerability potential, risk management, and corresponding mitigation activities. For proper drought preparedness and mitigation, it is necessary to monitor drought-effective variables (meteorological, hydrological, agricultural, and social) for the purpose of reliable predictions and impact assessments with the least-expected hazard consequences. In the previous chapters, the necessary arsenal for effective drought vulnerability, hazard, and mitigation elements were explained in a quantitative manner with numerical examples.

In general, drought as defined by the International Negotiating Committee of Convention to Combat Desertification is the naturally occurring phenomenon that exists when rainfall is significantly below recorded normal levels, causing serious hydrological imbalance that adversely affects the land resources production system (Dambe, 1997). Recently, it can be clearly recognized that the rhythm of the changes in the natural and social environments has become more rapid, complex, and permanent. Drought generally results from a combination of natural factors that can be enhanced by anthropogenic influences. The primary cause of any drought is deficiency in precipitation, and in particular, its intensity, occurrence time, and distribution. All these processes influence everyday and future human life more directly compared to previous decades. Dryness and drought are the result of special interactions between natural and social environments. Man and society play active as well as passive roles in this process, which influences the global development of a region. In the last few decades it became also obvious that drought effects can cause damages (Naginders and Kundzewich, 1997).

As already explained in "chapter: Introduction," droughts have natural, physical, and social aspects and dimensions. Drought risks have two parts: probabilistic risk based on an event such as precipitation for meteorological drought, runoff for hydrological drought, and harvesting risk for agricultural droughts; and also vulnerability of society to the event incurring losses. In general, each drought-stricken area has different meteorological, climatological, hydrological, and social features; therefore, no two droughts should be expected to have the same impact and risk. Vulnerability and risk are interrelated such that high vulnerability level means that a population is at risk for negative drought impacts. Hence, for risk assessment concerning the society, the fundamentals of vulnerability in the drought-stricken area must be well understood and methodological steps must be implemented with joint contributions of responsible authorities and the general public. Unfortunately, there are also differences among scientists about the existence and intensity of drought; therefore, the manager or decision maker may be disappointed about such disagreements. This is also due to differences in views between the drought managers and scientists, because the former are interested in drought impact on the society, the economy, and drought mitigation works in general, whereas the latter are more physically oriented and interested in drought modeling and predictions. For successful consequences, it is necessary to encourage dialog, mutual understanding, and integration between these two ends. In practical treatments, most often the administrators, managers, and decision makers do not care much about scientific projects and their conclusions.

In a region, as a buffer during drought periods, an increase in storage possibilities also increases drought resilience due to extra water storage. The resilience can be augmented through an efficient reservoir operation, which has been functional, for instance, in Istanbul City, by means of conjunctive management ("see chapter: Introduction"). Legal agreements between political jurisdictions (central and local authorities) concerning the amount of water to be delivered from one jurisdiction to another impose legal requirements on water managers to maintain flows at certain levels. It is necessary to reduce possible conflict and tension between different consumers. Effective government policy is the most important ingredient for the management of legal aspects during water resources management implications in a region. Governments provide relief from the immediate impact of drought through livestock support and provision subsidization or free food.

This chapter provides definitions of basic terminology and meanings and explanations of various hazard and risk management, vulnerability, mitigation, and combat procedures. After the proper terminology id established, the bulk of this chapter is about drought risk assessment, management, and possible applications on the basis of qualitative and quantitative approaches.

7.2 BASIC DEFINITIONS

For better understanding of any disastrous situation the manager, decision maker, and stakeholders should be aware of distinctive meanings among the following disaster terminologies that are also common in any drought hazard case. The definitions are arranged in alphabetical order.

1. *Acceptable risk:* Acceptable risk is a level of any injury or loss from a disastrous situation that is considered to be tolerable by a society or authority in view of the social, political, and economic cost-benefit analyses. In any uncertain study, the solution cannot be achieved absolutely without error. Hence, a level of vulnerability is considered to be "acceptable" and balancing factors such as cost, equity, public input, and the probability of drought.

2. *Adaptation:* Adaptation is a physical or behavioral characteristic that has developed to allow a society to achieve better survival in its environment. In the context of this chapter, it also means adjustment to present and expected drought effects in the social system so that beneficial opportunities can be moderated by human intervention.

3. *Capacity:* In general, capacity is the ability to receive, hold, or absorb something. It is the combination of all strengths, attributes, and resources available within a community, society, or organization that can be used against disaster during the mitigation process. It is not only physical capacity such as infrastructure and institutions, but also human experience, knowledge, skills, management practices, and administrational regulations.

4. *Crisis management:* Crisis management is the process by which an organization deals with a major event that threatens to harm stakeholders or the general public. Here, it is concerned with the management of large-scale industrial and environmental disasters. Crisis management deals with drought responses and actions without any preliminary preparedness plan. Generally, it includes ineffective, poorly coordinated, and ultimately individual or governmental initiatives.

5. *Disaster:* Disasters often follow natural hazards. For instance, a disastrous drought causes calamitous, distressing, or ruinous effects in such scales that they threaten to disrupt critical functions of a society for a period long enough to significantly harm it or cause its failure. Disaster is the unprecedented alteration of a currently normal situation in a community as a result

of physically hazardous events that interact with vulnerable social conditions leading to widespread adverse human, material, economic, or environmental effects. In such a situation, it is necessary to provide an immediate emergency response to satisfy critical human needs, which may also need external support for recovery to a normal state.

6. *Disaster risk and reduction:* Disaster risk is the disaster occurrence likelihood over a certain time period, again due to hazardous physical events as specified in the previous item. It is equal to the multiplication of hazard and vulnerability. According to the UN Office for Disaster Risk Reduction, it aims to reduce the damage caused by natural hazards like droughts, earthquakes, floods, and cyclones through an ethic of prevention.

7. *Disaster risk management:* A management effort that includes the processes after evaluating the disaster and its risk, design, implementation, strategy evaluation, policy making, and specification of measures in order to improve disaster risk appreciation and reduction. In such a management effort, continuous preparedness improvement promotion and recovery possibilities should also be considered for additional human security, property, life quality, resilience, and sustainable development. Incorporation of disaster risk management into development plans can help to lower the impact of drought disasters on property and human lives.

8. *Drought contingency plan:* A contingency plan is a process that prepares an organization to respond coherently to an unplanned event. It identifies specific actions that can be taken before, during, and after a drought occurrence to mitigate some of the resulting impacts and conflicts. Frequently, these actions are triggered by a monitoring system. For instance, public water systems must have a contingency plan ready in case of drought or similar water shortages.

9. *Drought impact:* In general, impact means a place open to assault and difficult to defend against a natural hazard. In the context of this chapter, it is a specific effect of drought on the environment and human activities with undesirable consequences or outcomes.

10. *Drought impact assessment:* Drought impact assessment is the process of looking at the magnitude and distribution of drought effects.

11. *Exposure:* Exposure implies adversely affected people, livelihoods, environmental services and resources, infrastructure, or economic, social, or cultural assets. For instance, the mental health impact of drought is poorly quantified and a relationship exists between distress and environmentally based measures of drought, which is referred to as drought mental exposure (O'Brien et al., 2014).

12. *Hazard:* Hazard is a dangerous phenomenon, substance, human activity, or condition that may cause loss of life, injury, or other health impacts, property damage, loss of livelihoods and services, social and economic disruption, or environmental damage. It represents future threats to the society such as natural drought events. Any precaution against the hazard should not depend only on scientific methodologies, but additionally the

experience of the society should also be taken into consideration. In the context of this chapter, drought hazard is a threatening event that would make supply inadequate to meet demand.

13. *Mitigation:* Mitigation is the act of lessening (reducing) the effect or impact of any natural disaster such as drought. In other words, it is the act of a condition or less-severe consequence. For this purpose, short- and long-term actions, programs, or policies should be implemented in advance of drought, or during its various stages it should be reduced to the degree of risk to people, property, and productive capacity.

14. *Preparedness:* Preparedness is a mitigation action prior to the occurrence of disaster. It corresponds to the state of being prepared or readiness against a disaster such as drought. Predisaster activities are designed to increase the level of readiness or improve operational capabilities for responding to a drought emergency.

15. *Resilience:* Resilience is the ability of a system with its component parts to anticipate, absorb, accommodate, or recover from the effects of a hazardous event in a timely and efficient manner. During this process, preservation, restoration, or improvement in essentially basic infrastructure and its components must be ensured for the safety of the community or society. Policy makers are facing increasing calls to consider the resilience of communities that rely on ecosystem goods and services, and the resilience of natural systems themselves. These calls are in response to increasing threats to communities in general from external factors such as environmental (possibly associated with climate change), social (reductions in natural resources), and economic (changes in local and regional economic conditions) hazards. Unfortunately, most communities have had little experience in explicitly managing for resilience.

16. *Response:* Response is a congregation in reply to the officiant, which is the drought occurrence in this case. It includes a set of actions taken immediately before, during, or directly after a drought event to reduce impacts and improve recovery for the next drought occurrence. Response measures are important parts of drought preparedness as one part of a more comprehensive mitigation strategy.

17. *Risk:* Risk is a potential of losing something or a valuable source. This takes into consideration negative consequences of any event such as the drought phenomenon based on numerical data treatment by probabilistic methodologies. The consequence of any risk is the total loss, which is referred to as the potential losses, say, due to drought occurrence, duration, and areal coverage. It is exposure to the chance of injury or loss; a hazard or dangerous chance. It is concerned with the potential adverse effects of drought as a product of both the frequency and severity corresponding to vulnerability.

18. *Risk analysis:* Risk analysis is the study of the underlying uncertainty of a given course of action, which are the drought impacts in this book. Risk analysis refers to the uncertainty of forecasted future drought probability,

which depends on statistical analysis in order to determine the probability of a drought success or failure leading to possible future economic losses. It also corresponds to the process of identifying and understanding the relevant components associated with drought risk as well as the evaluation of alternative strategies to manage the risk.

19. *Risk management:* Risk management is the identification, assessment, and prioritization of risks followed by systematic application of available resources so as to minimize, monitor, and control the probability. Risk management is the opposite of crisis management, where a proactive approach is taken well in advance of drought so that mitigation can reduce drought impacts, and the relief and recovery decisions are made in timely, systematic, coordinated, and effective manners during a drought.

20. *Transformation:* Transformation refers to various alterations in the fundamental attributes of a societal system (value attributes, regulations, legislative or bureaucratic regimes, financial institutions, and technological systems).

21. *Vulnerability:* Vulnerability means adversely affected predisposition and it is one of the main problems to be dealt with in any natural disaster such as a drought. It can also be defined as defenselessness, insecurity, exposure to risk and difficulty in coping with them through adaptation and mitigation. Vulnerability should also describe how people, buildings, and infrastructure are susceptible to damage that may be created by drought hazard or by any other hazard type. Impacts are related to vulnerability, which is related to population, assessments, activities, or environmental characteristics that are susceptible to drought effects. In a region the degree of vulnerability depends on the environmental and social characteristics. The vulnerability can be measured by the ability to anticipate, cope with, resist against, and recover from drought.

22. *Vulnerability assessment:* Vulnerability assessment is concerned with the processes of identification, quantification, prioritization, and ranking different vulnerability types in a system. Vulnerability assessment is also concerned with drought-related impacts by taking these into consideration within the framework of identification and prediction. In addition to drought impact, other adverse social, economic, and environmental conditions may also contribute to vulnerability; hence, the overall assessment may be achieved.

7.3 GOALS AND OBJECTIVES

The development of goals and objectives for drought mitigation helps the community envision how the prepared plans can be implemented before and during an actual drought event. For the planning process to result in a more resilient community, the goals and objectives should be known transparently by different stakeholders and responsible authorities jointly. Long-term, consistent guidelines for the adoption and implementation of effective policies and strategies are necessary for developing a resilient community (Tang et al., 2008; Burby, 2005; Nelson and French, 2002).

Any multihazard mitigation plan should include community expectations about the drought mitigation procedures toward a common and successful goal, which would help to identify the desired outcomes related to the community's ability to mitigate against, prepare for, respond to, and recover from a given drought threat or disaster. Objectives should be more concrete than goals in that they should provide specific, measurable, and achievable statements and advice (Tang et al., 2008). They should be directly related to specific activities, implementation, operating procedures, and preparedness measures by taking into consideration available resources with additional required resources during a drought. Goals are directly related to drought mitigation and preparedness including primarily water consumption reduction requirements (Vickers, 2005; Wade, 2000), public education related to the vulnerability (Wilhite, 2000; Burby et al., 2000), and property protection (FEMA, 2008; Godschalk, 2003).

In the front line of any drought mitigation plan, water conservation is the main concern. To achieve such a successful plan, it is necessary to establish water conservation programs for the community to implement significant water consumption reductions. It is well known that for proper protection and, hence, reduction in hazard impacts and disaster losses, public education is a key component in any drought hazard mitigation, which can result in empowerment for residents to enact personal conservation efforts.

7.3.1 Policies, Strategies, and Tools

The bulk of any drought mitigation plan lies on the foundation of specific policies, tools, and strategies for realizing the stated goals and objectives. Policies and strategies aim at what will be necessary to help the core capabilities to remain intact during a disaster situation. Specific strategies and policies may include regulations, protective policies, land-use restrictions, and incentives (Tang et al., 2008).

Among the most significant tools related to drought is the development of an early warning system related to local timely drought impacts (Fontaine et al., 2012; Gupta et al., 2011; Knutson, 2008; Wilhite et al., 2000). These systems should include soil moisture, streamflow, snowpack, reservoir levels, and groundwater monitoring (Steinemann et al., 2005). An early warning system in place enables communities to take early actions, which have proven to be more effective in drought impact combat and mitigation studies (Vickers, 2005). Specific to drought, policies and tools that are directed at the agricultural sectors have particular importance (Rockstrom, 2003). Potential preparedness and mitigation measures related to the agricultural sector include improved soil and water management practices, in addition to adaptive tilling practices, intercropping, crop insurance, and agricultural irrigation standards (FEMA, 2013; Rockstrom, 2003; Wilhelmi and Wilhite, 2002).

Two types of water conservation approaches are effective in practical applications: behavioral approaches that depend on human reactions before and during an actual drought, and remedial solutions through technological gadgets.

Among the human behavior-dependent activities are short shower durations, reduction in lawn watering frequencies and periods, reduced car washing, and water fountain turn offs. On the other hand, technological aids include low-flow fixtures, water system audits, and improvements in the irrigation systems such as the use of dripping water systems instead of surface (furrow, flood, or level basin) irrigation systems. Among other relevant methods are water, rainfall, and runoff harvesting (RH) systems, voluntary water restriction incentives, and water conservation pricing ("see chapter: Climate Change, Droughts and Water Resources"). In the Kingdom of Saudi Arabia, rainfall and RH projects and their actual implementations reduce the stress on potable water (Şen et al., 2011). Schmidt and Garland (2012) discuss water reuse as a form of redundancy in hazard planning; this redundancy is achieved by introducing a water source that was not previously available. The use of water restrictions is a common tool used by municipalities to address short-term water shortages (Knutson, 2008; Kenney et al., 2004). Mandatory water restrictions prove to be more effective in water consumption reduction.

During the last 5–6 decades, diesel and electric motors led to water abstraction systems that can pump groundwater out of major aquifers faster than drainage basins replenishment and recharge. This has led to permanent loss of groundwater storage (aquifer) capacity coupled with water-quality deterioration, ground subsidence, and social problems. In many places, presently as well as expected in the future, food production will be threatened by this phenomenon. This is also one of the drought-triggering factors in excessive (more than safe yield) groundwater withdrawal without proper planning and management strategies (Şen, 1995, 2014; and "chapter: Climate Change, Droughts and Water Resources").

Understanding and improving of drought hazard and subsequent mitigation studies at micro and macro scales are possible by enhancing drought preparedness and mitigation efforts in all aspects of life. It is necessary to emphasize greater understanding and description of both the physical features of the hazard and the social factors that influence societal vulnerability. Any society that lacks risk-based drought management programs and early warning systems is under the threat of the worst effective drought impact. In such cases, drought crisis management activities are the only alternative to combat the drought impact. Not only should managers, decision makers, and scientists be aware of drought hazards and possible consequences, but importantly society should be, too. Lack of predrought preparedness is characterized by delayed crisis response in the postdrought setting. This situation will have negative impacts and result in a long period of recovery. In order to avoid such a worst-case scenario, it is necessary to have a drought-resilient society, a risk-based drought policy, preparedness plans, and proactive mitigation strategies. It is part of a long-term management strategy directed at reducing societal vulnerability to drought. An early warning system integrates a wide range of physical and social indicators and their implementation.

7.4 DROUGHT WATCHES SYSTEMS AND RELIEF

Early warning helps to inform individuals, institutions, and the community at large about drought conditions and the necessary mitigation measure implementations. In such a system, the following components are necessary in an integrated manner.

1. Hydrometeorological variables must be monitored through an automated network of instruments.
2. Automatic monitoring of water resources availability also must be affected for reserve (surface reservoir and groundwater aquifer) assessments.
3. Data from the previous two steps must be collected, stored, and treated through suitable methodologies at a research center, where drought indices and spatiotemporal evolution of droughts can be achieved through effective software such as GIS.
4. After the treatment of all available information, the output information must be provided to anyone concerned with drought danger through e-mails, Internet accesses, mobile phones, and other media.
5. Necessary training and seminar programs must be arranged by experts for the local population, and relevant brochures and circulars must be printed and distributed for public awareness.
6. A public campaign must be started for making people aware of the importance of a drought monitoring service.

Different drought-related alternative criteria may have crisp, probabilistic, or fuzzy information content. In a multicriterion approach, each one of these information sources must be considered in a joint manner for the best mitigation alternative. Some of this information needs to enter a prediction model, but others may support the outcomes of such models linguistically toward more refined solutions.

Drought mitigation programs can be visualized along two branches as short- and long-term measures. Among the short-term activities are the supplements from groundwater resources (balanced recharge and abstraction plans), programs and plans for water scarcity management, and natural supports such as a terrible event that causes a lot of damage or harm.

Also among the short-term solutions are the use of rather marginal water sources such as ponds, relaxation of water-quality limitations, and the efficient use of water resources. In case of local droughts, it is also among the short-term implications to reduce the water supply at such rates that the consumer cannot feel the loss physiologically. Short-term demand reduction may be effected by means of reductions in municipal consumption, annual crop restrictions, and implementation of mandatory rationing.

As for the long-term mitigation instruments, it is possible to count at least the revision and improvement of the current water distribution and management system, water transportation from nearby or far distances (see the case for Istanbul City in "chapter: Introduction"), cases of agriculture modernization of irrigation systems (instead of traditional watering systems better to introduce sprinkler or even dripping system alternatives), waste water treatment and reuse activities, local and small dam construction not only for water supply purposes but also for groundwater recharge possibilities, and RH especially from floods and flash floods. Additionally, seawater desalinization plant construction is for long-term drought mitigation purposes. On the demand reduction side, long-term activities include municipal network design for rainwater and wastewater management, industrial water recycling, and especially reduction in irrigation consumption through modern irrigation techniques and crop pattern alternatives.

To cope with drought, it is necessary to shift emergency management to a proactive approach. The experience gained throughout the years about droughts and their watch systems represent an effective tool for implementing such a proactive approach. Lessons drawn from previous analyses about short- and long-term drought mitigation measures in a water supply system confirm that a certain method may be capable to describe multiple societal viewpoints and stakeholder interests to foster the decision-making process about drought vulnerability and mitigation tasks.

In all drought mitigation work, not only is water supply augmentation a top priority, but also demand reduction is equally important. As for the long-term strategic solutions, and increase in the water supply can be achieved through the construction of additional impoundments, water transfer in addition to wastewater reuse, and desalination plant construction. On the other hand, in drought mitigation for supply increase and demand reduction purposes the following short- and long-term policies may be followed:

1. Drought contingency plans based on scientific and technological aspects.
2. Water resources allocation based on quality classifications.
3. Implication of economic policies.
4. Temporary reallocation of water resources.
5. Rehabilitation and education programs.
6. Tax relief and public aid.

Recent climate change implies dry climate belts around the equator will shift toward the polar regions by 225–250 km (this chapter). Therefore, it is necessary for many countries, especially those in the subtropical regions, to plan convenient planting patterns under different climate scenarios from now into the future. After long years' experience, Wilhite et al. (1996) suggested the following points for drought combat and mitigation activities:

1. After obtaining reliable, timely information about a drought situation, its effects must be predicted and the conclusions must be passed to convenient authorities so that they can take necessary precautions. For success, it is necessary to have drought characteristics' treatment based on the real meteorological measurements.
2. Effective evaluation methodologies must be empowered with new scientific information and technologies. To inform the decision makers' final decisions, it is necessary to develop new drought solution methodologies. For this purpose, the decision makers must grasp the importance of drought

and its intensity for pondering drought impact on agriculture, and accordingly, for taking the necessary precautions.
3. All knowledge and information must be gathered in the same center and the decisions from this center must be applied locally and regionally. For the necessary preliminary assistance and aid, the plans must be prepared beforehand and the necessary studies must be continuously updated with the arrival of new information,
4. For reaching drought targets through the previous studies, there must not be delays in assistance programs and they must be announced with proper management plans and administered by a professional center.

7.4.1 Early Warning System

Although an understanding of underlying vulnerability is essential to understand the risk of drought in a particular location and for a particular group of people, a drought early warning system should be designed to identify negative trends and thus to predict both the occurrence and the impact of a particular drought and to elicit an appropriate response (Buchanan-Smith and Davies, 1995). Numerous natural indicators of drought should be monitored routinely to determine drought onset, end, and spatial characteristics. Severity must also be evaluated continuously in frequent time steps and internationally, some of the most prominent challenges with developing drought monitoring and early warning systems, which have been noted, are given below (World Meteorological Organization, 2006).
1. Meteorological and hydrological data networks are often inadequate in terms of the density of stations for all major climate and water supply parameters. Data quality is also a problem because of missing data or an inadequate length of record.
2. Data sharing is inadequate between government agencies and research institutions, and the high cost of data limits their application in drought monitoring, preparedness, mitigation, and response studies.
3. Information delivered through early warning systems is often too technical and detailed, limiting its use by decision makers.
4. Forecasts are often unreliable on a seasonal time scale and lack of specificity, reducing their usefulness for agriculture and other sectors.
5. Drought indices are sometimes inadequate for detecting the early onset and end of drought ("see chapter: Basic Drought Indicators").
6. Drought monitoring systems should be integrated, coupling multiple climate, water, and soil parameters and socioeconomic indicators to fully characterize drought magnitude, spatial extent, and potential impact.
7. Impact assessment methodologies, a critical part of drought monitoring and early warning systems, are not standardized or widely available, hindering impact estimates and the creation of regionally appropriate mitigation and response programs.
8. Delivery systems for disseminating data to users in a timely manner are not well developed, limiting their usefulness for decision support.

Some factors that can provide the basis for recognizing drought at an early stage include the following points (United Utilities, 2008):

1. A high probability of sources failing to meet demand or to refill sufficiently.
2. Storage in reservoirs below control curve levels. Rapid weekly decline in stocks (or slow rise in stocks during winter) of key reservoirs.
3. Low and declining river flows at river sources resulting in abstraction being limited.
4. Significant reductions in the output from spring and groundwater sources.
5. Significant declines in groundwater levels as measured at key observation boreholes in major aquifers.
6. Magnitude and duration of peak demands significantly higher than normal for the time of year.
7. Rainfall amounts significantly below average for periods of 2 months or longer durations.

The following points are suggested by Wilhite (2005) for early drought warning precautions:

1. *Data networks:* Inadequate station density, poor data quality of meteorological and hydrological networks, and lack of networks on all major climate and water supply indicators reduce the ability to accurately represent the spatial pattern of these indicators.
2. *Data sharing:* Inadequate data sharing between government agencies and the high cost of data limit the application of data in drought preparedness, mitigation, and response.
3. *Early warning system products:* Data and information products are often too technical and detailed. They are not accessible to busy decision makers, who, in turn, may not be trained in the application of this information for decision making.
4. *Drought forecasts:* Unreliable seasonal forecasts and the lack of specificity of information provided by forecasts limit the use of this information by farmers and others.
5. *Drought monitoring tools:* Inadequate indices exist for detecting the early onset and end of drought, but the standardized precipitation index was cited as an important new monitoring tool to detect the early emergence of drought ("see chapter: Basic Drought Indicators").
6. *Integrated drought/climate monitoring:* Drought monitoring systems should be integrated and based on multiple physical and socioeconomic indicators to fully understand drought magnitude, spatial extent, and impacts.
7. *Impact assessment methodology:* Lack of impact assessment methodology hinders impact estimates and the activation of mitigation and response programs.
8. *Delivery systems:* Data and information on emerging drought conditions, seasonal forecasts, and other products are often not delivered to users in a timely manner.
9. *Global early warning system:* No historical drought database exists and there is no global drought assessment product that is based on one or two

key indicators, which could be helpful to international organizations, non-governmental organizations, and others. As has now been well documented, early warning alone is not enough to improve drought preparedness (Buchanan-Smith and Davies, 1995). The key is whether decision makers listen to the warnings and act on them in time to protect livelihoods before lives are threatened. There are many reasons why this is often the missing link. For example, risk-averse bureaucrats may be reluctant to respond to predictions, instead waiting for certainty and quantitative evidence. This invariably leads to a late response and to hard evidence that the crisis already exists.

7.5 DROUGHT MITIGATION PLANNING HISTORY AND OBJECTIVES

The content of drought mitigation effective planning should include require-ments and specifically should have a clear understanding of the drought phe-nomenon and how it can be accounted for as a part of the mitigation process (Hayes et al., 2004). This typically is manifested in drought plans that rely on crisis response techniques rather than striking a balance between crisis response and risk management (Wilhite et al., 2000; Knutson et al., 1998).

Drought differs from other natural hazards in three main ways. First, drought is a creeping phenomenon with no clear onset or regression (Tannehill, 1947). In many cases there is a failure to identify a developing drought until there are tan-gible impacts like dead or dying crops or an insufficient supply of water to meet the regular water demands. As a result there have been many calls for the devel-opment of a drought early warning system (Wilhite and Smith, 2005; Wilhite and Svoboda, 2000; Lohani and Loganathan, 1997). The first four priorities in drought mitigation are expressed according to their significance as

1. The planning process,
2. Hazard identification and risk assessment,
3. Mitigation strategies,
4. Plan updates, evaluations, and implementation.

These items are interrelated and not completely independent from other drought mitigation activities. The planning process, hazard identification, and risk assessment can be achieved by a strong factual basis establishment for the plan-ning area, where both numerical and verbal data should be collected and treated together for an optimal approach to planning. For effective drought mitigation, one should look at population, rainfall and temperature, water demand and sup-ply trends, land use, and vegetation cover within the subject area. Among the mitigation strategies are goals and objectives for the planning area as well as specific strategies, tools, and policies that may be used to achieve the stated goals and objectives.

Communities must develop a firm factual basis for the planning process. This is especially true for events such as drought that can have devastating impacts (Tang et al., 2008). A factual basis can be established by conducting an analysis of the population in the planning area. This analysis must include an inventory of vulnerable populations so that mitigation actions can be developed (CDC, 2010). As related to drought, vulnerable populations include the agriculture sector, tourism, animal husbandry, and communities with insufficient water supplies (Schmidt and Garland, 2012; Gupta et al., 2011; CDC, 2010). Establishing a clear definition for the threat is a critical element in establishing a factual basis. The identification of past events is essential in the process of managing the threats that the community currently faces, as doing so aids in the development of possible mitigation strategies (FEMA, 2008, 2013; Fontaine et al., 2012; Gupta et al., 2011). Establishing a factual basis for a plan by analyzing temporal past events ("see chapter: Temporal Drought Analysis and Modeling") as well as identifying and mapping potential hazards ("see chapter: Regional Drought Analysis and Modeling") provide parameters that can guide decisions regarding future events.

Drought hazard identification and risk analysis is a very complex issue due to the temporal and extensive spatial variability effects. In the study of such risk analysis, concerned communities need to consider the drought frequency and duration in addition to potential impacts and specific vulnerabilities (Wilhite et al., 2007; Geringer, 2003). In some cases planners fail to assess objectively the drought hazards a community faces, typically by underestimating the potential threat (Newkirk, 2001).

Drought manifestation progression is a key component of drought preparedness from the community viewpoint, which is accomplished by defining drought levels or classifications and the triggers that will be used in drought progression identification (Fontaine et al., 2012; Knutson, 2008; Steinemann and Cavalecanti, 2006; Steinemann et al., 2005; Wilhite et al., 2000). Communities should plan not only for drought progression levels but also drought regression levels/criteria. The identification of the community's water sources is essential in understanding the potential threat that the community faces (Fontaine et al., 2012). Monitoring the community's water supply leading up to drought helps in mounting a timely and effective response (CDC, 2010; Wilhite and Svoboda, 2000). In doing so, attention should be given to water supply as well as concerns with water quality or water system capacity (Gupta et al., 2011; Wilhite, 2000).

The composition of local land use is another factor that should be analyzed and understood when planning for any natural hazard type (Burby et al., 2000; Burby, 2005; Nelson and French, 2002). Land-use concerns specially related to drought include the vulnerability of agricultural lands, tourism-dependent areas, recreational areas, and other urban features such as parks and tree canopies.

In addition to codes addressing interior improvements, landscape standards can be enacted that are helpful in reducing water consumption (FEMA, 2008, 2013; Schmidt and Garland, 2012; Vickers, 2005). These standards can be used to require high-efficiency irrigation heads (Schmidt and Garland, 2012; Knutson, 2008) and rainwater harvesting (Gupta et al., 2011; CDC, 2010). Enhanced building and landscape standards represent impacts that can be made by all residents in the community, which is essential in developing a resilient community (Wardekker et al., 2010).

Reinforcing and improving infrastructure represent another area that can be addressed with specific strategies and tools. In 2000, it was estimated that water supply systems experienced 10–20% of water loss that was unaccounted for (Lahlou, 2001). This water loss was a result of water theft, unapproved users, and unmetered uses (eg, firefighting), but the most significant source of water loss was leakage. Water leakage can be addressed through the auditing of water delivery systems (Knutson, 2008; Vickers, 2005). There are many benefits to auditing water systems that include reduced property damages, more responsible consumption of resources, and improved public relations between public water utilities and water customers (Lahlou, 2001). Water audits can be offered to homeowners as well as farms.

Public education programs are useful in addressing drought awareness and informing the public in how they can have an impact on the situation. Communities may consider developing a website that can be used to keep community members informed as the drought condition progresses or regresses (Wade, 2000; Wilhite et al., 2000).

7.5.1 Drought Mitigation

Drought mitigation and combat require restrictions on various social, economic, technological, and anthropogenic activities, and on the other hand, their consequences are on the cost, damage, and deterioration of human activities in every aspect of life. To reduce the actual impacts and losses during the next drought, it is necessary to be on alert-based preparedness under the light of mitigation activities and plans. Different mitigation strategies can be applied depending on the type of drought. For instance, solutions to agricultural drought epochs are possible by expanding the irrigation water facilities and, especially, by increasing the efficiency of existing systems so that crops can be grown that require less water. In many irrigation networks, less than half the water actually benefits crops and the rest is lost through seepage from unlined canals, evaporation and runoff from poorly applied water, and poor management that fails to deliver water to crops at the right time and in suitable quantities (Gleick, 1993). Further developments in the irrigation projects are not without problems. If necessary precautions and measures are not taken in a timely way, such developments may lead to other irrigation problems such as waterlogged and salinized lands, groundwater pollution, and hence, reduction in the use of water sources and destruction of aquatic habitats.

To combat possible drought occurrences in a region, the following simple but effective mitigation measures should be considered continuously. The basic idea underlying these activities is the maintenance of sustainable and balanced ecological systems.

1. Currently available irrigation systems should be improved by careful managing and the addition of new developments.

2. Pasture and cropping pattern of plant developments must be restructured for better efficiency.
3. Afforestation should be enhanced with soil and water conservation purposes.
4. Agronomic practices should be adapted suitably to regional agricultural activities.
5. Drinking water supply provisions should be considered.
6. Rural communications must be developed.
7. Small but local self-sufficient marginal farms and farmers should be supported.
8. Solutions to agricultural drought epochs are possible by expanding the irrigation water facilities and, especially, by increasing the efficiency of existing systems, so that crops can be grown with less water requirements.
9. Groundwater resources are the major water reservoirs for irrigation and agricultural productions in many arid lands of the world. They are the most extensive sources and their quality is very suitable for irrigation purposes. Therefore, in the long run, in order to combat agricultural droughts by considering effective mitigation plans, more than the surface water supplies, the subsurface water supplies are also significant. Even in the Arabian Peninsula deserts, local agricultural production, today, is possible with the help of well technology abstraction of deep groundwater and by radial sprinklers to grow crops for local and regional uses (Şen, 2014). Unfortunately, today even in the developed countries, the groundwater abstraction rates for irrigation exceed the natural recharge and due to overpumping the groundwater levels drop several meters per year. This will also increase economically the rising cost of the groundwater.
10. After the 1970s there was a slowing rate of increase in the irrigable land area in the world. Due to either pollution or overexploitation of local and nearby water resources for irrigation, there appears to be new pressure to transfer water away from agricultural areas. On the other hand, future threats such as global climatic change complicate the rational management of agriculture ("see chapter: Climate Change, Droughts and Water Resources"). Another significant mitigation toward the droughts is the maximum benefit from rainfall harvesting or rain-fed farming.

Water scarcity requires appropriate combat and mitigation plans. For this purpose, relevant drought impacts need to be ranked according to their risk priorities by consideration of environmental, social, and economic aspects of the region. Such activities provide a fundamental opportunity to decide on the appropriate mitigation movements in order to reduce short- and long-term drought risks. As explained in the previous chapters, it is necessary for an effective mitigation process to consider the qualitative and quantitative characteristics and consequences based on prediction procedures. Accordingly, the necessary preparedness and emergency measurements can be developed and implemented in an efficient manner.

The environmental, social, and economic drought impacts are not dependent scientifically on the drought duration, magnitude, and intensity only, as explained in earlier chapters, but also to a significant extent on the environmental, social, and economic vulnerability of the region. An effective mitigation cannot be achieved with numerical data only but linguistic data in terms of logical rules are also necessary for a successful end with the least loss (Şen, 2013).

Although there is an increasing alertness against drought hazards and several initiative networks, risk assessment strategies, monitoring-based prediction and mitigation with their further improvements are necessary for effective drought combat. Generally, existing strategies are reactive rather than proactive with effective prediction work. There is a continuous need for drought planning, monitoring, appropriate planning, and mitigation strategies not only through the scientific methodologies, but additionally with the support of central and local administration, people awareness, and stakeholder shares. For successful achievements, a regional drought mitigation plan is essential, which should be dependent on the factual data of the area. In fact, country drought mitigation plans prepared by central governments are the most essential documents on which different local administrations within the same country can base their local strategies.

Unfortunately, risk-reduction activities are not treated as an integral part of water resources management because most often environmental, social, and economic aspects are not taken into consideration. Multiple droughts and inadequate political and institutional capacities are among the major uncertain drawbacks in efficient mitigation plan preparation. However, in the preparation of efficient drought mitigation planning the following major points are helpful.

1. At least monthly water resources planning must be effected based on the supply and demand elements through a water balance (budget) equation.
2. The share of each water consumption sector (domestic, agricultural, and industrial) must be considered in terms of preference.
3. For an effective and reliable care of the two previous steps, a team of experts for drought mitigation must be on alert based on numerical scientific methodologies and local linguistic knowledge and information.
4. Interactive network and convenient information distribution systems should be developed through local educational seminars, alertness, and reliable information distribution by means of social or any other media.

In an integrated drought planning and response team effort toward an effective mitigation plan, two different but simultaneously functioning branches of tasks should exist. The first branch should deal with prodrought activities including drought mitigation plan research and development cooperation with related departments. This branch should depend on available drought policy or develop such a general policy for the region under its responsibility. The drought plan preparation, purpose, and objectives should be ordered according to their mitigation significance in case of actual drought. This branch should also update the current information automatically by inputting new numerical and linguistic knowledge and information. Prior to drought occurrence the same branch should identify research and development needs in the region for better improvement of drought policies and plans. Drought area political and scientific aspects should be intermingled through an effective synthesis for improved drought management strategies. The outputs of such work can be shared with the public through presentation of current drought management goals and as feedback the public may also suggest some logical rule revisions in the current drought management work. In this way, public opinion is taken into consideration and in the meantime, the public is alerted for the coming drought occurrence. To publicize the drought activities, it is among the duties of this branch to propose training programs for all stakeholders. After the drought occurrence, the same branch experts with additional experience should continue for the next drought preparation. All these activities and functions can be achieved through reliable drought predictions provided that there are effective monitoring, measurement, and knowledge and information gathering and processing units.

The second branch of expert activities for drought mitigation should be functionally in operation during the drought occurrence. This branch's duties are first to settle water share conflicts during the drought, to assist all the concerned parties and agencies that work for the drought offset, and also to coordinate all the assessment activities by different agencies for a better information system and improved drought policy, which assists policy makers and decision makers. This branch can also provide anticipations and identifications of drought-related impacts from environmental, social, and economic points of view. Based on the earlier drought information, if any, this branch of experts can estimate past drought impacts and provide ready information to decision makers.

Interactive cooperation of these two branches for the same goal leads to successful drought plans, which during the drought occurrence may cause the least hazard to the stakeholders. On the other hand, drought mitigation should be planned according to short- and long-term measurements and planning.

The primary impact felt from any drought starts with water shortages; therefore, most of the drought planning and mitigation tasks focus first on water shortage reduction activities before, during, and after the drought offset. For this purpose, even during wet periods water resources management strategies must take into consideration the next drought occurrence, depending on the drought temporal and regional extents, magnitudes (total surplus and total deficit amounts), and drought intensities. Among their long-term activities, water resources planners and managers should always consider the possibilty of drought occurrences by taking into consideration meteorological dry spell occurrence possibilities during summer seasons at least for, say, 6–8 months, and this can be repeated with improvements each year. There is, however, less understanding related to how drought impacts society during the fall and winter seasons. This lack of understanding can lead to more intense water shortages and a less-prepared community when spring arrives. In the meantime, interannual drought possibilities may also be taken into consideration by knowing the drought characteristics and features of the area concerned under the light of scientific methodologies presented in the previous chapters. Additionally, for short-term actions the predrought branch task experts should be on alert to face an incoming drought event within the existing framework of current infrastructure and management strategic policies. Yevjevich et al. (1983) grouped the drought mitigation and alleviation combats into three categories.

1. Water supply oriented measures
2. Water demand oriented measures
3. Drought impact minimization measures.

As for water supply management, new water resources (surface, subsurface—groundwater, desalination plant) integration into the existing system has a top priority depending on the geographical and geological features of the region. Additionally, efficient use of existing water resources also helps to reduce the drought impact. At worst situations, conventional water resources benefits enter the drought mitigation strategic solution as marginal additions.

As regards the demand management side, one can rely on the water loss reductions such as the leakages from the water conveyance and distribution pipe–canal–culvert system. Water demand level modification provides extra water potential use in case of future drought occurrences. It is recommended to use low water consumption in industrial and domestic activities. In order to alleviate an agricultural drought situation, drought-resistant crop patterns are advised so as to save water for drought combat and hazard reduction. For general cases, during drought necessary water regulations and guidelines must be designed, published, and disseminated for a common rational usage of water resources. All these preparations for drought mitigation and their combined effectiveness assist in drought hazard reduction measure developments as a pre-mitigation stage. Effective incorporation and implementation of all the aforementioned drought reduction activities can be supported by environmental, agricultural, social, and economic activities. As for environmental improvements, not only groundwater quantity measurements, but also quality variations must be inspected both temporally and spatially. Wildlife must be preserved during the drought periods with the implementation of remedial solution suggestions. Different societal sectors need to be assessed sensitively against any drought impacts with the cooperation of the central government and local authorities based on their joint drought mitigation capacities, subsidies, and investment possibilities. Among the social and economic consequences, insurance systems and relief funds must be enlarged. Especially, trade activities can be directed toward the drought influence area. Finally, for agricultural drought impact reductions, it is possible to restrict agricultural and irrigation areas according to the food demand of the region through reasonable crop variety choices. In order to reduce drought impact on agricultural lands, activities on such lands can be concentrated on pieces of land where physical soil properties are more convenient by taking into consideration fertility improvements. Especially, the central government may guarantee crop yield insurance and damage compensation.

Effective adaptation measures are necessary by strengthening the resilience of the society concerned and natural ecosystems in addition to the technical measurements and scientific prediction methodologies to obtain accurate climate variation and change predictions based on effective scenarios (this chapter).

In many countries, there are reactive approaches against drought mitigation without rational, logical, and scientific benefits from past records, scientific methodologies, and experts or local experiences. As a result of this, there are shortcomings regarding effectiveness and natural resources sustainability. In many parts of the world, natural water resource replenishment decreases, and hence, water resource share per capita. Especially, irrational withdrawal of groundwater resources coupled with climate change, consumptive lifestyle, and population increase are among the most significant drought-triggering effects. Within the same country, there may be different drought-stricken areas and in such cases, it is possible to transfer water through pipes from relatively water-rich areas to potential drought areas. Accordingly, each country may develop drought strategic plans supported by the central government with local support. Even in the same region, different transborder water countries may collaborate for the common benefit to offset drought impacts in a better and joint manner than individual solutions. There may be bilateral, trilateral, or multilateral cooperation for joint optimum solutions. Individual or joint drought mitigation plans should be integrative and proactive with drought early warning system support, drought risk management, and impact assessment under the light of institutional arrangements and continuous research activities.

Recent years' experience with frequency and duration increases in drought events indicate that more care must be taken for future drought occurrences in trying to reduce drought hazard through proper mitigation. Accordingly, programmatic strategies help to decrease drought impact on a concerned community. In such a program, the following series of questions wait for answers:

1. To what extent do local hazard mitigation plans cover drought risks?
2. What are the drought causative components and indicators that should be given priority weight and close attention?
3. The drought planning efforts should not rely on traditional crisis management techniques. Hence, the basic question is, "Do adaptive risk management measures incorporate into planning approaches?"

With modern lifestyles many drought combat and mitigation techniques are lost to history. Although their effects are local and small in scale, at the time of water stress and shortage due to dry spells and consequent drought periods, they become effective in many ways. Although water conservation methods are rather old-fashioned, they are powerful drought mitigation tools that can save significant amounts of water reduce financial losses, improve public safety, and provide dependable security. Vickers (2001) stated that hundreds of hardware technologies and behavior-driven measures are available to boost the efficiency of water use, but when implemented and put into action, they can drive down short-term as well as long-term water demands.

During water stress or shortage periods, simple water conservation methods provide vital importance and show ways of efficiency implementations in response to drought. These simple but effective ways of water conservation play significant roles even in wet periods by storing water; giving the chance for economical savings by reducing infrastructure expansion. Hence, their benefit is not only temporal but also spatial. These approaches are very effective for combating multiyear drought.

Water conservation applications can be thought as short-term water shortage and drought alleviation solutions, but they are also very effective for long-term water resources management and operation. At places where the conservation methodology can be implemented, one should not think of them only for drought mitigation, but also as an integral part of the water supply system at all times. Recently, large-scale water conservation systems are taking place in many parts of the world, especially in arid and semiarid regions. Among the most used water conservation large-scale technologies are water desalinization and wastewater reclamation.

7.6 VULNERABILITY MANAGEMENT

Vulnerability analysis is important because it covers an area in which all stakeholders can contribute to the risk analysis studies. Vulnerability analyses require more time for areal coverage and concerned sectors. Additionally, communities within this area try to provide views about the risk.

The impact and severity of climate extremes (droughts, floods) as regards their infliction on the society, environment, and economy depend on exposure and vulnerability levels. The contribution of weather and climate effects on exposure and vulnerability are closely related in a directly proportional manner to consequent disaster losses. In particular, weak or insufficient vulnerability reduction prior to actual disaster occurrence causes an increase in risk management deficiencies. Environmental degradation, poor water resources development and management, unplanned urbanization, and governance insufficiencies may cause high exposure and vulnerability.

Any place in the world may be subject to some drought type (meteorological, hydrological, agricultural, socioeconomical, famine). The most dangerous drought may continue many years as multiyear drought, because precipitation or soil moisture fluctuations may have low frequency of occurrences. Droughts pose serious threats to water resources supply systems, irrigation, and agricultural activities.

Crop types may also increase the vulnerability, especially if the crops are of special species for water saving, and hence, increase the earnings. Water-resistant crops also help in drought resistance by providing a wider range of options for consideration in connection with delayed rainfall. In any drought case, it is well known that surface water resources (reservoirs, rivers, lakes) have much more vulnerability than aquifers (groundwater storages). This point must be taken into consideration in any water resources system management, especially at places where groundwater resources are a rather high percentage of the total available water resources. Drought hazard inflictions can be alleviated to a certain extent by allowing groundwater contribution into the overall management system.

Reduction or amplification of vulnerability and related disaster risk depend on global incorporation and mutual activity alternatives. The best solution would be local, national, and international activities' interconnectedness.

Not only temporal but more importantly spatial drought exposure and vulnerability are particularly important for planning, implementation, adaptation, and risk management strategies, which can reduce the risk level in the short term. For instance, groundwater storage, safe, and sustainable management offer immediate relief from the drought periods.

Efficient and reliable management work can reduce the vulnerability to a significant extent with comparatively less damage on the overall drought infliction system. Water resources management during a drought may have several components, such as identification of available water resources, their distribution, quality control, and sufficient supply to consumption locations.

In general, strategic preparedness aims at vulnerability reduction and preparation for appropriate responses to various drought effects. To achieve the best and optimum solution, it is necessary to have timely programs, adequate knowledge, and scientific information that enter into a manageable operational communication. During a drought event, the decision makers should benefit from the relevant routines and available information for the best support of the overall system. For this purpose, the necessary ingredients are knowledge-based weather, crops, and cultivation information, task sharing about "who does what" for efficient management strategies, and suitable contingency plans and response tactics.

Among the mitigation measures, the most important factor is the availability of public water supply storage and efficient distribution systems. They add versatility and reliability to local public water supplies. Other mitigation factors are rainwater and RH possibilities, which are used quite often in arid and semiarid regions such as the Kingdom of Saudi Arabia (Şen et al., 2011). The most extensive, rich, and significant recharges in arid regions are due to indirect recharge, which spreads floodwater over thousands of square kilometers on both sides along the main wadi channel (Şen, 2008). In many places, at times both direct and indirect recharges occur simultaneously. RH is among the indirect groundwater recharge possibilities and their calculation is comparatively easier due to the following points:

1. The Earth's surface area giving rise to indirect recharge such as the RH is smaller than the direct recharge; hence, the estimations are more reliable.
2. Due to smaller influence areas, the geometrical composition of indirect RH areas is more homogeneous.
3. As rainwater takes the simplest and shortest way to the groundwater storage area, its quality is more similar to rainwater composition than groundwater composition.

4. The water movement is almost vertical; hence, there is less probability for contact with lateral geological variations.

The methodological aspects should include a series of the following significant steps for the success of RH activities:

1. Identification of suitable rainfall harvesting locations within the region starting as a pilot study at the most drought-prone area.
2. Proposition of an effective rainfall harvesting design and its application in the field for actual performance and measurements.
3. Development of a suitable mathematical model with boundary and initial conditions.
4. Verification of rainfall harvesting and possible groundwater recharge and later exposition performances.
5. Generalization of all gained experience, methodology, technique, and modeling to other potential rainfall harvesting locations within the study area.
6. Assessment of rainwater harvesting storage through various techniques.
7. Climate change downscaling modeling and as a consequence to identify at each point within the study area rainfall patterns up to a certain future projection, preferably up to 2030 or 2050 ("see chapter: Climate Change, Droughts and Water Resources").
8. Delineation of monthly rainfall maps for various purposes such as flood, inundation, agriculture, urban population concentration, and migration.
9. Monitoring of groundwater level and quality fluctuations for groundwater mixture problem solutions, if necessary.

The direct recharge or runoff and indirect recharge through infiltration and following deep percolation lead to the formation of groundwater accumulations. These accumulations are formed on impervious layers and the groundwater is in the voids of the overlying permeable layer. Groundwater recharge estimation from rainfall harvesting is one of the most significant hydrological component calculations for further drought combat, especially in arid regions. To avoid undesirable vulnerability effects, the following precautious points must be taken into consideration:

1. Poverty reduction.
2. Better education and awareness.
3. Sustainable development.

It is also possible to mention among the exposures the following points, which support the risk reduction.

1. Asset relocation.
2. Weatherproofing assets.
3. Early warning systems.

In general, short-term vulnerability reduction supports long-term risk reductions. In any drought risk management, a sequence of processes such as monitoring, evaluation, learning, and innovation should be followed. Exposure and vulnerability play a significant role in the climate change severity impacts, which can be established in a better way by taking into consideration the effects of the extremes (droughts and floods). Adverse climate impacts may cause disasters with widespread damage and severe alterations in the normal functioning of communities or societies. Anthropogenic climate change, climate variability, and socioeconomic developments need close care; otherwise, they may lead to extreme climate events, exposure, and vulnerability. The exposure and vulnerability of a drought can be reduced through proper drought (disaster) risk management and adaptation to climate change with increasing resilience to the potential adverse impacts of climate extremes. However, whatever the means (scientific, technological, and administrational), it is not possible to get rid of the risk completely. Adaptation and mitigation activities are not independent from each other, but they support and complement each other in risk-reduction tasks of climate change and drought.

7.6.1 Water Resources Management

Among additional drought mitigation opportunities are long-term water supply design, planning, operation, and management; short-term drought contingency plans; additional public water supply sources development; conjunctive use of available water resources systems; water conservation awareness and encouragement through seminars and press media; groundwater recharge enhancement; runoff minimization and harvesting, reduction in agricultural water demand (about two-thirds of water consumption in the world goes to agricultural and food production facilities); and rainwater and RH, as explained above. More information about environmental hazards and their assessments can be obtained from the work by Smith and Petley (2009).

Short-term precautions against drought effects can be taken by local human responses, but long-term emergency situations enter the domain of societal activities and highly visible government intervention enters the circle of mitigation. The government starts to distribute various aid to drought-stricken areas including food and water rations. Long-term adjustments favor increasing the supply of water to meet anticipated demands; for instance, by building more storage reservoirs, which is the case for Istanbul City, Turkey ("see chapter: Introduction"). If such precautions are not considered and implemented prior to the next drought period, then the type and scale of problems may increase with greater demand on water and, consequently, risk assessment and management.

The existing water systems need continuous efficiency in maintenance and improvement in addition to management promotions for water supply and demand ends. If the strategy is to satisfy water demand, then more sustainable responses must be developed such as water recycling in urban areas, better irrigation practices, and the increasing use of drought-resistant crops. The following points are among the drought mitigation requirements, but this list can be augmented or the items can be reordered depending on the specific vulnerability and hazard characteristics of the region concerned.

1. Long-term water supply plans and management must be continuously updated with the possibility of short-term or proactive drought activity contingency plans.

2. As already explained in "chapter: Introduction" regarding Istanbul City water supply resources, additional water supply sources must be sought in an augmentative manner, if necessary, through the interconnections of additional sources from far distances.

3. Local administrators and authorities must be encouraged for water conservation by all means so as to spare water for the next drought duration.

4. Groundwater recharge possibilities must not be left outside the water resources system circuit, because in drought periods the most dependable water resources are from groundwater storage in the aquifers (Şen, 2014).

5. To avoid flood and inundation risks in urban areas, most often the streams, creeks, and banks are improved through casing (most often reinforced concrete), but the surface water that flows in these channels must not be discharged to seashores. These surface waters can be directed to hydrogeologically convenient depression locations so as to generate temporary artificial lakes and, hence, groundwater recharge for aquifer storage enrichment (Şen, 1995, 2014).

6. Vegetation cover and especially forest regions are carbon dioxide (CO_2) sink areas, but they are also rainfall sources with runoff reduction. All these activities are in the favor of groundwater recharge.

7. Because the most water consumption in any country or region is for agricultural activities depending on local water resources, various alternative water sources must be encouraged. For instance, groundwater resources must be preserved for drought duration consumption.

8. Rainwater must be harvested in any catchment system in addition to RH as is the case in the Kingdom of Saudi Arabia (Şen et al., 2011).

Compared to droughts and flash floods the consequences of extreme droughts may be more destructive environmentally, socially, and economically. The drought impact increases with the frequency of occurrence. For instance, successive drought periods after short wet spells render more damaging hazards on the society at large. The most dangerous effect of droughts is their impact on the food production and security of a nation and in the most extreme case they may end up with famine ("see chapter: Introduction"). However, in a well drought-managed society such undesirable effects do not occur easily. It has already been explained in "chapter: Introduction" that the occurrence and severity of extreme weather, including extreme droughts, can escalate in case of global warming or general climate fluctuations ("see chapter: Climate Change, Droughts and Water Resources").

During any drought management and operation one should also consider risk minimization parallel to beneficial value maximization in an optimum manner. During the drought management, if one crop fails then the experts should be able to suggest another crop pattern suitable for the present local meteorological, climatological, and hydrological conditions. Decisions should not be instantaneous and discrete, but should be processed in a continuous manner with updates. At each time, there is a set of risks and a corresponding set of options. The main task is to identify the least risk with the best course of option, which can be identified based on the available knowledge, such as weather statistics, storage operation, and cropping and cultivation options. Although one may think that these preparations can be updated during the drought occurrence, there will not be enough time for this.

Emergency water stocks, agricultural seeds, and pesticides for the purpose of food production may be considered among the preparedness activities. Contingency plans and strategical approaches cannot reach the best optimal solution without the consent of farmers and local range owners.

Among the preliminary standard measures in urban areas are support-by-awareness campaigns and enforcement in addition to mobilized support and support from the media. If the scale of drought disaster is very large, then international support can be called in.

Social vulnerability of droughts can be reduced, especially in the agricultural sector, by access to extraordinary credit and debt relief. In any part of the world, generally, any farmer has some experience about past drought effects on their crop yields, but many small-scale farmers risk losing their land in the event of two successive droughts, particularly in a context of escalating prices for agricultural land. In order to offset these situations, a national emergency fund can be generated, which may serve other disaster-related purposes as well.

In combat against drought disasters new technological developments can offer attractive benefits, but should be implemented gradually and with due regard to unexpected side effects in the area of drought.

7.6.1.1 ENGINEERING STRUCTURES AND MANAGEMENT

Engineering structures help to manage the water resources utilization according to supply and demand requirements in the best possible (optimum) manner. Different alternatives are developed and applied in the water sector over many years, but many of them do not take into account the climate change affects explicitly. However, in some countries water resources managers become aware of the climate change effects, which are bound to be rather significant in future decades, especially in the mid-latitudes and subtropical climate belts of the world. If reservoirs are full after a wet period, then a short-lived summer flood may not end up in a water impoundment; hence, the extra water may be lost into the sea or desert without any benefit and, consequently, drought may be caused as a result of prolonged lack of reservoir inflows. Therefore, droughts are not dependent on possible climate changes only but critically on the water resources system characteristics and especially on their managements. In- and off-stream consumptive and nonconsumptive water resources exploitations are expected to be affected in the long run.

There are several indicators of water resource stress, including the amount of water available per person and the ratio of volume of water withdrawal to potentially available water volume. When withdrawals are greater than 20% of total renewable resources, water stress is often a limiting factor on development, but withdrawals of 40% or more represent high stress (Falkenmark and Lindh, 1976). Similar water stress may be a problem if a country or region has less than 1700 m^3/year of water per capita. Simple numerical indices, however, give only partial indications of water resources pressures in a country or region because the consequences of water stress depend on how the water is managed (IPCC, 2007)

The potential impact of climate change on the hydrologic regime is a crucial question for water resources management. Potential change in hydrologic regime resulting from changed climate is an important topic in contemporary hydrology and water resources management (for further detailed information "see chapter: Climate Change, Droughts and Water Resources").

7.6.2 Water Harvesting and Management

Negative impacts of droughts on agricultural, societal, and environmental activities occur first as water stress and then as scarcity. In general, agricultural water consumption in any area covers about 70–75% of total demand. Population growth in an area causes diversion of some part of this water to domestic, industrial, and environmental uses. In cases of scarcity, water resources must be well managed with rationality. During drought periods, water demand is met mostly by groundwater resources and long-term drought continuations may damage the safe yield capacity of available aquifers and even may cause their mining, which is one of the most dangerous situations because in case of the next drought period, more serious damages and hazards may inflict environmental, social, and economic consequences on the society at large. Groundwater abstraction more than safe yield may damage also the water quality (Şen, 2014). Land-use practices leading to degradation may be intensified during drought periods and the managers and decision makers must take this point into consideration.

In steppe regions, low rainfall frequently accumulates through wadis in the form of surface flow as ephemeral water and largely this water is lost through direct evaporation or in salt sinks. However, to benefit from such ephemeral runoff, RH applications are becoming widely employed in the Kingdom of Saudi Arabia (Şen et al., 2011).

It is possible to talk about direct and indirect recharge possibilities. Direct recharge is the entrance of rainwater at the place of surface though without transformation to the surface runoff. Such areas are often in the upstream parts of drainage basins, where the rainfall occurs. However, in the case of local convective rainfall events direct recharge may also occur in any part of the basin. Indirect recharge is due to runoff water, which occurs mostly outside the rainfall influence area. The recharge types in arid regions together with the mechanism is presented by Lloyd (1986) and Şen and Al-Subai (2002).

As the indirect recharge area is along the wadi channel and may extend from the upstream to the outlet point at the downstream, its calculation is more difficult and needs additional care and preservation for estimation. The locations of indirect recharge occurrences are given as follows (Şen, 2008):

1. During the sheet overflow through distinct fractures, fissures, and solution cavities in hard rocks (igneous and metamorphic) and limestone (soft rock).

2. Depressions over the drainage basin area, which first appear as small lakes with surface water storage, then the stored water recharges the aquifer partially and the other part is lost as evaporation.

In addition to the abovementioned recharge types, there are also different causes and mechanisms that play a role in the recharge process. Among such dominant mechanical factors are atmospheric processes (temperature, evaporation, humidity, solar irradiation, and wind speed), surface processes (topography, morphology, runoff, depression dimensions, and vegetation) land use (agriculture, transportation roads, and urban areas), drainage pattern (streams, creeks, rivers, and subbasins), geology (soil type, rock type, and fracturing), and unsaturated zone (granular composition, thickness, and effective porosity). The role of these factors changes from humid to arid regions, where the basic mechanisms are indirect recharge in wadi channels, depressions, limestones and sabkhahs, volcanic rocks, sand dunes, and contact lines between different lithology. These factors affect the rainfall harvesting location in arid regions along wadi main channels. During extended periods of droughts without rainfall, different systems can serve such as cisterns and water haulages by truck from other sources.

Development of rainwater and RH facilities in a drainage basin helps to reduce storm water runoff and supply reuse needs. These large systems are based on the same principals as the traditional "rain barrels." The storage of rain water on the surface is a traditional technique, and various structures are in use such as underground tanks, ponds, check dams, weirs, and so forth in practical applications. Groundwater recharge from rainwater harvesting is a new concept and generally the following structures are in practical use:

1. *Pits:* Recharge pits are constructed for recharging the shallow aquifer. These are constructed as 1 to 2 m wide and to 3 m depth, which are backfilled with boulders, gravel, and coarse sand.

2. *Trenches:* Trenches are constructed when the permeable stream is available at shallow depth. They may have 0.5–1 m width, 1–1.5 m depth, and 10–20 m length depending on the water availability. These are backfilled with filter materials.

3. *Hand pumps:* Existing hand pumps may be used for recharging the shallow/ deep aquifers, if the availability of water is limited. Water should pass through filter media before its diversion into hand pumps.

4. *Recharge wells:* Generally, recharge wells have 100–300 mm diameter and are constructed for recharging the deeper aquifers. Water is passed through filter media to avoid choking of recharge wells.

5. *Recharge shafts:* Recharge shafts are for recharging the shallow aquifers, which are located below a clayey surface. Recharge shafts with 0.5–3 m in diameter and 10–15 m in depth are constructed and backfilled with boulders, gravel, and coarse sand.

6. *Lateral shafts with bore wells:* For recharging the upper as well as deeper aquifers, lateral shafts with 1.5–2 m width and 10–30 m length are constructed depending on the availability of water with one or two bore wells. The lateral shafts are backfilled with boulders, gravel, and coarse sand.

7. *Spreading techniques:* When permeable strata start from the top this technique is used. Spread the water in streams by making check dams, nala bunds, cement plugs, gabion structures, or percolation ponds.

In arid regions rainfall is sporadic with unpredictable storms and is mostly lost to evaporation. Only a small portion of the rainfall infiltrates into surface soil and hardly penetrates to aquifers through deep seepage.

Water harvesting (rainfall or runoff) practices are used to improve the situation and substantially to increase the portion of beneficial rainfall. One definition of water harvesting is the concentration of rainfall after runoff behind dams, in subsurface dams or large-scale ditches for groundwater recharge or direct haulage by local settlers for nearby agricultural activities. It is an ancient practice supported by a wealth of local innovative knowledge. Water harvesting is a means of providing additional water for human and animal consumption, domestic and environmental purposes, and agriculture. In an effective water harvesting implementation the following components are necessary:

1. The drainage area (wadi) collects all rainfall toward main and branch streams.

2. Among the possible storage facilities are small-scale surface dams and reservoirs, subsurface dams, cisterns, large-scale artificial ditches, or groundwater aquifers.

3. The target regions for exploitation of harvested water may include agricultural areas, plant or animal husbandry, or domestic use.

4. Additional facilities may be setup such as injection wells behind surface dams or scattered within the drainage basin at convenient locations, where the recharge to groundwater may take place in the shortest possible time (Şen, 2014).

The harvesting system should include surface or subsurface storage facilities ranging from an on-farm pond or tank to a small dam constructed across the flow of a channel with an ephemeral stream. The two most important problems are evaporation and seepage losses. The former must be reduced as much as possible, whereas the latter must be increased to the utmost possible level. To satisfy these two processes, the harvested water should be transferred from the reservoir into the soil as soon as possible after collection. The reservoir must be emptied prior to rainy seasons to accumulate as much water as possible.

7.6.3 Safe Yield Management

Based on previous experiences and the scientific information at hand, it is necessary to measure and compute the safe yield of the present water resources systems prior to drought infliction in the region. The safe yield can be defined as the maximum quantity of water that can be extracted consistently from an available water resources system without failure. The safe yield calculation can be achieved through a water balance model, which has been already mentioned in "chapter: Basic Drought Indicators," where only natural water balance has been considered. However, for drought water balance, the water supply and demand balance gains the most importance. In such water balance not only the natural resources, but also any type of other water sources must be entered into the balance equation, such as groundwater aquifers and desalination plant yield.

The safe yield calculations in reservoirs and lakes should be based on the original design documents by taking into consideration the minimum pool levels beyond which water extraction leads to a failure causing the drought effect. One should be cautious with dated safe yield calculations, but depend rather on the global calculations over a specific duration such as seasonal, annual, or multi-annual durations by consideration of possible average drought duration, say, 3 years. The safe yield may change over time due to siltation, changes in the reservoir use, and also in water quality. It is recommended that the water balance calculations for safe yield should be repeated with the records of new hydrometeorological data. The safe yield calculation for groundwater resources needs numerical data about the aquifer hydrogeological parameters, the maximum pumping rate for each well in the drought-stricken area, and the well-rated maximum yield (Şen, 2014).

In any safe yield calculation, the water quality must also be taken into consideration; otherwise, quantity-related water balances may lead to undesirable drought disasters and hazards. In rivers, streams, and reservoirs minimum flows must be maintained to protect fish and wildlife. Groundwater overpumping has the potential to cause poor quality groundwater intrusion to already good quality water areas (Şen, 2014).

7.7 RISK ANALYSIS MANAGEMENT

The risk analysis method has diverse approaches based on various concepts, because such analyses are employed in many disciplines with convenient terminology. Technical calculations alone are not enough; risk perception and communication must also be provided to the population in the drought-prone area. Unfortunately, during any risk analysis the vulnerability is overlooked or vice versa. Risk analysis cannot be thought of as without hazard and vulnerability components. It includes probabilistic and deterministic (scenario) facets, loss and damage estimations, risk indexing and risk mapping, multirisk analysis, and cost-benefit analysis (Benjamin and Cornell, 1970). The results of risk analysis should enter the thoughts of the local authorities and decision makers in planning and developing mitigation policy of the region. For instance, in land-use allocation various alternatives can be obtained by considering drought or any other hazard possibility enough to reach a set of best allocation strategies. Risk analysis is not equivalent to hazard analysis; the latter needs information from the former. In general, risk analysis is based on analyzing vulnerability only with respect to physical systems. Increasing vulnerability, exposure, or severity and frequency of climate events increase disaster risk. On the other hand, disaster risk management and climate change adaptation can influence the degree to which extreme events translate into impacts and disasters.

Drought risk association in any area is a product of the exposure to the natural hazard and the vulnerability of the society to a drought event. In addition to the vulnerability definition given in Section 7.2, it is also the characteristics and circumstances of a community, system, or asset that make it susceptible to the damaging effects of a hazard. Conduction of risk assessment is necessary to better understand the drought hazard and to identify the factors and processes concerning who and what is most at drought risk and why?

Drought risk reduction strategies can be achieved in the best possible way provided that there is a harmonious cooperation between several stakeholders such as political authority, high-level engagement, strong institutional framework, and appropriate governance based on the environmental, social, and economic characteristics of the area. The local drought risk reduction studies should be an integral part of national strategies involving a wide range of stakeholders, community and civil society organizations, regional and subregional organizations, multilateral and bilateral international bodies, the scientific community, the private sector, and the media.

Drought impact assessment should try to measure any change or actual drought event by identifying various drought-triggering causes. Such an assessment should begin with the identification of direct drought consequences, which may include crop yield reductions, livestock losses, and depletion in reservoir water levels. After the identification of these main (direct) factors, other secondary causes can be brought out, which are more socially oriented. Hence, at the end of the drought experts hold the reasons for drought impacts but cannot yet identify the underlying reasons for these impacts.

To combat drought events, it is necessary to consider interdisciplinary causes and analyses by means of expert teamwork derived from different disciplines related to the drought phenomenon. Providing available data, information, and methodology to this team may help them produce a convenient and efficient mitigation procedure based on drought risk. In all these studies, not only drought-related statistical information but also environmental, social, and economic factors must also be taken into consideration. The collection of complete information leads to efficient planning and strategic solutions; otherwise, any shortfall in information or perspective could lead to results that fall far short of planning goals.

Vulnerable social conditions are under the effect of weather, meteorological, and climate factors, which may deviate from the long-term averages in their temporal behavior covering a range of area; hence, drought disaster risk is associated with these uncertain conditions. Drought disaster losses and risks have had a tendency to increase in recent years due to extravagant lifestyles, demographical changes, and climate change effects ("see chapter: Climate Change, Droughts and Water Resources"). Although economic disaster losses are higher in developed countries, fatalities are higher in developing countries.

Extreme events in weather, meteorological, and climate conditions vary across regions; each region with its unique vulnerability potential and exposure to drought hazards together with effective risk management and adaptation indicating the factors contributing to exposure and vulnerability. One can count among the main risk factors more variable rainfall, population growth, ecosystem degradation, and poor health and education systems. On the other hand, risk management and adaptation efforts include improved water management, sustainable farming practices, drought-resistant crops, and effective drought forecasting methodologies ("see chapters: Temporal Drought Analysis and Modeling; Regional Drought Analysis and Modeling; Spatio-Temporal Drought Analysis and Modeling").

Whether a drought event becomes a disaster depends on the local vulnerability under a set of stresses, among which water stress plays the main role. The resulting risk can be reduced by various coping strategies, which are usually related to withdrawals from standby asset bases and resources, and involve diversification of activities so as to relieve stress on each part. Only those communities with appropriate social networks can reduce drought risk to a desirable level and then cope with the incoming drought event. Cooperation among various approaches and methods enable detailed understanding among different experts in a drought combat team that works for risk reduction and final mitigation activities. Vulnerability and risk of drought may increase if interventions do not focus on the local context. Poor infrastructure and services in addition to inadequate and poor planning are bound to lead to unplanned settlement expansion into marginal areas that may increase the vulnerability of communities to drought impacts. The drought risk can be reduced by the effective communication of information with the support of the available early drought warning system.

On the other hand, emergency response by local, governmental, and voluntary parties plays a major role in significant drought risk reduction. Drought risk reduction requires expert practitioners, concerned governmental departments, and development organizations for prioritizing and assessing possible potential investments. To achieve this task, cost-benefit analysis is a widely used tool, but with limited applicability to drought disaster risk management. Timely communications based on prepared plans and information are early sources for drought risk-reduction agents. National planning concerning each region within the country also includes responsibilities for further risk-reduction procedures. Such risk reductions can be achieved by changing risks through proper adjustments according to standards and their convenient applications. The establishment of national safety nets or wider social protection programs is increasingly important as a way to avoid poverty after a disaster.

The objective of the risk analysis is to develop a broader understanding of drought analysis procedures and methods for use by practitioners and policy makers in future drought mitigation efforts by providing drought risk predictions, which help to take the necessary precautions prior to drought occurrence. In disaster management, drought risk assessment provides useful information for planners and decision makers for their individual and joint work toward hazard reduction. Any local representative or decision maker should be aware of the basic concepts, methods, and tools to assess hazard, risks, and losses associated with droughts. For this purpose, the basic statistical, probabilistic and, if possible, stochastic methodological techniques must be known by at least some team members. Such basic information is very essential in any drought risk mitigation and disaster risk-reduction studies. It is also significant to consider ethical details through advanced or specific methodologies for understanding the fundamental processes in drought risk assessment and management. Basic assumptions in drought risk analyses are explained for judgment and appreciation of drought information in order to make the necessary and convenient warnings and measurements for future drought risks.

For efficient management against drought impacts a risk-based approach is recommended, where risk depends on the drought occurrence probability with economic or other consequences. For example, Wilhite and Knitson (2008) noted that the principles of risk management can be promoted by adaptation of the following points:

1. Encouragement of the improvements and applications of seasonal and shorter-term forecasts ("see chapters: Temporal Drought Analysis and Modeling; Regional Drought Analysis and Modeling; Spatio-Temporal Drought Analysis and Modeling; Climate Change, Droughts and Water Resources")
2. Development of integrated monitoring and early warning systems and associated information delivery systems.
3. Development of preparedness plans at various levels of government and local administrative systems.
4. Adaptation of mitigation actions and programs collectively by each party concerned for drought hazards.
5. Generation of a safety emergency response programs network that ensures timely and targeted relief actions.
6. Provision of an organizational structure that enhances coordination within and among levels of government with the contribution of stakeholders.

The theory and methodology of drought risk assessment are a part of the broader context of disaster risk assessment. Prior information about the drought risk gives administrators and decision makers the chance to create basic perception, communication, and preparedness facilities. Knowing the drought risk provides decision makers with the opportunity for land allocation through proper activities and land-use planning. After reviewing the end products of scientific drought risk analysis, the local administrators may then convey the practical information to local people so as to make them aware of the future drought possibilities and preparation against drought occurrences. This task provides a domain for drought risk reduction in the region. As drought risk assessment procedures come from a set of different expertise, their practical rules as concept, terminology, methodology and approach can be conveyed to the people, who are subject to possible future drought occurrence.

A significant question in practice is whether risk and hazard analyses are the same and have the same contents? Although they are related to each other, they are also quite different in objectives. The former is concerned more with scientific assessments of drought disaster analysis, whereas the latter is based on the end products of risk analysis, and it is more concerned with vulnerability analysis. For safety against any risk danger, hazard should be treated distinctively and in more detail.

Risk is an inherent part of daily life and it has many different concepts in different contexts. This section focuses on the processes and methods related to quantifying drought disaster risk through probabilistic and statistical methodologies. Potential drought hazards can be appreciated after a comprehensive risk analysis. Risk (based on physical aspects) and hazard (based more on social and economic aspects) provide the necessary tools for decision making. An effective risk management and hazard mitigation can be achieved by incorporating also the stakeholders in the local or regional risk management processes. Hence, the necessary basic and practical information can be disseminated among individuals, communities, and decision makers alike. In order to make effective drought risk analyses and hazard as well as vulnerability work in an area, the following points are worth considering:

1. Uncertainty notions such as probability, statistics, stochastic processes, and fuzzy approaches.
2. Risk perception with reliability ingredients.
3. Disaster risk definition and basic concepts.
4. Practical understanding of risk impacts within a dynamic system including hazard, vulnerability, adaptation, and mitigation possibilities.
5. Multihazard conceptions, exposure, susceptibility, and resilience possibilities.
6. Practical applications of risk and hazard analyses.
7. Spatiotemporal extent of risk and hazard analyses.
8. Coupling among risk, hazard, adaptation, vulnerability, and mitigation against possible drought occurrence expectations.
9. Extraction of practical and logical verbal rules from scientific methodologies for disaster managers and decision makers.
10. Preparation of risk and hazard graphs, figures, and maps and their updates every 5–10 years according to the drought vulnerability expectation in the region.

As has already been explained in the previous chapters, there are different methodologies for risk analysis and assessments. Of these methodologies, some are very involved scientifically and precise, while others provide better understanding of possible drought risk and hazard assessments. For effective drought risk analysis, the following points help to create reliable information:

1. Basic probability concepts such as independence and mutual exclusiveness principles ("see chapters: Temporal Drought Analysis and Modeling; Regional Drought Analysis and Modeling; Spatio-Temporal Drought Analysis and Modeling").
2. Basic statistical concepts for average drought parameter assessments and their usage in drought prediction models.
3. For effective planning, the concept of return period and probability of risk relationship based on future planning short-, medium-, or long-term planning prospects.

4. Various convenient scenarios for probable cases and their analyses with future outcomes concerning vulnerability and mitigation evaluations.
5. Meteorological, hydrological, social, and economic situations and their uncertainty sources with possible contributions to overall risk analyses.
6. Calculation of different risk sources and their ranking according to impact significance, and finally, their weighted average risk amount.
7. Various risk loss assessments including cost, benefit, and their frequencies with impact on the society at large.

A large majority of disaster risk-related activities are concerned mainly with the assessment of risk and hazards based on technical methodologies and experts. Such an approach is appropriate but does not provide utilization by the disaster-affected community. However, even the technical knowledge and information should be translated into nontechnical linguistic terms so that local people can understand the risk concepts and severity of the drought phenomenon.

In any risk management, first of all the concerned technical and administrational staff should identify the location, geographic coordinates, cities, towns, agricultural land, military locations, and also should describe the drought impact possibilities in terms of damages, population displacement, economic losses, and so on, if possible, from the previous drought areas in the same or similar regions. Additionally, it is recommended to describe a drought event from the severity (magnitude), frequency (probability of occurrence or return period), and the areal extent of the drought-inflicted region points of view ("see chapters: Temporal Drought Analysis and Modeling; Regional Drought Analysis and Modeling; Spatio-Temporal Drought Analysis and Modeling; Climate Change, Droughts and Water Resources").

On the other hand, each drought-prone region must be described in terms of possible population vulnerability. To appreciate the hazard type, available information must be provided concerning the physical, social, and environmental vulnerability types. Depending on the expert view and previous experiences, the responsible committee can suggest briefly a few suitable measures against the drought hazard so as to reduce the vulnerability for the region prior to the next drought.

For an effective drought management in an area, it is necessary to have an inventory of past drought events, which provide not only scientific information, but also environmental, social, and economic knowledge. For this purpose, the responsible authority for drought management should seek answers to the following questions:

1. What were the starting and ending dates of previous drought events?
2. What were the drought durations, magnitudes (total deficits), and intensities?
3. What were the average return periods of the drought durations?
4. What were the losses incurred to different economic sectors?
5. What were the contribution rates from the central and local governmental and other agencies?
6. What was the major cause of drought in the meteorological or hydrological sense?

7. What were the extreme deviations from the long-term seasonal, annual, or multiannual averages?
8. Was there any contribution from climate change? If yes, then at what percentage?
9. How were the water resources managed before and during the drought events?
10. What was the extent of drought areal coverage?
11. What were the water management practices?
12. Which steps were taken in order to reduce the drought risks prior to drought occurrences?
13. What type of mitigation and vulnerability were implemented during past drought events?
14. Was there an early drought warning system? At which levels?
15. What type of water resources improvements were made for drought mitigation?
16. Has there been interregional or international aid? At which levels?
17. Has there been any public alertness and education prior to drought events?
18. How reliable were the drought prediction procedures? What means were used such as numerical weather forecasting, radar tracing, and early warning system and satellite images?
19. What were the hydrological prediction methodologies and their success rates?
20. Has there been any sharing of water management with neighboring regions?

7.8 DISASTER MANAGEMENT

Proper risk analysis is a prerequisite for subsequent hazard disaster management. Only accurate risk analysis outputs should be inputs for disaster and hazard risk management programs. Risk management and communication are intimately related to hazard disaster management and their harmonious consideration provides effective risk-reduction possibilities. Fundamental questions about disaster management cannot be answered prior to risk analysis, because early warning signals and preparedness of people for disaster damage depend on sound and reliable information on future risk possibility with its timing, location, and extent. Not only the managers or decision makers, but also the stakeholders must be informed jointly about the end outputs from the risk analysis without sophisticated formulations, procedures, or numbers but with their linguistical interpretations instead. This verbal information must be in the form of logical rule bases as "IF…(causative factors)…THEN…(consequence)…" statements only (Şen, 2013). The rule base helps each one concerned about drought impact with preparedness, mitigation, planning, operation, maintenance, and possible recovery actions and plans.

Those who are interested in drought mitigation should try to better understand the risk elements, hazard, and vulnerability. Hazard impact must not be concentrated on a location only but it must be visualized temporally, spatially, and socially. Unfortunately, vulnerability is not a well-defined concept and it may have differences depending on the vulnerable area or region, but it must be regarded as lack of resilience, susceptibility, and as a product of exposure. One must not depend on the physical dimension of the hazard and vulnerability based on the risk analysis calculations only, but their social, economic, traditional, cultural, institutional, and other related aspects must not be forgotten. The dependence of vulnerability on hazard must be taken into consideration dynamically during any drought mitigation work. Vulnerability should be appreciated by taking into consideration various elements including the target society, sector (domestic, industrial, agricultural), and economic consequences.

If the question is which methodology to depend on for the risk analysis, there is not a clear-cut answer because there are various approaches such as probabilistic, statistical, stochastic, deterministic, and empirical. It is advised that the final dependable result should consider various results through the statistical probability distribution function (PDF) as explained in "chapter: Basic Drought Indicators". In an integrated risk assessment leading to effective vulnerability and mitigation, one should consider understanding of basic concepts such as return period, exceedance probability (EP) curves, and expected or average expected losses in probabilistic risk analysis approach framework.

In general, probabilistic risk analysis is preferable for mitigation purposes because it includes the impacts of all the potential hazard sources and also the numerical uncertainty in the estimates. The probabilistic approach also provides a more reliable decision-making framework. One must not forget inclusion of nonnumerical uncertainty that can be gathered from the local people as for the drought features. This information may be fuzzy in content, but it may help to set up a rule base in a rational, logical, and science philosophical manner (Şen, 2013).

In drought and any other natural disaster management scenario studies based on model outputs and rule base information, a convenient set of scenarios can be visualized through the computer programs or available efficient software. In practice, scenario analyses are useful for emergency preparedness, operation, and management. Different scenarios allow managers and decision makers to visualize possible risk pictures and to use risk parameters as inputs for various planning, adaptation, and mitigation activities. As mentioned before, subjective information (linguistical), personal or institutional experiences, and knowledge accumulation in a community help to incorporate this information into effective planning, adaptation, and mitigation tasks. As for the integrated drought approach, it is useful to consider a set of effective points that are rather simple but significant.

1. Droughts occur temporally and spatially due to significant reduction of precipitation as the start of meteorological drought, which may continue to trigger hydrological drought and subsequent agricultural and socioeconomic droughts ("see chapter: Introduction"). The possibility of this sequence of drought types must be considered in an effective disaster (drought) management study.

2. Natural drought severity impact on society depends on vulnerability, especially on different sectors and accordingly, a concise and rational mitigation implementation must be planned, which helps for the preparedness of the community.

3. An integrated water resources management system must be revised under the possible effect of drought severity by revising water resources storage (surface and groundwater) opportunities, demand pattern adjustments, operation rules, reviews, and so on.

4. To face the risk, a shift may be very effective from a reactive approach (emergency assistance) to a proactive approach.

Among the reactive approach steps, one can mention the following points as necessary measures:

1. Prior to, during, or after a drought occurrence the available water resources and any other potential source must be monitored for objective rational decision making.

2. Even though the drought initiation time cannot be determined accurately, once the drought spell effects are felt, to face the drought safely, measures must be identified.

3. After some time has passed during the drought period, all available facilities, instruments, and methodologies must be implemented for the balance of the drought.

On the other hand, for a proactive drought mitigation approach the following two categories must be considered as planning and monitoring and implementation stages.

1. Planning stage

    a. Prior to any drought effect, available water resources should be estimated for short-term, medium-term, and especially long-term utilizations. The necessary comparison must be made for each term between the supply and demand patterns.

    b. Risks concerning water storage (dams, subsurface dams, and groundwater such as aquifers) and drought impact must be assessed under the prevailing circumstances.

    c. With the possible approach of the next drought period, long-term water resources integrated management must be implemented within the drainage basin or region, with possible impacts and their current situation assessments. This implementation must be cared for especially during the drought period.

    d. Prior to drought occurrence, short-term measurements must be taken of any water resources management planning, design, operation, management, and maintenance.

    e. Again prior to drought occurrence, the necessary and convenient short-term measurement definitions must be completed concerning drought contingency planning and its implementation preparations.

2. Monitoring and implementation stage

    a. Hydrometeorological variables (precipitation, temperature, evaporation, soil moisture, etc.) monitoring and treatments for dependable information digestion and extraction must be continued before, during, or after each drought occurrence.

    b. Water resources reserves must be cared for all the time during drought or normal conditions.

c. Water crisis management plans must be revised, refined, and implemented sometime after the drought starts. At this stage, warning and then the necessary alarms must be given for the water crisis.

d. With the extensive continuation of drought, natural disaster declaration must be announced so that each individual and institution can take their share. This is a stage for the start of water emergency.

e. Finally, a drought contingency plan must be implemented at full scale and it should continue after the start of drought is felt. Early drought warning system notices must be taken into consideration through an effective communication system so that the warnings can reach concerned stakeholders, in particular, and the general population.

Example 7.1 Return Period Calculation and Expected Cost

For a monthly drought with a probability occurrence of 0.01, what is the return period? It has already been explained in "Chapter 2" that the return period is equal to the reverse of the occurrence (exceedance) probability of an event. Hence, the return period is $1/0.01 = 100$ month, which is about $100/12 = 8.33$ years, or to be on the safe side one can take the result as 9 years. This is the expectancy of a drought to appear within every 9-year period. It is possible to generalize the expected loss, E_L, definition as follows:

$$E(L) = p \times L \tag{7.1}$$

where p is the probability of exceedance and L is the amount of loss. It is also possible to write the expected loss equation in terms of return period, R, simply as follows:

$$E(L) = \frac{L}{R} \tag{7.2}$$

Furthermore, the previous drought event is associated with a cost, say, US$5 million, and in the future for the same return period the expected loss can be calculated from Eq. (7.1) as $5 \times 106 \times 0.01 = 50,000$ USD.

Example 7.2 Expected Opportunity Loss

If the annual probability of dry spell occurrence in year i is p_i, with associated loss, L_i, then the overall total expected loss, E_T, is given as

$$E_T(L) = \sum_{i=1}^{n} p_i L_i \tag{7.3}$$

In Table 7.1 the dry spell occurrence probabilities and the losses are given for 12 cases. Calculate the total and average expected loss values.

TABLE 7.1 Drought Occurrence Probabilities and Losses

Drought Number	Annual Dry Spell Occurrence Probability, p_i	Loss, L_i (million USD)	Expected Loss (USD)
1	0.002	2.5	5000
2	0.005	1.5	7500
3	0.010	1.0	10,000
4	0.020	0.5	10,000
5	0.030	0.3	9000
6	0.040	0.2	8000
7	0.050	0.1	5000
8	0.050	0.08	4000
9	0.060	0.07	4000
10	0.070	0.05	3000
11	0.090	0.05	4000
12	0.100	0.03	3000
Total			72,500

The EP for a given level of loss, $E(L_i)$, can be determined by calculating

$$E(L_i) = P(L > L_i) = 1 - P(L \leq L_i) \tag{7.4}$$

or during the time duration of i years this expected loss becomes

$$E(L_i) = 1 - \prod_{j=1}^{i} (1 - p_j) \tag{7.5}$$

The resulting EP is the annual probability that the loss exceeds a given value. Eq. (7.5) indicates that 1 minus the probability that all the other events remain below this value. This notation can be written out, for example, for the first three events in Table 7.1 as

$$E(L_3) = 1 - [(1 - p_1) \times (1 - p_2) \times (1 - p_3)]$$
$$= 1 - [(1 - 0.002) \times (1 - 0.005) \times (1 - 0.010)] = 0.0169$$

Example 7.3 Average Annual Loss Damage

One can also calculate average annual loss damages, which is the area under the EP curve that is equal to the sum of the loss rates for each event. This means that average annual loss can be calculated as the sum of all probabilities times their associated losses. Consideration of the values in Table 7.1 leads to the average loss of $72,000 million.

Example 7.4 Exceedance Probability Calculation

The empirical EP curve can be constructed on the basis of historical observations if there are a sufficient number of records, say 10–15. Theoretical EP can be adopted as one of the theoretical PDFs (gaussian, Pearson, Gumbel, Gamma, etc.) that fits the data best and it helps to predict the "extreme" ("high") values given a probability of exceedance, which is equivalent with the risk and also the reverse of the return period. In general, the convenient PDF for EP has two extreme parts: one on the left for "low" and the other on the right for "high" values. Depending on the purpose, the probability predictions are sought in one of these parts. These parts include less certain situations and, therefore, conclusions based on the extreme parts of the PDF particularly include extremes, which should be depended on with caution. In case of droughts, most often "low" values are sought and, therefore, the left part is used.

EP curve for any, say, rainfall data can be constructed by executing the following steps:

1. Sort the available rainfall data values in an ascending order from the smallest to the largest.
2. Determine the number of available data as n.
3. In the ordered data the rank is denoted by m, where the smallest amount has the least rank as $m = 1$ and the greatest one has the rank $m = n$. Others will have ranks in between these two values.
4. Empirical probabilities, P_m, must be attached to each one of the ordered data (rainfall in this case) according to the rank, m, as in Eq. (2.24).
5. Hence, there are two sequences as the ordered data and corresponding empirical probability data. If the probabilities are shown on the vertical axis and the ordered data on the horizontal axis, their plots appear in the form of scatter points, shown as stars in Fig. 7.1.

   One can notice that the scatter points have a regular decrease as the rainfall values increase. This means that the extreme events will have less probability of occurrence, whereas low and medium rainfall events have greater probability of occurrences.
6. It is now time to try and find the best fitting theoretical PDF to the scatter points. After several trials, theoretical Gamma PDF has been fitted to the scatter points through MATLAB® software.
7. First the valid parameters of the given data are calculated through the software statement in MATLAB as,

$$\textbf{parameter} = \textbf{gamfit(data)}$$

Here "data" implies rainfall values. This software yields the parameters (shape, α and scale, β) as in Fig. 7.1, $\alpha = 2.9221$ mm and $\beta = 6.9143$.

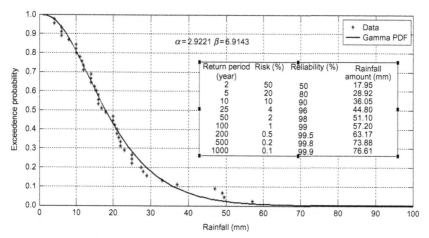

The table within the figure:

Return period (year)	Risk (%)	Reliability (%)	Rainfall amount (mm)
2	50	50	17.95
5	20	80	28.92
10	10	90	36.05
25	4	96	44.80
50	2	98	51.10
100	1	99	57.20
200	0.5	99.5	63.17
500	0.2	99.8	73.88
1000	0.1	99.9	76.61

$\alpha = 2.9221 \ \beta = 6.9143$

FIG. 7.1 Exceedance probability (EP) curve.

8. The empirical scatter data figure has a rainfall change range from 0 to 100 mm and, accordingly, the MATLAB statement is defined for the horizontal axis variable as

$$x = 0:001:100$$

which means that for theoretical calculations, the counter will start from 0 and continue by increment, 0.001, until the maximum value of 100 mm is reached.

9. The theoretical Gamma PDF values, y, are calculated again by another MATLAB statement, which includes previously calculated parameters as

$$y = \text{gamcdf}(x, 2.9221, 6.9143)$$

The $1 - y$ values are the theoretical Gamma EPs and their plot against x appears as a continuous line in Fig. 7.1.

10. To calculate risk amounts, it is first necessary to select a set of design durations (design period), which are adopted as 2-, 5-, 10-, 25-, 50-, 100-, 200-, 500-, and 1000-year durations. The corresponding probabilities to this set of selected design periods are $1/2 = 0.500$, $1/5 = 0.200$, $1/10 = 0.100$, $1/25 = 0.040$, $1/50 = 0.050$, $1/100 = 0.010$, $1/200 = 0.005$, $1/500 = 0.002$, and $1/1000 = 0.001$, respectively.

11. The theoretical rainfall values that correspond to these probabilities, p, can be obtained by means of the following MATLAB statement:

$$\text{AnnualRainfall} = \text{gaminv}(p, 2.9221, 6.9143)$$

Hence, the rainfall amounts in the last column of the table in figure are obtained accordingly.

7.9 DROUGHTS RISK CALCULATION METHODOLOGY

Although many areas are prone to drought risks, unfortunately the drought appearances have vague, imprecise, uncertain, and most of the time fuzzy behaviors not only for the general population, but also for the experts in the subject. It is difficult to find an objective, unique definition for droughts, and current definitions are based rather on expert and professional concepts ("see chapter: Temporal Drought Analysis and Modeling"). Unlike aridity, drought is a temporary phenomenon that is a permanent feature of the local climate. Seasonal aridity (well-defined dry season) must be separately examined apart from drought.

It is important for quantitative studies to know the basic frequency, intensity, duration, and areal coverage of drought risks from the assessments of the previous measurements, observations, and experiences ("see chapters: Temporal Drought Analysis and Modeling; Regional Drought Analysis and Modeling; Spatio-Temporal Drought Analysis and Modeling"). Provided that there are not foreseeable damages on human activities, drought severity is not significant practically. For instance, continuous droughts in desert areas are not effective because local people are accustomed to living under the prevailing conditions.

7.9.1 Probabilistic Risk and Safety Calculations

Fig. 3.2 presents a hydrometeorological series of measurements, which provides a simple basis for drought description and calculation. The probabilistic risk, $P(R)$, can be defined as the nonexceedance probability of the standardized hydrologic variable, x, over the standard threshold level, x_o (see Fig. 3.2). If the standardized sequence of rainfall, runoff, or soil moisture record is x_1, x_2, ..., x_n, then the risk probability can be defined generally as

$$P(R) = P(X_1 \leq X_o, X_2 \leq X_o, ..., X_n \leq X_o) \tag{7.6}$$

On the other hand, the safety probability, $P(S)$, is defined notationally as the complementary event.

$$P(S) = 1 - P(R) \tag{7.7}$$

7.9.1.1 DEPENDENT AND INDEPENDENT PROCESSES

The simplest form of serial dependence in any hydrometeorological process can be modeled by first-order Markov process given by Eq. (3.11) with complete probabilistic description around x_o level on the basis of the three basic transition probabilities, $p = P(x_1 > x_o)$, $P(x_i > x_o | x_{i-1} > x_o)$, and $P(x_i \leq x_o | x_{i-1} \leq x_o)$. In "chapter: Temporal Drought Analysis and Modeling," Section 3.6.1, $P(x_i > x_o | x_{i-1} > x_o)$ is defined as the first-order autorun coefficient, r. Hence, the explicit forms of the transitional probabilities are similar to Eq. (3.26) as follows:

$$P(x_i > x_o | x_{i-1} > x_o) = r \tag{7.8}$$

$$P(x_i > x_o | x_{i-1} \le x_o) = \frac{p}{q}(1 - r) \tag{7.9}$$

$$P(x_i \le x_o | x_{i-1} > x_o) = 1 - r \tag{7.10}$$

$$P(x_i \le x_o | x_{i-1} \le x_o) = 1 - \frac{p}{q}(1 - r) \tag{7.11}$$

$P(x_i \le x_o | x_{i-1} \le x_o)$ can be written explicitly similar to Eq. (3.32) as

$$P(x_i \le x_o | x_{i-1} \le x_o) = q + \frac{1}{2\pi q} \int_0^\rho e^{-z^2/2(1+z)} \left(1 - z^2\right)^{1/2} dz \tag{7.12}$$

where z is the standardized variable according to Eq. (2.1). The numerical solutions for various ρ and q values are given already in "chapter: Temporal Drought Analysis and Modeling," Section 3.6.1, and Table 3.4. For the first-order Markov process, it is possible to factorize Eq. (7.6) implicitly in terms of the transitional probability given in Eq. (7.11) as

$$P(R) = P(x_1 \le x_o) \prod_{i=2}^n P(x_i \le x_o | x_{i-1} \le x_o) \tag{7.13}$$

and explicitly as follows:

$$P(R) = q \left[1 - \frac{p}{q}(1 - r) \right]^{n-1} \tag{7.14}$$

which for independent processes, $\rho = 0.0$ and $r = p$, reduces to

$$P(R) = q_n \tag{7.15}$$

This expression has already been derived and presented by Yeh (1970).

Example 7.5 Simple Risk Calculation

At this stage a question might be asked as to how does the serial dependence affect the simple risk and safety? Let a decision maker want to know the risk probability for the next $n = 10$ duration with wet spell probability, $p = 0.30$ ($q = 1 - p = 0.70$, dry spell probability), provided that the system has a first-order Markov dependence structure with $\rho = 0.5$.

If the risk is calculated with the assumption that the process is independent, then from Eq. (6.15) one can calculate that $P(R) = 0.70^{10} = 0.0282$ or $P(S) = 1 - 0.0282 = 0.9718$. However, in case of dependence with $\rho = 0.5$ and, correspondingly, by use of Table 3.5, one can see that $r = 0.795$; hence, the substitution of the relevant values into Eq. (7.14) yields the safety probability $P(R) = 8.583 \times 10^{-4}$ or $P(S) = 1 - 8.583 \times 10^{-4} = 0.9991$. This simple

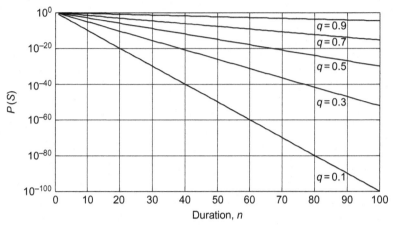

FIG. 7.2 Safety straight line.

correlation proves that an increase in ρ decreases risk and, in turn, increases the safety as a result of which the duration and the cost for drought disaster both decrease. This simple calculation indicates the significance of the serial correlation coefficient in the assessment of risk (safety) for drought planning.

On the other hand, the corresponding safety probability can be found from Eq. (7.7) by consideration of Eq. (7.14) as

$$P(S) = 1 - q\left[1 - \frac{p}{q}(1-r)\right]^{n-1} \tag{7.16}$$

For an independent case $r = p$, this equation reduces to its simplest form,

$$P(S) = 1 - q_n \tag{7.17}$$

This expression yields to a set of straight lines on semilogarithmic graph paper for a set of q values as in Fig. 7.2. This graph is very useful in practical applications of water resources systems design for calculating the nonoccurrence probability of threshold value, x_0, given the level of safety and the expected life (return period) of the project as n.

Example 7.6 Design Variable Magnitude

Let a planner be interested in designing his or her project for $n = 10$ years with $P(S) = 0.90$, which is to say that in the long-run the project will be subject to risk probability, $P(R) = 0.10$. What is the nonexceedance (dry spell) probability?

It is possible to find the nonexceedance probability either from Fig. 7.2 or from Eq. (7.17) as $q = [1 - P(S)]^{1/n} = (1 - 0.90)^{1/10} = 0.7943$, with corresponding probability of occurrence, $p = 1 - 0.7943 = 0.2057$. The planner may be able to calculate the magnitude of standardized design value, x_s, corresponding to $p = 0.2075$ by adopting a suitable PDF to the data set at hand. For example, if the underlying PDF has standard normal behavior with mean, $\mu = 0$, and standard deviation, $\sigma = 1$, then corresponding to $q = 0.7943$, the standard variable, x_s, can be found from normal PDF tables as 0.8214. In general, the actual design variable, X_0, can be calculated as follows provided that the data mean and standard deviation values are known:

$$X_0 = \mu + x_s \sigma$$

Example 7.7 Safety Calculation

Sometimes, it is necessary to know the safety of an already existing engineering structure. For instance, if the structure has been designed originally for an expected life of $n = 30$ units (month, years) with $q = 0.99$, after its completion the safety probability is $P(S) = 1 - 0.99^{30} = 0.2603$; that is, the safety is about 26% and, therefore, the corresponding risk is 74%.

If the design value has not been exceeded for the first 20 years, then the safety probability of the same structure for the remaining 10 years will be $P(S) = 1 - 0.99^{10} = 0.0956$ (9.56%), which is considerably smaller than the original safety. The risk probability is $P(R) = 1 - 0.0956 = 0.9044$ (90.44%). An important conclusion is that as the number of years without any damage increases, the risk over the remaining life of the structure is also bound to increase.

7.9.1.2 RETURN PERIOD AND RISK

For risk assessment, it is necessary first to calculate the probability of return period, T_r, exceedance, $P(T_r \geq j)$, once over the life period, j. The calculation of this probability is given by Feller (1967) for independent process as

$$P(T_r \geq j) = q^{j-1} \qquad (7.18)$$

or

$$P(T_r = j) = pq^{j-1} \qquad (7.19)$$

On the other hand, the expected (arithmetic average) return period value can be obtained as

$$E(T_r) = \sum_{j=1}^{\infty} jP(T_r = j) = p\sum_{j=1}^{\infty} jq^{j-1} = \frac{1}{p} \qquad (7.20)$$

This expression indicates that there is an inverse relationship between the average return period and probability of exceedance (wet spell), p. The numerical solutions of return period probability distribution are given in Table 7.2 for a set of average return periods, $E(T_r)$, by consideration of Eqs. (6.18) and (7.20).

One can understand from this table that over a long period of years, 25% of the intervals between drought duration equal to or greater than the 100-year return period is less than about 30 years, while an equal number is in excess

TABLE 7.2 Independent Processes Return Period Theoretical Distribution

Average Return Period $E(T_r)$	Actual Return Period T_r Exceeded Various Percentages of Time, $P(T_r \geq j)$						
	0.01	0.05	0.25	0.50	0.75	0.95	0.99
2	7.64	5.32	3.00	2.00	1.41	1.07	1.01
5	21.64	14.42	7.21	4.10	2.28	1.23	1.04
10	14.71	28.43	14.16	7.58	3.73	1.48	1.09
30	136.84	89.36	41.89	21.44	9.48	2.51	1.29
100	459.21	299.07	138.93	69.97	29.62	6.10	2.00
1000	4603.86	2995.23	1386.60	692.80	288.53	52.53	11.11

during 139 years. In other words, for 25% risk (75% safety), drought duration will not be exceeded within the next 30 years by consideration of a 100-year return period.

The probability of risk can be obtained in terms of return period from Eq. (7.15) by making use of Eq. (7.20) as

$$P(R) = \left[1 - \frac{1}{E(T_r)}\right]^n \tag{7.21}$$

It is possible to calculate average (expectation) return period from this expression after the necessary algebraic arrangements as

$$E(T_r) = \frac{1}{1 - [P(R)]^{1/n}}$$

This expression is valid for risk probability, and its solution for a set of record numbers is given in Table 7.3. An inspection of this table shows that there is a 1% chance that the average return period of the maximum event occurrence in this 10-year record is as low as 1.66.

On the other hand, the safety probability as a complementary event to Eq. (7.21) can be written readily as follows:

$$P(S) = 1 - \left[1 - \frac{1}{E(T_r)}\right]^n \tag{7.22}$$

The necessary tables and graphs for the application of this expression have been presented by Gupta (1973).

TABLE 7.3 Average Return Periods

	Risk Probability, P(R)				
Record Number, n	**0.01**	**0.25**	**0.50**	**0.75**	**0.99**
2	100.00	4.00	2.00	1.33	1.01
5	249.25	9.2	4.13	2.35	1.19
10	498.00	17.89	7.26	4.13	1.66
20	995.49	35.26	14.93	7.73	2.71
60	2985.50	104.80	43.80	22.10	7.00

For a dependent drought case similar to Eqs. (7.18) and (7.19) the theoretical probability of the return period can be obtained from Eq. (7.16) as

$$P(T_r \geq j) = q \left[1 - \frac{p}{q}(1-r) \right]^{j-2} \tag{7.23}$$

or

$$P(T_r = j) = P(T_r \geq j) - P(T_r \geq j+1) = p(1-r) \left[1 - \frac{p}{q}(1-r) \right]^{j-2} \tag{7.24}$$

These two expressions reduce to Eqs. (7.18) and (7.19) for an independent process case with $r=p$. Hence, the return period, T_r, which is the expected value (average) of the random variable, can be derived after the necessary algebra leading to

$$E(T_r) = \frac{q^2}{\left[1 - \frac{p}{q}(1-r) \right] p(1-r)} \tag{7.25}$$

This expression reduces to Eq. (7.20) for $r=p$. It shows that in the case of dependent variables the return period is not a function of probability of exceedance, p, only, but also of the autorun coefficient, r, which is explicitly related to ρ (Table 3.5). The numerical solutions of the return period probability are given in Table 7.4 for $\rho=0.2$.

A comparison of this table with Table 7.2 reveals that the return period averages are generally greater than the independent process case. By consideration of return period, the risk and safety probabilities for a dependent process can be written as

TABLE 7.4 Markov Process ($\rho = 0.2$) Return Period Theoretical Distribution

Average Return Period $E(T_r)$	Actual Return Period T_r Exceeded Various Percentages of Time $P(T_r \geq l)$								p	r
	0.01	0.05	0.25	0.50	0.75	0.95	0.99			
2	8.83	6.02	3.21	2.00	1.29	0.88	0.80	0.500	0.5640
5	24.14	16.00	7.88	4.37	2.32	1.13	0.92	0.200	0.2818
10	48.39	31.80	15.20	8.06	3.88	1.44	1.00	0.100	0.1681
30	143.95	93.97	43.99	22.47	9.88	2.54	1.26	0.033	0.0810
100	471.22	306.88	142.53	71.75	30.35	6.21	2.00	0.010	0.0352
1000	4629.65	3012.00	1394.36	697.68	290.14	52.54	11.09	0.001	0.0065

$$P(R) = q \left[\frac{q^2}{E(T_r)p(1-r)} \right]^{n-1} \tag{7.26}$$

and

$$P(S) = 1 - q \left[\frac{q^2}{E(T_r)p(1-r)} \right]^{n-1} \tag{7.27}$$

respectively. The solutions of these expressions can be achieved for given pairs of p and r values leading to numerical results in Table 7.5.

There are different methodological approaches, procedures, and algorithms for risk management. For instance, Kates and Kasperson (1983) suggested three steps for the risk assessment.

1. Determination of what type of dangerous situation may arise and, accordingly, risk calculations must be carried out by considering damage levels.
2. Each potentially dangerous cause can be calculated by probability principles.
3. Social losses caused by different events can be assessed by risk calculations.

7.10 DROUGHT DURATION–SAFETY CURVES

Feller (1951) presented an objective definition of drought based on run-lengths by consideration of a truncation level, X_o, along a measurement sequence, X_1, $X_2, ..., X_n$ (Fig. 3.4). Feller (1967) digitized surplus and deficit states into +1 and −1, each with probability of occurrence, p, and nonoccurrence, q, respectively. The probability, $P_i(L=j)$, of critical drought duration (maximum drought duration), L, in a sample of length i with duration j has already been derived in "chapter: Temporal Drought Analysis and Modeling," Section 3.7.1. Eq. (3.68) helps to find the cumulative PDF of the critical drought duration by successive summation of the probabilities. The plot of critical drought duration on the horizontal axis versus the safety probability results in a set of curves, which are referred to as drought duration–safety curves.

One of the basic questions is the determination of critical drought duration, L, in a given sample length, n (for instance, the planning horizon for a water structure) at a certain demand or capacity level, p, and risk probability, $P(R)$, or safety probability, $P(S)$. Answers to such questions can be achieved by the graphical representation of Eq. (3.68). In Appendix 3.1, MATLAB software is given for this purpose. Fig. 7.3 presents drought duration–safety curves at $p = 0.5$ level.

If the demand level is $p = 0.90$ the answer to the same questions are presented in Fig. 7.4.

TABLE 7.5 Markov Process ($\rho=0.2$) Risk and Safety Values

p	q	r	E(Tr)	Economic Life (Return Period) of the Project, n									
				n=10		n=20		n=30		n=50		n=100	
				Safety	Risk	Safety	Risk	Safety	Risk	Safety	Risk	Safety	Risk
0.33	0.67	0.414	3.26	0.031	0.969	0.001	0.999	0.000	1.000	0.000	1.000	0.000	1.000
0.25	0.75	0.334	4.34	0.078	0.992	0.006	0.994	0.000	1.000	0.000	1.000	0.000	1.000
0.20	0.80	0.282	5.43	0.134	0.866	0.018	0.982	0.003	0.997	0.000	1.000	0.000	1.000
0.10	0.90	0.168	10.72	0.375	0.625	0.158	0.842	0.060	0.940	0.008	0.992	0.000	1.000
0.05	0.95	0.104	21.14	0.615	0.385	0.400	0.600	0.234	0.766	0.089	0.911	0.008	0.992
0.03	0.97	0.083	35.22	0.772	0.228	0.561	0.439	0.434	0.566	0.237	0.763	0.056	0.944
0.01	0.99	0.035	102.5	0.906	0.094	0.822	0.278	0.745	0.255	0.613	0.687	0.375	0.625

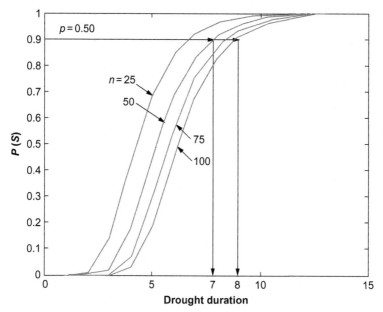

FIG. 7.3 Drought duration–safety relationship ($p=0.5$)

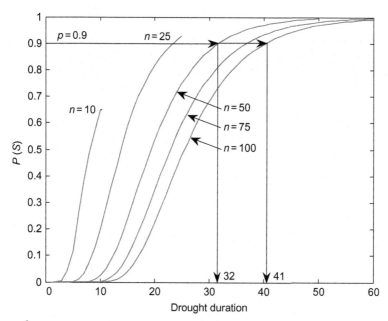

FIG. 7.4 Drought duration–safety relationship ($p=0.9$)

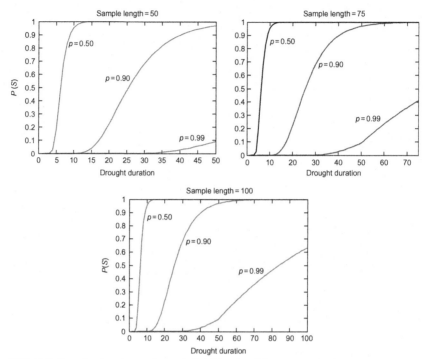

FIG. 7.5 Drought duration–safety relationships for different sample sizes.

Example 7.8 Critical Drought Duration Calculation

If the design life of a water project at a place is 50 years, what is the duration of a critical drought for meeting $p=0.50$ (50% of the demand) at 10% risk level?

Here, $n=50$, $p=0.50$ and safety probability $P(S)=1-0.10=0.90$. Entrance to Fig. 7.3 on the vertical axis at 90% safety level and consideration of $n=50$, the critical drought duration can be read on the horizontal axis as $L=7$ years. The same question for 100-year life has the answer as $L=8$ years.

Similar questions can be answered at $p=0.90$ level from Fig. 7.4, which yields $L=32$ years and $L=41$ years for $n=50$-year and $n=100$-year water project durations, respectively.

On the other hand, Fig. 7.5 presents different drought duration–safety curves for a set of demand (truncation) levels. In this figure project durations are labeled as sample lengths.

For dependent processes the drought duration–safety curve formulations are already presented in "chapter: Temporal Drought Analysis and Modeling," Section 3.7.2, for dependent Bernoulli trials. Numerical solutions of these equations are achieved for different sample lengths through software given in Appendix 3.2 in MATLAB form. The solutions are presented graphically in Fig. 7.6 for truncation (demand) level $p=0.7$ ($q=0.3$). It is now possible to read possible critical drought durations in a dependent process by use of this graph.

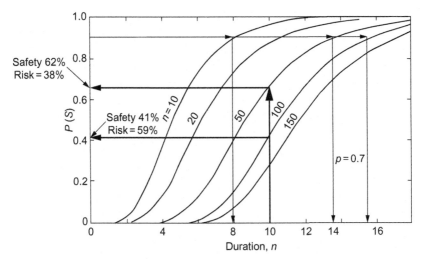

FIG. 7.6 Dependent process duration–safety relationships

Example 7.9 Average Critical Drought Duration Calculation

At a location of planned water structure, flow series abide by the Bernoulli trials. The life span of the structure is considered to be 10-, 50-, or 100-year. It is also desirable that each one of these plans meet the demand at probability level of $p=0.70$. Calculate the average critical drought duration at $P(S)=0.90$ level.

In Fig. 7.6 duration is shown by n, where all curves are valid for $p=0.70$. The critical drought duration corresponding to 90% safety (10% risk) level can be found as corresponding horizontal axis values, which give 8, 14, and 16 years, respectively.

Example 7.10 Critical Drought Areal Coverage Calculation

Within the next 50-year duration, what is the risk of observing a 10-year duration critical drought in the same area as in the previous example?

For the answer, first, 10-year critical drought duration is fixed on the horizontal axis in Fig. 7.6. A perpendicular line from this point reaches $n=50$ years curve and then the horizontal arrow from this intersection shows safety probability as $P(S)=0.62$ and $P(R)=0.38$ on the vertical axis. The answer to the same question for 100-year life duration can be found as $P(S)=0.41$ and $P(R)=0.59$.

7.11 WEATHER MODIFICATION

Among the drought mitigation procedures, some countries tried to depend on weather modification (cloud seeding) for rainfall augmentation; hence, additional surface water storage impoundments occur in surface dams as has been the case during the drought period in Istanbul City from 1990 to 1994. Research type cloud seeding operations have been conducted in many countries for the last 50 years with the hope to augment water resources in a watershed. Early studies stated that the rainfall increments in a cumulus cloud varied from 5% to 30 % (Bethwaite et al., 1966). Woodley and Solak (1990) reported an increase of about 17% over watersheds after a seeding program near San Angelo, Texas. Also, considerable work has been done at the University of Washington on the "storm-type" problem in weather modification experiment (Hobbs and Rangno, 1978; Rangno, 1979).

Warner and Twomey (1956) mentioned two different operations, namely, static and dynamic seeding. The dynamic seeding programs are used frequently in practical applications and they are related to latent heat release due to particle growth in the updraft region of a cloud. In the static seeding operations the vertical development of the cloud due to dynamic effects is assumed to be negligible.

One of the many perplexing problems in natural or artificial effects on meteorological events is the statistical analysis of experimental data. Problems are introduced by nonnormality of the hydrometeorological variables. Various statistical procedures for the assessment of external effects suffer from random conclusions due to scarce and uncertain data. In order to better assess the external effects, especially in the cloud seeding subject, many programs have been conducted since the early 1960s (Henderson, 1966; Smith, 1967; Grant and Mielke, 1967; Mooney and Lunn, 1969; Chappell et al, 1971). After reviewing many of the cloud seeding programs, Elliott (1986) concluded that it is not possible to unequivocally determine the amount of influence that may arise due to external effects.

The physical control of the entire cloud seeding procedures has not yet been achieved fully. The assessments of the experimentally obtained rainfall data from any seeding experiment can be achieved by various statistical methods. Simple regression applications are used between the seeded and unseeded rainfall data of target (seeding) and control (nonseeding) areas. It is assumed that there is a linear relationship between the rainfall amounts in control and target areas before and after the seeding experiments. Hence, there are two straight lines, one for seeded and the other for unseeded rainfall data. In case of an increase due to the seeding operation the slope of the straight line from seeding application is expected to be greater than the long-term relationship slope of unseeded rainfall quantities (Fig. 7.7).

The difference between the seeded and unseeded lines is attributed to the effects of seeding with a degree of confidence determined by the scatter of historical data. Although Woodley and Solak (1990) suggested the use of historical regression technique in the seeding assessments, they included many procedural assumptions such as the normality, linearity, and homoscedasticity, which are rather difficult to have in any cloud seeding operation. The main drawbacks of the regression technique are as follows:

1. There must be a high correlation between past rainfall in the target and control areas.

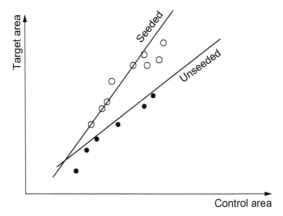

FIG. 7.7 Target and control area rainfall amounts.

2. There must be enough data available for the confirmation of the relationship.
3. The regression technique necessitates long-year data of seeding experiment (at least for 10 years), for the confirmation of the cloud seeding success or failure.

Although the regression method seems very reasonable, it is open to the serious criticism that if the patterns of storms during the seeding period differ appreciably from the long-term average pattern on which the historical relationships are based, then the latter cannot be used to predict accurately the rainfall amounts in the target area and, consequently, the assessment of seeding effects may be most misleading (Neyman and Scott, 1967). Randomized seeding technique is used, which requires a similar weather pattern over the target area and, accordingly, the clouds are seeded or unseeded. In the seeding case, if the cloud formation and structure are not different than for the unseeded case, and the cloud seeding is effective, then the rainfall amount is expected to be greater during seeded days. The rainy days are grouped separately for seeded and unseeded cases for the same target area stations. Rainfall average, $\bar{R}_n$, and standard deviation, S_n, are calculated for each station in the target and control areas. The seeding assessment is achieved by considering the differences for $\bar{R}_n$ and S_n values between two groups (Dennis et al., 1975). These differences are checked on the basis of a certain significant level and then the decision is made as to rainfall increment or not.

None of the aforementioned evaluation techniques are applicable in the Istanbul City cloud seeding experiment because all of them are dependent on the gaussian PDF of the rainfall amounts. However, daily rainfall amounts in the Istanbul area are distributed according to log-normal PDF. Besides, one cannot measure the efficiency factor through these methods.

For this reason the statisticians suggested that either an acceptable form of randomization should be incorporated into the operation from the beginning of the cloud seeding experimentation or the use of other short-term cloud seeding-effect calculation procedure, namely, double ratio method (DRM), which provides an answer whether the seeding operations carried out in short spans of time (daily) are effective or not (Dennis et al., 1975; Woodley and Solak, 1990). The way the double ratio technique has been applied so far in the cloud seeding literature has serious drawbacks by not taking into account the natural variability of the effectivity coefficient in the target and control areas. Gabriel and Rosenfeld (1990) presented the results of the second Israeli randomized rainfall enhancement experiment for alternative targets by employing various statistical breakdowns of the rainfall data. It was earlier noticed that during the Israeli I project by Gabriel (1967) and Gabriel and Baras (1970) the ratio statistics were practically sensitive to cloud seeding data. Hence, analysis of Israeli II mainly used ratios. Gabriel and Rosenfeld (1990) used mean and median values in the statistical analysis.

It is suggested in this section that the use of the most frequently occurring rainfall amount (ie, mode value) is more plausible in cloud seeding evaluations. Although daily rainfall data are broken down into months and seasons, the proposed procedure herein concentrates on finding the DRM frequency distribution for any desired period. On the other hand, although a frequency distribution of the DRM is applied before and during the seeding period, herein the frequency distribution of the double ratio values without seeding for any desired period of successive days are used, but seeding day rainfall amounts are employed individually without any statistical treatment whether they fall within the statistically acceptable region of seeding increment or not. Herein, DRM is employed and a new method is developed for the assessment of cloud seeding evaluations based on the Chebyshev's inequality.

7.11.1 Frequency Double Ratio Method

In the cloud seeding assessments regression analysis, double and root ratio methods, randomized experiments and bayesian approaches are among the statistical treatments, but the most widely used is the DRM. The reason being that in the regression approach long-year (at least 10 years) cloud seeding data are necessary, whereas DRM requires only a number of daily cloud seeding data, preferably at least 30 days for reliable applications.

To investigate differential effects of seeding for different areas and under different meteorological conditions, one can ask the question with regard to the entire experiment, "Does the cloud seeding affect rainfall?" To answer this question in cloud seeding evaluations the most frequently used method is the DRM given by Dennis et al. (1975) and Woodley and Solak (1990) as

$$E = \frac{(R_t/R_c)_s}{(R_t/R_c)_{us}} \tag{7.28}$$

where E is referred to as the "effectivity coefficient" and R indicates total rainfall amount with subscripts t, c, s, and 'us' that refer to target area, control area, seeded period, and unseeded period, respectively. The numerator is the ratio of total target area daily rainfall to control area daily rainfall during any desired short durations of 3 years. On the other hand, the denominator is also the ratio of unseeded daily rainfall totals. It is preferable to take the longest rainfall records possible for calculating the ratio in the denominator. In general, the possible longest record is restricted by the available longest past records of rainfall within the study area considered. The use of past records for quantification of the natural variability does not create any problem during the procedure. However, if possible at least 10-years' data are needed for effective use of Eq. (7.28) in practical applications. Generally, numerator should be calculated from short-term records, whereas denominator from comparatively long-term records. The numerator term corresponds to sampling of various short durations from the denominator long-term records. The total rainfall values during subperiods in Eq. (7.28) are never equal to zero and, therefore, the ratio of infinity does not occur.

There are fundamental pitfalls in the use of the DRM in practice, as has been noticed during actual cloud seeding operations. Although most often the DRM formulation yields values equal to 1, even without seeding its value may deviate from 1 due to the inherent random fluctuation character in natural daily rainfall amounts. Some portion of these deviations might originate from possible measurement errors or heterogeneities in the rainfall records' behavior. Prior to the application of the DRM any heterogeneity should be discarded by some convenient technique. Unfortunately, in practice, any upward deviation from 1, for instance, say, for effectivity coefficient 1.20, the deviation of 0.20 is interpreted as percentage increase due to the cloud seeding operation. It cannot be considered as rainfall increment until inherent random errors in daily rainfall series are isolated from this deviation. Hence, it is necessary to develop a simple criterion to discern the natural deviation from the forced deviation by cloud seeding. The global deviation, D_G, can be written in terms of effectivity coefficient as

$$D_G = E - 1 \tag{7.29}$$

in which D_G may assume positive or negative random values. In any cloud seeding operation, the global deviation is expected to have two subdeviations as naturally inherent deviation, D_N, and seeding deviation, D_S. Hence,

$$D_G = D_N + D_S \tag{7.30}$$

Consequently, the net change (increment or decrement) due to cloud seeding only can be calculated as

$$D_S = D_G - D_N \tag{7.31}$$

The right-hand side terms in this expression can be calculated from available data. It is obvious that Eqs. (7.29)–(7.31) are linear and, therefore, the arithmetic averages for small samples and theoretical expectations of both sides yield $\bar{D}_G = \bar{E} - 1$; $\bar{D}_G = \bar{D}_N + \bar{D}_S$; and $\bar{D}_S = \bar{D}_G - \bar{D}_N$, respectively. In calculating the net effect of cloud seeding according to the above equations the following procedure must be applied:

1. Calculate the frequency diagram of the effectivity coefficient from available past observations. In this case, D_N becomes equal to D_G. Subsequently, calculate the natural deviation $D_N = E - 1$ and its frequency distribution for any desired daily period (5 day, 10 day, 20 day, etc.). Hence, for each period there will be a unique frequency distribution for D_N and, consequently, it is possible to calculate its statistical summary values such as the arithmetic mean, standard deviation, mode value, and so on. This distribution function is referred to as the "frequency DRM," the application of which is furnished in the following section. Let us denote the mean value of this frequency function by $\bar{D}_N$ and so if the frequency diagram is symmetrical, $D_N = 0.0$. Otherwise, for positively (negatively) skewed distributions the value is greater (less) than zero.

2. Apply the formula given in Eq. (7.28) for target and control areas, but this time using seeded daily rainfall totals in the numerator, whereas the denominator remains as it was for the unseeded case. This calculation gives a global deviation and it reflects the effects of rainfall resulting from either seeding or seeding with natural rainfall. The calculation will result in a single D_G value only.

3. The net effect of seeding can be calculated as a difference given in Eq. (7.31) or as the average.

Example 7.11 Cloud Seeding Application

The results of the cloud seeding experiment in Istanbul City, Turkey, described by Omay et al. (1993), have been reanalyzed in an attempt to discuss any difference in the effects of seeding on different daily rainfall totals. They did not present scientifically based cloud seeding evaluation techniques except a simple application of double ratio formulation, which led to a single value. As explained above, such a single value is subject to random error at least partially and, therefore, cannot be considered as a reliable figure in cloud seeding assessment.

The cloud seeding project was performed over surface water reservoirs that supply domestic and industrial water to Istanbul City. The static cloud seeding operation was used during the whole project period. Depending on the radar reflectivity factor between 20 and 0 dBZ; liquid water content greater than 0.05 g/m^3; and ice particle count less than 100 n/L, silver iodide solution was injected into the suitable clouds, which were determined on the basis of radar and aircraft measurements. The seeding agent was delivered into the clouds by airborne generators or flares. Conventional radar with a digital video indicator and processor was used for effective radar reflectivity factor and nowcasting for seeding availability in the cloud. Silver iodide seeding began from aircraft in 1991 and continued from time to time depending on the availability of representative clouds through 1992 and 1993. The seeding operations were carried out on the European and Asian sides of Istanbul City as shown in Fig. 7.8.

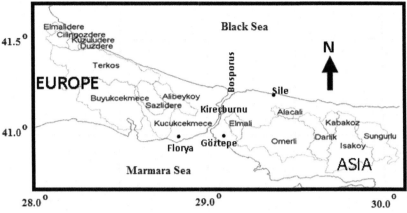

FIG. 7.8 Control and target areas.

On the basis of cloud availability, in some days, the target area was taken on the European or Asian side, and accordingly, the control area was considered on the other side of the city. In the selection of control and target areas similar behaviors in one of these two parts of the city were considered on the basis of topography, meteorology, and climatology. However, during some of the seeding operations both the target and control areas were taken on the same side of the city. The choice of alternative target and control area positions depended on the direction from which the seedable cloud was coming. The main purpose was to reduce drought vulnerability and hence to provide additional water to the other water reservoirs located in the vicinity (Fig. 7.8).

The seeding experiment was divided into periods whose exact lengths depended on the existing meteorological conditions. During each period, one or two of the reservoirs were considered as target area with seeding and the other(s) were used as control area. In order to establish a basic guidance for the cloud seeding effectiveness measurement, DRM was applied first to rainfall stations located on both sides of the city with at least 10 years of daily records. In each station the record length started from 1937 and extended up to 1993, inclusive. The characteristics of these stations are presented in Table 7.6.

TABLE 7.6 Characteristics of Daily Rainfall Stations

Continent	Station Name	Mean (mm)	Standard Deviation (mm)
Europe	Florya	1.67	4.79
	Kireçburnu	2.21	6.55
Asia	Göztepe	1.47	4.22
	Şile	2.41	6.25

Based on these daily rainfall records target and control areas were located on each side of the city, which were referred to as the European stations (Florya and Kireçburnu) and Asian stations (Şile and Göztepe). It is possible to apply Eq. (7.28) for different periods between the target and control areas prior to any seeding operation. For this purpose, various 10-year period total daily rainfall amounts were substituted in the denominator up to year 1960 and for the remaining period till to the present time subsamples of 20-, 30-, 45-, and 60-day periods were considered for the numerator in Eq. (7.28). In these manners, sampling distributions of the natural deviation as $\bar{D}_N = \bar{E} - 1$ for these subsample lengths were calculated and they led to the following conclusions:

1. Invariably all the frequency distributions were positively skewed, which implies that the arithmetic average value was greater than the mode value. Hence, a dilemma arose as to whether to adopt the arithmetic mean or the mode value as a reference to the cloud seeding experiment assessment. In different previous experiments, like in Istanbul City, the arithmetic mean value was adopted as a reference value. However, herein the mode value was used due to the following reasons.
2. Natural deviation, D_N, coincided with the maximum frequency (ie, mode value).
3. Increase in the subsample length caused the frequency diagram to become more symmetrical and, consequently, the mean and mode values became closer to each other. Table 7.7 shows the difference (deviations) between mean and mode values for different sample lengths at various stations.
4. Natural deviations were either positive or negative. This is tantamount to saying that even without cloud seeding naturally observed daily rainfall amounts might be misinterpreted according to Eq. (7.28) as rainfall increment or decrement. This point indicates that the major drawback in the use of classical double ratio or square ratio methods is the use of only a single value.
5. Provided that the mean, $\bar{D}_N$, and standard deviation, σ_{D_N}, are known the probability of a DRM result to fall within $\pm k\sigma_{D_N}$ bounds around its mean is approximately given in the form of Chebyshev inequality (Feller, 1967).

$$P[(\bar{D}_N - k\sigma_{D_N}) \le D_N \le (\bar{D}_N + k\sigma_{D_N})] \ge 1 - \frac{1}{k^2} \qquad (7.32)$$

where k is an integer number as $1, 2, 3, \ldots$ It is obvious that as the sample size increases, there appears decreases in the upper, $U_U = \bar{D}_N + k\sigma_{D_N}$, and lower,

TABLE 7.7 Mean and Mode Differences

Target/Control	Subsample Size (days)			
	20	30	45	60
Asia/Europe	0.259	0.122	0.110	0.067
Europe/Asia	0.996	0.437	0.316	0.211

$U_L = \bar{D}_N - k\sigma_{D_N}$, confidence limits. Table 7.8 indicates the necessary results for two standard deviations ($k=2$) in the case of different sample sizes. After the cloud seeding, if the calculated D_G value according to Eq. (7.33) falls outside the upper and lower limits as shown in this table, only then may a decision be taken regarding rainfall increment due to the applied seeding operation. Furthermore, the amount of seeding effect is found by considering limits for rainfall increase as

$$D_S = D_G - U_U \quad (D_G > U_L) \tag{7.33}$$

and for rainfall decrease as,

$$D_S = U_L - D_G \quad (D_G < U_L) \tag{7.34}$$

According to Omay et al. (1993), the global deviation due to cloud seeding operations over Istanbul City during the 3-year period from 1991 to 1993, inclusive, are summarized in Table 7.9.

Comparison of D_G values in this table with the relevant confidence intervals in Table 7.8 by considering 44 and 62 days as 45 and 60 days, respectively, one can see that they all remain within the upper and lower limits. This is tantamount to saying that deviations even after cloud seeding are within the sampling errors. Hence, the final objective decision was that there had not been any significant rainfall increment due to cloud seeding operation in Istanbul City. However, Omay et al. (1993) have treated global deviations in Table 7.9 directly as

TABLE 7.8 Confidence Limits at 10% Significance Level

	Number of Seeding Days							
	20		30		45		60	
Target/Control	U_L	U_U	U_L	U_U	U_L	U_U	U_L	U_U
Göztepe/Florya (Asia/Europe)	−2.22	2.61	−1.20	1.42	−1.03	1.21	−0.83	0.93
Kireçburnu/Şile (Europe/Asia)	−2.71	3.61	−2.30	3.23	−0.97	2.56	−1.23	1.76

TABLE 7.9 Global Deviation, D_G

Year	Seeding Day Number	Asia/Europe	Europe/Asia
1991	44	0.42	0.28
1992	62	0.48	0.25
1993	20	0.20	0.25

increments, which is a rather very deterministic approach and does not take into account random components originating from sampling errors as a result of physical behavior of cloud composition, in concordance between the meteorological conditions within the cloud and the amount of injection material, uncertainties in atmospheric conditions, wind direction, and synoptic conditions over Istanbul City.

Finally, it is not possible to support that there was real rainfall increment because as shown above the deviations are within the confidence interval of no seeding daily rainfall records. Hence, it is not possible to use weather modification (cloud seeding) method for drought hazard mitigation because this method is not reliable and yet does not have a sound scientific basis for rainfall increase or as a vulnerability approach alternative.

REFERENCES

Benjamin, J.R., Cornell, C.A., 1970. Probability, Statistics and Decisions for Civil Engineers. McGraw-Hill, New York, 684 pp.

Bethwaite, F.D., Smith, E.J., Warburton, J.A., Heffernan, K.J., 1966. Effects of seeding isolated cumulus clouds with silver iodide. J. Appl. Meteorol. 5, 513–520.

Buchanan-Smith, M., Davies, S., 1995. Famine Early Warning and Response—The Missing Link. Intermediate Technology Publications, London.

Burby, R.J., 2005. Have state comprehensive planning mandates reduced insured losses from natural disasters? Nat. Hazards Rev. 6 (2), 67–81.

Burby, R.J., Deyle, R.E., Godschalk, D.R., Olshansky, R.B., 2000. Creating hazard resilient communities through land-use planning. Nat. Hazards Rev. 1 (2), 99.

Center for Disease Control (CDC), 2010. When Every Drop Counts Protecting Public Health during Drought Conditions: A Guide for Public Health Professionals. In: National Center for Environmental Health, Division of Emergency and Environmental Health Services, Centers for Disease Control and Prevention, Atlanta, GA.

Chappell, C.F., Grant, L.O., Mielke, P.W., 1971. Cloud seeding effects on precipitation intensity and duration of wintertime orographic clouds. J. Appl. Meteorol. 10, 1006–1010.

Dambe, D.D., 1997. Agrometeorological inputs in measures to alleviate the effect of drought and to combat desertification. WMO-TD, No. 836, Geneva.

Dennis, A.S., Koscielski, A., Cain, D.E., Hirsch, J.H., Smith, P.L., 1975. Analysis of radar observations of a randomized cloud seeding experiment. J. Appl. Meteorol. 14, 897–908.

Elliott, R.D., 1986. Review of wintertime orographic cloud seeding. In: Braham, R.R. (Ed.), Precipitation Environment—A Scientific Challenge. In: Meteorological Monographs, vol. 21. American Meteorological Society, Switzerland, pp. 87–101.

Falkenmark, M., Lindh, G., 1976. Water for a Starving World. Westview Press, Boulder, CO.

Federal Emergency Management Agency (FEMA), 2008. In: Local Multi-Hazard Mitigation Planning Guidance. Department of Homeland Security, Washington, DC.

Federal Emergency Management Agency (FEMA), 2013. In: Mitigation Ideas: A Resource for Reducing Risk to Natural Hazards. Department of Homeland Security, Washington, DC.

Feller, W., 1951. The asymptotic distribution of the range of sums for independent random variables. Ann. Math. Stat. 22, 427–432.

Feller, W., 1967. An Introduction to Probability Theory and Its Application. John Wiley and Sons, New York, 509 pp.

Fontaine, M.M., Steinemann, A.C., Hayes, M.J., 2012. State drought programs and plans: survey of western US. Nat. Hazards Rev. http://dx.doi.org/10.1061/(ASCE)NH.1527-6996.0000094.

Gabriel, K.R., 1967. The Israeli artificial rainfall stimulation experiment: statistical evaluation for the period 1961–1965. In: Proceedings of the Fifth Berkeley Symposium on Mathematical Statistics and Probability, Berkeley, CA, pp. 91–113.

Gabriel, K.R., Baras, M., 1970. The Israeli rainmaking experiment 1961–67. Final statistical tables and evaluation. Technical report. Department of Statistics, The Hebrew University, Jerusalem, Israel, 47 pp.

Gabriel, K.R., Rosenfeld, D., 1990. The second Israeli rainfall stimulation experiment: analysis of rainfall on both target area. J. Appl. Meteorol. 29, 1055–1067.

Geringer, J., 2003. The future of drought management in the states. Spectrum J. State Gov. 23–28.

Gleick, P.H. (Ed.), 1993. Water in Crisis: A Guide to the World's Fresh Water Resources. Oxford University Press, New York.

Godschalk, D.R., 2003. Urban hazard mitigation: creating resilient cities. Nat. Hazards Rev. 4 (3), 136–143.

Grant, L.O., Mielke Jr., P.W., 1967. A randomized cloud seeding experiment at climax, Colorado 1960–1965. In: Proceedings of the Fifth Berkeley Symposium in Mathematical Statistics and Probability, vol. 5. University of California, California.

Gupta, K.K., 1973. Eigen problem solution by a combined storm sequence and inverse iteration technique. Int. J. Numer. Methods Eng. 7, 17–42.

Gupta, A.K., Tyagi, P., Sehgal, V.K., 2011. Drought disaster challenges and mitigation in India: strategic appraisal. Curr. Sci. 100, 1795–1806.

Hayes, S.C., Wilson, K.G., Gifford, E.V., Bissett, R., Piasecki, M., Batten, S.V., Byrd, M., Gregg, J., 2004. A preliminary trial of twelve-step facilitation and acceptance and commitment therapy with polysubstance-abusing methadone-maintained opiate addicts. Behav. Ther. 35 (4), 667–688.

Henderson, T.J., 1966. A ten year non-randomized cloud seeding program on the Kings river in California. J. Appl. Meteorol. 5, 697–702.

Hobbs, P.V., Rangno, A.L., 1978. A reanalysis of the Skagit cloud seeding project. J. Appl. Meteorol. 17, 1661–1666.

IPCC, 2007. IPCC Fourth Assessment Report Working Group I Report "The Physical Science Basis" Cambridge University Press, New York.

Kates, R.W., Kasperson, J.X., 1983. Comparative risk analysis of technological hazards (a review). Proc. Natl. Acad. Sci. U. S. A. 80, 7027–7038.

Kenney, D.S., Klein, R.A., Clark, M.P., 2004. Use and effectiveness of municipal water restrictions during drought in Colorado. J. Am. Water Resour. Assoc. 40 (1), 77–87.

Knutson, C.L., 2008. The role of water conservation in drought planning. J. Soil Water Conserv. 63 (5), 154–160.

Knutson, C., Hayes, M., Phillips, T., 1998. How to Reduce Drought Risk. A guide prepared by the Preparedness and Mitigation Working Group of the Western Drought Coordination Council. Lincoln, Las Vegas, USA.

Lahlou, S., 2001. Functional aspects of social representations. In: Deaux, K. (Ed.), Representations of the Social: Bridging Theoretical Traditions. Blackwell, Oxford, UK, pp. 131–146.

Lloyd, J.W., 1986. A review of aridity and groundwater. Hydrol. Process. 1 (1), 63–78.

Lohani, V.K., Loganathan, G.V., 1997. An early warning system for drought management using the palmer drought index. J. Am. Water Resour. Assoc. 33 (6), 1375–1386.

Mooney, M.L., Lunn, G.W., 1969. The area of maximum effect resulting from the Lake Almanour randomized cloud seeding experiment. J. Appl. Meteorol. 66, 264–273.

Naginders, S.S., Kundzewich, Z.W., 1997. Sustainability of water resources under increasing uncertainty. In: Proceedings of Rabat Symposium S1, April 1997, IAHS Publication Number 240.

Nelson, A.C., French, S.P., 2002. Plan quality and mitigating damage from natural disasters: a case study of the Northridge earthquake with planning policy considerations. J. Am. Plan. Assoc. 68 (2), 194–207.

Newkirk, R.T., 2001. The increasing cost of disasters in developed countries: a challenge to local planning and government. J. Conting. Crisis Manag. 9 (3), 159–170.

Neyman, J., Scott, E.L., 1967. Planning an experiment with cloud seeding. In: Proceedings of the Fifth Berkeley Symposium on Mathematical Statistics and Probability, Berkeley, CA, pp. 327–350.

O'Brien, L.V., Berry, H.L., Coleman, C., Hanigan, L.C., 2014. Drought as a mental health exposure. Environ. Res. 131, 181–187.

Omay, E., Incecik, S., Şen, O., 1993. Istanbul rainfall increment project. Final report (in Turkish).

Rangno, A.L., 1979. A reanalysis of the Wolf Creek Pass cloud seeding experiments. J. Appl. Meteorol. 18, 579–605.

Rockstrom, J., 2003. Resilience building and water demand management for drought mitigation. Phys. Chem. Earth Parts A/B/C 28 (20–27), 869–877.

Schmidt, D.H., Garland, K.A., 2012. Bone dry in Texas: resilience to drought on the upper Texas gulf coast. J. Plan. Lit. 1–12.

Şen, Z., 1995. Applied Hydrogeology for Scientists and Engineers. Taylor and Francis Group, CRC Press, Boca Raton, 496 pp.

Şen, Z., 2008. Wadi Hydrology. Taylor and Francis Group, CRC Press, Boca Raton 347 pp.

Şen, Z., 2013. Philosophical, Logical and Scientific Perspectives in Engineering. Springer, New York, 260 pp.

Şen, Z., 2014. Practical and Applied Hydrogeology. Elsevier, Amsterdam, 406 pp.

Şen, Z., Al-Subai, K., 2002. Hydrological considerations for dams sitting in arid regions: a Saudi Arabian study. Hydrol. Sci. J. 47 (2), 173–186 (Journal des Sciences Hydrologiques).

Şen, Z., Al Al-Sheikh, A., Al-Turbak, A.S., Al-Bassam, A.M., Al-Dakheel, A.M., 2011. Climate change impact and runoff harvesting in arid regions. Arab. J. Geosci. 6 (1), 287–295.

Smith, E. J., 1967. Cloud seeding experiments in Australia. In: Berkeley Symp. on Math. Statist. and Prob. Proc. Fifth Berkeley Symp. on Math. Statist. and Prob., vol. 5, Univ. of Calif. Press, pp. 161–176

Smith, K., Petley, D.N., 2009. Environmental Hazards: Assessing Risk and Reducing Disaster, fifth ed. Routledge, New York.

Steinemann, A.C., Hayes, M.J., Cavalcanti, L.F.N., 2005. Drought indicators and triggers. In: Drought and Water Crises: Science, Technology, and Management Issues, vol. 86, CRC Press, p. 71.

Steinemann, A.C., Cavalcanti, L.F.N., 2006. Developing multiple indicators and triggers for drought plans. J. Water Resour. Plann. Manag. 132 (3), 164–174.

Tang, Z., Lindell, M.K., Prater, C.S., Brody, S.D., 2008. Measuring tsunami planning capacity on U.S. Pacific Coast. Nat. Hazards Rev. 9 (2), 91–100.

Tannehill, I.R., 1947. Drought, its causes and effects. Soil Sci. 64 (1), 83.

United Utilities, 2008. Report on adaptation under the Climate Change Act. Water PLC, pp. 126–135.

Vickers, A., 2001. Handbook of water use and conversion: Homes, Landscapes, Bussiness, Industries, Farms. Amherst, Massachusetts., USA: WaterPlow Press.

Vickers, A., 2005. Managing demand: water conservation as a drought mitigation tool. In: Wilhite, D.A. (Ed.), Drought and Water Crises: Science, Technology, and Management Issues. Dekker/CRC Press, New York.

Wade, B., 2000. Tapping into water shortage solutions. Am. City County. http://americancityandcounty.com/mag/government_tapping_water_shortage/.

Wardekker, J.A., de Jong, A., Knoop, J.M., van der Sluijs, J.P., 2010. Operationalising a resilience approach to adapting an urban delta to uncertain climate changes. Technol. Forecast. Soc. Change 77, 987–998.

Warner, J., Twomey, S., 1956. The use of silver iodide for seeding individual clouds. Tellus 8, 453–459.

Wilhelmi, O.V., Wilhite, D.A., 2002. Assessing vulnerability to agricultural drought: a Nebraska case study. Nat. Hazards 25 (1), 37–58.

Wilhite, D.A., 2000. Drought preparedness and response in the context of Sub-Saharan Africa. J. Conting. Crisis Manag. 8 (2), 81–92.

Wilhite, D.A., 2005. Drought and Water Crises: Science, Technology, and Management Issues. CRC Press, Taylor & Francis Group, Boca Raton, 406 pp.

Wilhite, D.A., Knitson, C.L., 2008. Drought management planning: conditions for success. Options Mediterr. Ser. A 80, 141–148.

Wilhite, D.A., Smith, M.B., 2005. Drought as hazard: understanding the natural and social contex. In: Wilhite, D.A. (Ed.), Drought and Water Crises. Science, Technology and Management. Taylor and Francis, Boca Raton, pp. 3–29.

Wilhite, D.A., Svoboda, M.D., 2000. Early warning Systems for Drought Preparedness and Drought Management. (Proc. Expert Group Meeting, Lisbon). World Meteor. Organiz., Geneva, pp. 1–21.

Wilhite, D.A., Beadle, C.L., Worledge, D., 1996. Leaf water relations of *Eucalyptus globulus* ssp. globulus and *E. nitens*: seasonal, drought and species effects. Tree Physiol. 16, 469–476.

Wilhite, D.A., Hayes, M.J., Knutson, C., Smith, K.H., 2000. Planning for drought from crisis to risk management. J. Am. Water Resour. Assoc. 36, 697–710.

Wilhite, D.A., Svoboda, M.D., Hayes, M.J., 2007. Understanding the complex impacts of drought: a key to enhancing drought mitigation and preparedness. Water Resour. Manag. 21 (5), 763–774.

WMO, World Meteorological Organization, 2006. Drought monitoring and early warning: concepts, progress and future challenges. WMO-No. 1006.

Woodley, W.L., Solak, M.E., 1990. Results of operational seeding over the watersheds of San Angelo, Texas. J. Weather Modif. 22, 1–17.

Yeh, W.-G., 1970. Reservoir management and operations models: a state-of-the-art review. Water Resour. Res. 21 (12), 1797–1818.

Yevjevich, V., Da Cunha, L., Vlachos, E., 1983. Coping with Droughts. Water Resources Publications, Littleton, CO.

Index

Note: Page numbers followed by *f* indicate figures, *b* indicate boxes and *t* indicate tables.

Printed in the United States
By Bookmasters